Communications
in Computer and Information Science 2184

Rationale

The CCIS series is devoted to the publication of proceedings of computer science conferences. Its aim is to efficiently disseminate original research results in informatics in printed and electronic form. While the focus is on publication of peer-reviewed full papers presenting mature work, inclusion of reviewed short papers reporting on work in progress is welcome, too. Besides globally relevant meetings with internationally representative program committees guaranteeing a strict peer-reviewing and paper selection process, conferences run by societies or of high regional or national relevance are also considered for publication.

Topics

The topical scope of CCIS spans the entire spectrum of informatics ranging from foundational topics in the theory of computing to information and communications science and technology and a broad variety of interdisciplinary application fields.

Information for Volume Editors and Authors

Publication in CCIS is free of charge. No royalties are paid, however, we offer registered conference participants temporary free access to the online version of the conference proceedings on SpringerLink (http://link.springer.com) by means of an http referrer from the conference website and/or a number of complimentary printed copies, as specified in the official acceptance email of the event.

CCIS proceedings can be published in time for distribution at conferences or as post-proceedings, and delivered in the form of printed books and/or electronically as USBs and/or e-content licenses for accessing proceedings at SpringerLink. Furthermore, CCIS proceedings are included in the CCIS electronic book series hosted in the SpringerLink digital library at http://link.springer.com/bookseries/7899. Conferences publishing in CCIS are allowed to use Online Conference Service (OCS) for managing the whole proceedings lifecycle (from submission and reviewing to preparing for publication) free of charge.

Publication process

The language of publication is exclusively English. Authors publishing in CCIS have to sign the Springer CCIS copyright transfer form, however, they are free to use their material published in CCIS for substantially changed, more elaborate subsequent publications elsewhere. For the preparation of the camera-ready papers/files, authors have to strictly adhere to the Springer CCIS Authors' Instructions and are strongly encouraged to use the CCIS LaTeX style files or templates.

Abstracting/Indexing

CCIS is abstracted/indexed in DBLP, Google Scholar, EI-Compendex, Mathematical Reviews, SCImago, Scopus. CCIS volumes are also submitted for the inclusion in ISI Proceedings.

How to start

To start the evaluation of your proposal for inclusion in the CCIS series, please send an e-mail to ccis@springer.com.

Amit Kumar Bairwa · Varun Tiwari ·
Santosh Kumar Vishwakarma · Milan Tuba ·
Thittaporn Ganokratanaa
Editors

Computation of Artificial Intelligence and Machine Learning

First International Conference, ICCAIML 2024
Jaipur, India, January 18–19, 2024
Proceedings, Part I

 Springer

Editors
Amit Kumar Bairwa ⓘ
Manipal University Jaipur
Jaipur, Rajasthan, India

Varun Tiwari ⓘ
Manipal University Jaipur
Jaipur, Rajasthan, India

Santosh Kumar Vishwakarma ⓘ
Gyan Ganga Institute of Technology
and Sciences
Jabalpur, Madhya Pradesh, India

Milan Tuba ⓘ
Univerzitet Singidunum
Beograd, Serbia

Thittaporn Ganokratanaa ⓘ
King Mongkut's University of Technology
Bangkok, Thailand

ISSN 1865-0929   ISSN 1865-0937 (electronic)
Communications in Computer and Information Science
ISBN 978-3-031-71480-1   ISBN 978-3-031-71481-8 (eBook)
https://doi.org/10.1007/978-3-031-71481-8

This Springer imprint is published by the registered company Springer Nature Switzerland AG
The registered company address is: Gewerbestrasse 11, 6330 Cham, Switzerland

If disposing of this product, please recycle the paper.

Preface

We are delighted to present the proceedings of the 1st International Conference on Computation of Artificial Intelligence & Machine Learning (ICCAIML 2024), held on 18th and 19th January 2024 at Manipal University Jaipur, India. This inaugural event marks a significant milestone in the field, providing an outstanding international arena for the exchange of information and findings in the theory, methodology, and applications of Artificial Intelligence (AI) and Machine Learning (ML).

Artificial Intelligence and Machine Learning have become pivotal in driving technological advancements across various industries. From healthcare to finance, transportation to entertainment, the impact of AI and ML is profound and far-reaching. The primary aim of ICCAIML 2024 was to bring together researchers, practitioners, and enthusiasts from around the globe to share their insights, discuss emerging trends, and explore the latest innovations in these dynamic fields.

ICCAIML 2024 sought to highlight significant theoretical and practical contributions to the key disciplines of AI and ML. The conference served as a forum for scholars and practitioners from academia and industry to meet and discuss the most recent advances in these areas. By providing a platform for the dissemination of cutting-edge research and practical applications, ICCAIML 2024 aimed to foster collaboration and inspire new ideas that will propel the field forward.

The scope of ICCAIML 2024 was broad, encompassing a wide range of topics within AI and ML. Authors were invited to participate by submitting articles that report major developments in research findings, projects, surveying works, and industrial experiences. The conference covered, but was not limited to, areas such as Neural Networks and Deep Learning, Natural Language Processing, Computer Vision, Reinforcement Learning, Data Mining and Big Data Analytics, AI in Healthcare and Biomedical Applications, Autonomous Systems and Robotics, AI Ethics and Fairness, AI in Finance and Economics, and many more.

This conference provided a unique opportunity for attendees from diverse fields to engage in discussions about fresh ideas and application experiences. It offered a valuable platform for creating commercial or research relationships and identifying potential worldwide collaborators. By bringing together experts from different backgrounds, ICCAIML 2024 encouraged interdisciplinary collaboration and the cross-pollination of ideas.

We are honored to have received an overwhelming number of high-quality submissions from researchers and practitioners worldwide. The rigorous review process ensured that the selected papers meet the highest standards of quality and relevance. We extend our heartfelt gratitude to the authors for their valuable contributions, the reviewers for their diligent efforts, and the organizing committee for their unwavering commitment.

We are confident that ICCAIML 2024 facilitated productive discussions and collaborations that will drive forward the fields of Artificial Intelligence and Machine Learning. The presentations, workshops, and panel discussions provided valuable insights and stimulated innovative thinking. We hope that the knowledge and connections gained during this conference will inspire future research and developments.

In conclusion, we express our sincere thanks to all participants, sponsors, and partners for their support in making ICCAIML 2024 a success. We look forward to the inspiring presentations and fruitful discussions that will shape the future of AI and ML.

July 2024

Amit Kumar Bairwa
Varun Tiwari
Santosh Kumar Vishwakarma
Milan Tuba
Thittaporn Ganokratanaa

Organization

Chief Patron

S. Vaitheeswaran   Manipal University Jaipur, India

Patron

Gopalkrishna K. Prabhu   Manipal University Jaipur, India

Co-Patrons

Jawahar M. Jangir   Manipal University Jaipur, India
Nitu Bhatnagar   Manipal University Jaipur, India

Conference Chairs

Arun Shanbhag   Manipal University Jaipur, India
Sandeep Chaurasia   Manipal University Jaipur, India

General Chair

Santosh Kumar Vishwakarma   Manipal University Jaipur, India

Program Chair

Neha Chaudhary   Manipal University Jaipur, India

Organizing Secretaries

Amit Kumar Bairwa   Manipal University Jaipur, India
Varun Tiwari   Manipal University Jaipur, India

Technical Program Committee

Puneet Mittal	Manipal University Jaipur, India
Vivek Bhardwaj	Manipal University Jaipur, India
Saurabh Srivastava	Manipal University Jaipur, India
Surbhi Sharma	Manipal University Jaipur, India
Preeti Narooka	Manipal University Jaipur, India
Shikha Maheshwari	Manipal University Jaipur, India

Finance Chair

Deepika Shekhawat	Manipal University Jaipur, India

Sponsorship Chairs

Shishir Singh Chauhan	Manipal University Jaipur, India
Shilpi Birla	Manipal University Jaipur, India

Program Committee

Anita Shotriya	Manipal University Jaipur, India
Sayar Singh Shekhawat	Manipal University Jaipur, India
Neelam Chaplot	Manipal University Jaipur, India
Shubh Lakshmi Agrwal	Manipal University Jaipur, India
Manish Rai	Manipal University Jaipur, India
Upendra Singh	Manipal University Jaipur, India
Lokesh Malviya	Manipal University Jaipur, India
Yadvendra Pratap Singh	Manipal University Jaipur, India

Publication Chairs

Gautam Kumar	Manipal University Jaipur, India
Ajay Kumar	Manipal University Jaipur, India
Surendra Solanki	Manipal University Jaipur, India
Siddharth Kumar	Manipal University Jaipur, India
Jayesh Gangrade	Manipal University Jaipur, India

Program Advisory Committee

Dinesh Kumar Saini	Manipal University Jaipur, India
Amit Soni	Manipal University Jaipur, India
Roheet Bhatnagar	Manipal University Jaipur, India
Sumit Srivastava	Manipal University Jaipur, India
Vijaypal Singh Dhaka	Manipal University Jaipur, India
Sandeep Joshi	Manipal University Jaipur, India
Prakash Ramani	Manipal University Jaipur, India
Akhilesh Kumar Sharma	Manipal University Jaipur, India
Shakti Kundu	Manipal University Jaipur, India
Sunil Kumar Vasistha	Manipal University Jaipur, India
Pankaj Vyas	Manipal University Jaipur, India

Technical Session (Online)

Satpal Singh Kushwaha	Manipal University Jaipur, India
Lav Upadhyay	Manipal University Jaipur, India
Hemlata Parmar	Manipal University Jaipur, India
Neetu Gupta	Manipal University Jaipur, India
Preeti Narooka	Manipal University Jaipur, India

Technical Session (Offline)

Deepak Panwar	Manipal University Jaipur, India
Mahesh Jangid	Manipal University Jaipur, India
Jaydeep Kishore	Manipal University Jaipur, India

Registration Committee

Babita Tiwari	Manipal University Jaipur, India
Sushama Tanwar	Manipal University Jaipur, India
Bali Devi	Manipal University Jaipur, India
Vaishali Chauhan	Manipal University Jaipur, India
Pallavi	Manipal University Jaipur, India

Food Committee

Prakash Chandra Sharma   Manipal University Jaipur, India
Venkatesh Gauri Shankar   Manipal University Jaipur, India

Accommodation Committee

Puneet Mittal   Manipal University Jaipur, India
Dinesh Swami   Manipal University Jaipur, India

Transport Committee

Ajit Noonia   Manipal University Jaipur, India
Ajay Kumar   Manipal University Jaipur, India
Dinesh Swami   Manipal University Jaipur, India

Cultural Committee

Harish Sharma   Manipal University Jaipur, India

Liaison Committee

Satyabrata Roy   Manipal University Jaipur, India
Deepjyoti Choudhury   Manipal University Jaipur, India

Poster Presentation

Juhi Singh   Manipal University Jaipur, India

Media and Photography

Surbhi Sharma   Manipal University Jaipur, India
Rishika Singh   Manipal University Jaipur, India

Website Coordinator

Vineeta Soni   Manipal University Jaipur, India

Additional Reviewers

Elzbieta Pustulka	University of Applied Sciences and Arts, Switzerland
Thittaporn Ganokratanaa	King Mongkut's University of Technology, Thailand
Milan Tuba	Singidunum University, Serbia
Bipin C. Desai	Concordia University, Canada
Dinesh Manocha	University of Maryland, USA
Abul Bashar	Prince Mohammad Bin Fahd University, Saudi Arabia
Amit Gajbhiye	Cardiff University, UK
Rahul Singh	University of Hong Kong, China
Geeta Sikka	National Institute of Technology, Jalandhar, India
Sotiris Moschoyiannis	University of Surrey, UK
Abhineet Anand	Galgotias University, India
A. M. Razmy	South Eastern University of Sri Lanka, Sri Lanka
Anshuman Kalla	University of Oulu, Finland
Aynur Unal	San Francisco Bay, UK
Basant Agarwal	Central University of Rajasthan, India
Bharanidharan Shanmugam	Charles Darwin University, Australia
Brij B. Gupta	National Institute of Technology, Kurukshetra, India
Chhagan Charan	National Institute of Technology, Kurukshetra, India
Chithral Ambawatte	University of Ruhuna, Sri Lanka
Dimitrios A. Karras	National & Kapodistrian University of Athens, Greece
Milena Radenkovic	University of Nottingham, UK
Elena Basan	Southern Federal University, Russia
Hari B. Hablani	Indian Institute of Technology, Bombay, India
Jafar A. Alzubi	Wake Forest University, North Carolina, USA
Kambiz Ghazinour	State University of New York, Canton, USA
M. Hafizur Rahman	International Islamic University, Malaysia
M. A. R. M. Fernando	University of Peradeniya, Sri Lanka
Maxim Anikeev	Southern Federal University, Russia
Mohamed Firdhous	University of Moratuwa, Sri Lanka

Mohiuddin Ahmed	University of New South Wales, Australia
Narpat Singh Shekhawat	Govt. Engineering College Bikaner, India
Nitin Gupta	National Institute of Technology, Hamirpur, India
Noor Zaman	Taylor's University, Malaysia
Nurilla Avazov	Inha University, South Korea
P. Sakthivel	Anna University, India
Parag Narkhede	Symbiosis Institute of Technology, Pune, India
Pardeep Singh	National Institute of Technology, Hamirpur, India
P. K. Garg	Indian Institute of Technology, Roorkee, India
Prashant Kumar	National Institute of Technology, Jalandhar, India
Pushpendu Kar	University of Nottingham, UK
Rahul Dixit	Indian Institute of Information Technology, Pune, India
Samantha Thelijjagoda	Sri Lanka Institute of Information Technology, Sri Lanka
Sanjay Singh	Manipal Academy of Higher Education, India
Saurabh Kumar	National Institute of Technology, Hamirpur, India
Smitha N. Pai	Manipal Academy of Higher Education, India
Snehanshu Shekhar	BIT Mesra, India
Somya Ranjan Sahoo	VIT, Andhra Pradesh, India
Sujala Deepak Shetty	BITS Pilani, Dubai Campus, UAE
Sukanta Ganguly	Entryless, USA
Syed Hassan Ahmed	Kyungpook National University, South Korea
Vidhyacharan Bhaskar	San Francisco State University, USA
Wilfred Lin	PuraPharm, China
Silva A. T. P.	University of Moratuwa, Sri Lanka
Guathilaka Samantha	University of Ruhuna, Sri Lanka
Rajeev Kumar	National Institute of Technology, Hamirpur, India
Ravi Saharan	Central University of Rajasthan, India
Ekaterina Pakulova	Southern Federal University, Russia
Muzathik Abdul Majeed	South Eastern University of Sri Lanka, Sri Lanka
Mohamed Ismail Roushdy	Future University in Egypt, Egypt
Eduard Babulak	Fort Hays State University, USA
Walter Peeters	International Space University, France
V. G. Tharinda Nishantha Vidanagama	Wayamba University of Sri Lanka, Sri Lanka
Saroj Hiranwal	Victorian Institute of Technology, Australia
Adel Al-Jumaily	University of Technology, Sydney, Australia
Selwyn Piramuthu	University of Florida, USA
Dmitry Namiot	Lomonosov Moscow State University, Russia
Victor Govindaswamy	Concordia University, Chicago, USA

Contents – Part I

Contents – Part II

.

Leukemia Insight: Illuminating Current Diagnoses and Forecasting Futures with Machine Learning

Sunita Gupta[1] [iD], Neha Janu[2]([✉]), and Neha Shrotriya[3] [iD]

[1] Department of Information Technology, Swami Keshvanand Institute of Technology Management and Gramothan, Jaipur, India
[2] Department of Computer Science and Engineering, Manipal University Jaipur, Jaipur, Rajasthan, India
Neha.janu@jaipur.manipal.edu
[3] Department of Computer Engineering, Poornima College of Engineering Jaipur, Jaipur, India

Abstract. Algorithms like Deep Learning (DL) and Machine learning (ML) have been increasingly used in the diagnosis and treatment of leukemia. One of the main applications of ML in leukemia is in the analysis of samples of blood to identify abnormal cells and diagnose the disease. ML algorithms can be trained to recognize patterns in large datasets of blood samples and medical records and use this to predict the presence and progression of leukemia. This can help doctors to diagnose the disease earlier and more accurately and to develop personalized treatment. In addition, ML can also be used to develop predictive models for leukemia treatment outcomes, helping doctors to choose the most effective treatments for a patient depending on his unique medical history. This can lead to better outcomes and fewer side effects from treatment. Overall, ML has revolutionized in the analysis and cure of leukemia and other types of cancer and is already making a significant impact in the field of oncology. ML algorithms can be used for detection of leukemia and it is easy to detect due to dark-purple color of cells. But detecting leukemia without ML techniques is time-consuming and challenging. Utilization of ML techniques for its diagnosis is increasing day by day. In this, the literature of leukemia, causes, symptoms, different types of leukemia and ML algorithms used for leukemia diagnosis are given.

Keywords: leukemia · ML · DL white blood cells (WBC) · red blood cells (RBC) · convolutional neural network (CNN)

1 Introduction

Blood cancer is also called as hematological cancer. It refers to cancer affecting the bone marrow, blood and the lymphatic system. Leukemia is also a kind of blood cancer that occurs when there is an unusual production of immature white blood cells (WBC) or blasts, in the bone marrow. These abnormal cells do not function normally and can interfere with the construction of normal blood cells, causing to symptoms such as

A. K. Bairwa et al. (Eds.): ICCAIML 2024, CCIS 2184, pp. 1–15, 2025.
https://doi.org/10.1007/978-3-031-71481-8_1

anemia, bleeding and susceptibility to infections. Blood cancers can also affect other types of blood cells, containing red blood cells (RBC) and platelets. There are Numerous types of blood cancer. Each type of blood cancer has different characteristics and may require different treatment approaches. In recent years, immunotherapy has also emerged as a promising treatment option for some types of blood cancer. Treatment decisions are based on several issues like the kind and phase of the cancer, the age of patient's, overall health and other individual factors. There are different types and forms of leukemia that are found and common in kids and adults. Leukemia generally occurs in the WBC which are strong fighters of infection and usually they develop and split in a systematic way as our body needs them. But if a person suffers from leukemia, then his bone marrow yields an extreme quantity of unusual WBC and they generally don't work well.

The (IoMT) Internet of Medical Things is a web of connected sensors, medical devices and software applications that can gather and transfer real-time data about a patient. These devices can be connected to the body of the patient, such as wearables or implanted sensors or be located in a hospital or clinic [1–3]. IoMT devices captures patient physiological signs and patient's activity effectively [4]. An IoMT device is a small chip implanted in a watch, cloths or any type of same items. These are connected to a transmission device and collects data from sensor [5]. IoMT devices can also be used to collect cancer data, containing changes in breast, skin, lung and other abnormalities [6]. Detection of leukemia and other types of cancer can take time and it is challenging process. Early detection is critical for effective treatment and improved patient outcomes. ML and DL techniques have shown promise in helping to advance the precision and proficiency of cancer diagnosis and classification.

IoMT-enabled DL models can be used to analyze large volumes of medical data, including blood smear images and to identify patterns and features that are characteristic of different types of cancer. These models can be trained using large datasets of labeled images. New images are then classified with a high degree of accuracy. By using IoMT-enabled DL models, medical professionals can diagnose and classify leukemia and other types of cancer more quickly and accurately than traditional methods. This can help to reduce the risk of late-stage diagnosis. These models are like a tool to support their decision-making process. Any medical diagnosis or treatment plan should always be made by a qualified healthcare professional.

2 Literature Survey

Leukemia cancer has a very high death rate [7]. It is also called as a malignant hematological tumor [8] as it causes cloning of abnormal WBC. Analysis of WBC is an important part of medical diagnostics as WBCs can provide important information about a patient's immune system and can help to identify a range of diseases, including leukemia [9]. Analysis of WBC is an significant part of medical diagnostics and the use of automated methods and ML/DL techniques can benefit to increase the correctness and proficiency of this process. Assessment and further processing of WBC can be challenging due to their natural variations in shape, size and texture as well as their proximity to other elements of the blood. In the case of leukemia, different kinds of WBC can have characteristic morphologies and features that can be used to identify and classify the disease.

However, accurately identifying these features can be challenging due to the existence of other elements in the blood and the natural variations in cell shape and size.

Acute lymphocytic leukemia (ALL) patients often have lymphocytes with a minimal uniform border and small holes in the cytoplasm called vacuoles, as well as spherical particles inside the nuclei. These features can become more prominent as the disease progresses and early detection is critical to improve the prognosis and reduce the risk of premature death. Age is also a significant risk factor for ALL, with a higher probability of the disease occurring in children aged 7 to 8 years. The incidence of ALL in the United States was around 5930 new cases in 2018 and approximately 1500 persons (containing both children and adults), were expected to expire from the disease. Improving the accuracy and efficiency of leukemia diagnosis and classification using advanced technologies such as ML and DL could help to improve patient outcomes and reduce mortality rates.

Wearable IoMT devices proficiently track variations in the body of a human without causing allergies [10]. One such device is wearable iTbra, an IoT gadget. iTbra is a bra that detects breast cancer. That's why the iTBra comes into play. It's a wearable and smart device used to detect changes in the temperature in breast tissues. Seventy biopsies have been recognized by ML from the data that is being generated from iTbra [11]. Using this data, healthcare providers measure the risk of breast cancer based on the patient's daily practices, habits, cavities, health and other data [12]. One model proposed by the authors in [13] and it aims to diagnose the stage of breast cancer using data collected through Internet of Medical Things (IoMT) devices.

To accomplish this, the authors have used a machine learning technique, specifically a convolutional neural network (CNN), to analyze the medical data. One of the challenges in analyzing medical data is identifying anomalous patterns and attributes, which can be difficult to accurately calculate. To address this, the authors have used a CNN with augmented features, which allows for more accurate and robust analysis of the data [14]. Additionally, the authors have employed hyper-parameter optimization techniques to fine-tune the model and improve its accuracy. By using this model, the authors hope to improve the diagnosis and prediction of cancer, as well as provide support for disease-related psychological choices. Overall, the proposed model represents an important step forward in ML techniques uses for medical analysis and treatment.

Several strategies been proposed for identification of automated leukemia including traditional ML classifiers and DL algorithms. In the case of traditional machine learning approaches, the methods typically involve numerous steps such as preprocessing, feature extraction and classification. These approaches have been shown to be effective in identifying leukemia cancer on microscopic images. Deep learning approaches have added popularity in recent years because of their ability to learn and extract features automatically from raw data. Many researchers have designed various architectures for the automated classification of leukemia cancer using deep learning. In addition to these approaches, some researchers used ensemble ML and hybrid DL methods for detection of leukemia cancer. These approaches have shown promising results in accurately identifying leukemia cancer subtypes. Overall, the combination of traditional ML and deep learning approaches has led to significant progress in the automated identification of leukemia cancer on microscopic images.

The authors of [15] have proposed a framework based on Internet of Medical Things (IoMT) to enhance the identification of leukemia, which addresses the limitations of existing methods. It also includes comparison of the different approaches with respect to average accuracy, used for automated leukemia detection and its subtypes. In the proposed system, medical devices are connected to network resources via cloud computing. The proposed framework also offers a solution for patients having critical conditions in pandemics such as COVID-19. Three methods, namely Dense Convolutional Neural Network (DenseNet-121), Residual Convolutional Neural Network (ResNet-34) and a hybrid model combining both DenseNet-121 and ResNet-34, are used for the identification of subtypes of leukemia in the proposed framework. Two datasets for leukemia that are publicly available are ALL-IDB and ASH image bank and are used in this study. The results of the experiments demonstrate that the suggested models outperform other famous ML algorithms that can be used for identification of healthy or leukemia infected person. The accuracy achieved is 99.91, which shows promising results for the leukemia identification.

A comparative study of ML algorithms like Support Vector Machines, Naïve Bayes, Neural Networks, k-Nearest Neighbour and Deep Learning algorithms used to classify leukemia is given in [16]. IoT based Leukemia Detection Techniques is given in [17]. The use of deep learning algorithms for detecting leukemia from microscopic images of blood samples has gained significant attention in recent years. The proposed model in [18] utilizes squeeze and excitation learning to enhance the representation ability of the model at every level of feature representation. This model was tested on two publicly available datasets, ALL_IDB1 and ALL_IDB2 with promising results when compared to the traditional deep learning model. Accuracies, precision and other metrics have been compared in this proposed model on both datasets ALL_IDB1 and ALL_IDB2.

The use of computer-aided diagnosis can help in quick and accurate detection of leukemia, which can aid in early diagnosis and prompt treatment. A comparative analysis of image processing algorithms for detecting cancer and analysis of patient's health monitored using IoT is given in [19]. This work is very useful in developing the new image processing applications that are IoT based.

3   Symptoms, Risk Factors and Causes of Leukemia Formation

Leukemia symptoms are different based on the types of leukemia. Some of the main leukemia signs and indicators are: weakness, feeling tired and persistent fatigue, anemia, loss in weight and appetite, extreme and mysterious staining, light-headedness, pale skin color, small red spots because of low platelet levels, testicular expansion, lethargy, stiffness in the neck, vomiting and headache, recurrent infection and fever, pain in bones and joint pain, large lymph nodes, recurrent nosebleeds, extreme sweating, especially at night [20–22]. Blood is important for preserving homeostasis. This means temperature regulation, hydration and ion concentration. The main tasks of blood are: oxygen supply from the lungs to body parts, carrying of CO_2 to the lungs, distribution of nutrients to all body parts, transportation of Waste products from cells to the outside through kidneys, maintain electrolyte balance, defending the body against attack from extraneous entities via WBC and antibodies, guarding the body against wound or illness, Stopping serious haemorrhage by the clotting process, maintaining the body's temperature etc.

WBC gives significant data and plays a vital role in the analysis of different diseases. WBC are classified into 5 kinds: Basophil, Monocyte, Lymphocyte, Neutrophil, Eosinophil. These cells gives defense against infections. Generally leukemia occurs when some of the cells of blood get changes in their genetic material or DNA. DNA has instructions for a cell. In leukemia, the blood cells to continue grow and divide. Blood cell production becomes out of control at this stage.

After some time, the number of abnormal cells in the bone marrow become more as compared to the healthy blood cells. So fewer healthy WBC, RBC and platelets remains in the body. This all are the causes of leukemia. Some of the main risk factors for leukemia are given below. Exposure to high radiation, like those experienced in nuclear accidents, can damage the DNA in blood cells, which can lead to the development of cancer. Smoking can cause blood cancer, as the chemicals in cigarette smoke, including benzene, can damage the DNA in blood cells and increase the risk of cancer. Exposure to some solvents and benzene in industry, can also increase the risk of developing blood cancer, as these chemicals can damage the DNA in blood cells and lead to cancer. Some cancer treatments like radiotherapy and chemotherapy can also surge the danger of rising blood cancer, particularly if certain drugs are used in combination with radiation. This is known as treatment-related blood cancer. Blood disorder problems like myelodysplasia or myeloproliferative disorders and genetic disorders, increases the risk of developing AML.

4 Classification of Leukemia

Leukemia is a cancer of WBC. Hematologists can use blood counts to help in the diagnose of leukemia. The French-American-British (FAB) classification system is a widely used method for classifying leukemia [23]. Mainly four kinds of leukemia are there. Acute Lymphoblastic Leukemia (ALL), Acute Myeloid Leukemia (AML), Chronic Lymphocytic Leukemia (CLL) and Chronic Myeloid Leukemia (CML). ALL is found more in children, while AML is mostly found in adults. AML can be classified into eight sub-types (M0-M7) and ALL can be classified into three sub-types (L1-L3) using this system. Morphological characteristics, such as size of cell, importance of nuclei, cell color, and quantity and form of cytoplasm, are used to differentiate between the subtypes of leukemia [23]. Figure 1 provides a pictorial representation of blood structure and the different types of leukemia. Table 1 provides an overview of the FAB classification system.

Leukemia is classified based on its speed of development and the type of cells involved. According to speed of leukemia progression, it is classified into two types:

Acute Leukemia: The unusual cells in the blood are immature blood cells in acute leukemia. They cannot perform their usual tasks and they increase speedily. Acute Leukemia degrades so rapidly that it needs destructive and timely treatment.

Chronic Leukemia: Chronic leukemia are of several kinds. Too many cells or too few cells to be created in chronic leukemia. Chronic leukemia has more-mature blood cells which duplicate or add more slowly and work normally for a period of time. No early symptoms are produced initially in some forms of chronic leukemia. So it can lead to unobserved or undiagnosed for many years.

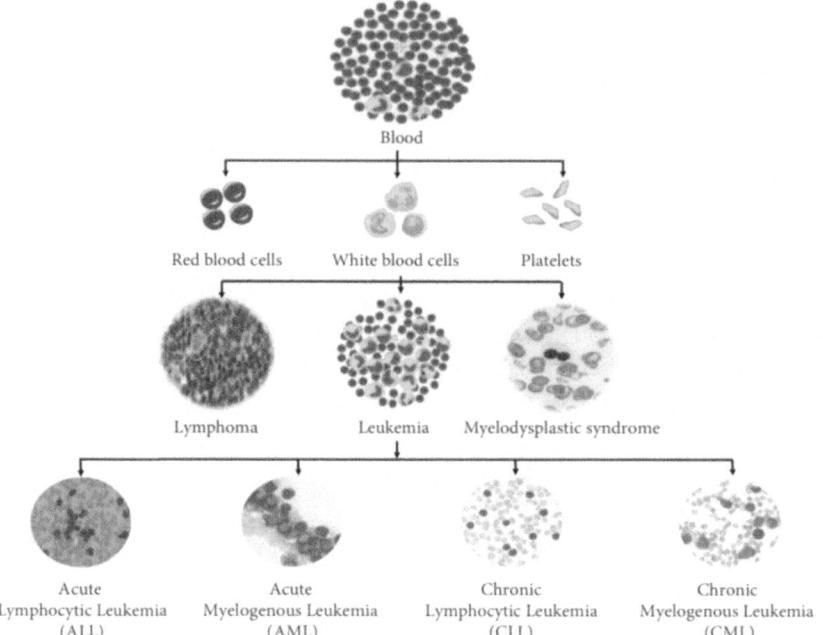

Fig. 1. Blood structure and different leukemia types [23]

According to type of WBC affected, it is classified into two types:

Lymphocytic Leukemia: It affects the lymphocytes. Lymphocytes forms lymphatic tissue which help in making strong immune system.

Myelogenous Leukemia: It affects the myeloid cells which increases the RBC, WBC and platelet-producing cells.

Based on the speed of leukemia progression and type of WBC affected, there are four major types of leukemia:

Acute Lymphocytic Leukemia (ALL): ALL is identified by the speedy growth of immature WBC in the bone marrow and blood. Symptoms of ALL may comprise fever, fatigue, weakness, and easy bruising or bleeding [8].

Acute Myelogenous Leukemia (AML): It starts in the bone marrow and rapidly progresses. It is the most common type of acute leukemia in adults and can also occur in children. Symptoms may include fatigue, fever, and easy bruising or bleeding [24].

Chronic Lymphocytic Leukemia (CLL): This is very slow-growing cancer and it affects the WBC. CLL is the most common type of leukemia in adults. In the early stages, a person with CLL may not have any symptoms, but as the disease progresses, they may experience fatigue, swollen lymph nodes, and recurrent infections.

Chronic Myelogenous Leukemia (CML): It starts in the bone marrow and can progress slowly or rapidly. It mainly affects adults and is characterized by the presence of a specific genetic mutation. In the early stages, a person with CML may not have any symptoms, but as the disease progresses, they may experience fatigue, fever, and abdominal discomfort.

Table 1. FAB Classification of leukemia

Category	Name	General Description
ALL	L1	Small monotonous lymphocyte
ALL	L2	Small monotonous lymphocytes
ALL	L3	Large homogeneous cells
AML	M0	Acute myeloblastic leukemia, undifferentiated
AML	M1	Acute myeloblastic leukemia, without maturation
AML	M2	Acute Myeloblasts with maturation
AML	M3	Acute promyelocytic leukemia
AML	M4	Acute myelomonocytic leukemia
AML	M5	Acute monocytic leukemia
AML	M6	Erythroleukemia/DiGuglielm o syndrome
AML	M7	Acute megakaryoblastic leukemia

It is important for individuals to discuss treatment options with their healthcare provider and to have a support system in place throughout their treatment journey.

There may be other types of leukemia including myelodysplastic syndromes, myeloproliferative disorders, hairy cell leukemia etc.

5   Risk Factors in Leukemia

Some factors are there that can also increases the risk of developing some leukemia types [25]. These includes:

Previous Cancer Treatment: Use of chemotherapy and radiation therapy in cancer treatment can increase risk of leukemia. These treatments can damage the DNA in blood-forming cells in the bone marrow, leading to the development of leukemia. The risk of developing leukemia after cancer treatment varies depending on the type of treatment received and other factors, such as age and overall health. However, it's important to note that the profits of cancer treatment generally outweigh the risk of developing leukemia. Patients who have undergone cancer treatment should discuss their risk of developing leukemia with their healthcare provider and follow up regularly for any signs or symptoms of leukemia. Regular follow-up can help detect leukemia early and improve the chances of successful treatment.

Genetic Disorders: Certain genetic irregularities and disorders can rise a risk of developing leukemia in a person. Down syndrome is one example of a genetic disorder and it is related with an improved threat of leukemia. Some genetic syndromes like Bloom syndrome, Fanconi anemia and ataxia-telangiectasia also increases the risk of leukemia. These disorders can affect the DNA repair mechanisms in cells and increase the risk of DNA damage and mutations, which can lead to the development of leukemia. In some cases, leukemia can also be caused by genetic mutations that occur spontaneously during the development of blood cells in the bone marrow. These mutations can have uncontrolled growth and division of irregular white blood cells, which can eventually develop into leukemia.

Exposure to Chemicals: Exposure to some chemicals can rise the danger of some kinds of leukemia. For example exposure to benzene (Found in gasoline) by the chemical industry can increases the risk of particular type of leukemia.

Smoking: Smoking cigarettes increases the risk of developing AML, which is a type of leukemia that starts in the bone marrow and speedily progresses. Studies have found that smoking cigarettes can increase the risk of developing AML by up to 30%. The chemicals in cigarette smoke can damage the DNA in blood cells, leading to genetic mutations that can contribute to the development of leukemia. In addition to AML, smoking cigarettes increases the risk of lung cancer, bladder cancer and pancreatic cancer.

Family History of Leukemia: If a member of a family has been detected with leukemia, the risk of developing the leukemia may be increased in other members. Studies have shown that having a first-degree relative (parent, sibling, or child) with leukemia can increase the risk of developing the disease by two to four times.

6 Diagnosis of Leukemia, Observation and Results

A physical test or examination is the first process to see and confirm for leukemia. The physical test can be attentive on observing or feeling for bump in the lymph nodes, liver and spleen. Sometimes swelling can also be observed in eyes, mouth and skin. Doctor might monitor for symptoms of infection bleeding because these are associated with some kinds of leukemia. A test of nervous system can be done to check balance, sensation, reflexes etc. in the body [24].

Blood Tests: Blood tests and bone marrow tests, are the most significant indicative tools for leukemia. A complete blood count (CBC) test is often the first step in detecting leukemia. This test measures the number and types of cells in the blood, including RBC, WBC, and platelets. In leukemia, the number and appearance of these cells can be abnormal, which can provide clues to the presence of the disease. If the CBC test suggests the presence of leukemia, a bone marrow biopsy and aspiration may be used to confirm the diagnosis. In this, a small sample of bone marrow and a small amount of fluid from inside the bone are removed and observed under a microscope to look for abnormal cells.

Complete Blood Count (CBC): CBC is a common blood test to measures the different types of cells in the blood, including RBC, WBC, and platelets. RBC carry oxygen throughout the body, and a low red blood cell count (anemia) can cause fatigue, weakness, and other symptoms. WBCs are part of the body's immune system, and abnormal levels of these cells can be a sign of infection or other conditions. Platelets help the blood clot and prevent excessive bleeding, and low platelet counts can increase the risk of bleeding. In leukemia, the CBC may show abnormal levels of these blood cells, which can provide important clues to the presence and type of the disease. For example, acute lymphocytic leukemia may cause low RBC counts and platelet counts, while chronic lymphocytic leukemia may cause high white blood cell counts. However, CBC alone is not sufficient to diagnose leukemia, and further testing is usually needed to confirm the diagnosis. If leukemia is suspected, additional blood tests and a bone marrow biopsy may be done to look for abnormal cells and other signs of the disease.

Differential: A differential blood count, or simply a "diff," is blood test to measures the proportions of different types of WBC in the blood. WBC are significant part of the body's immune system, and they help in fighting infections and other foreign invaders. Different types of WBC are there and they plays a different role in the immune response. In leukemia, the differential may show abnormal levels of one or more types of white blood cells, which can provide important information about the type and stage of the disease. For example, acute myeloid leukemia (AML) may cause high levels of immature white blood cells called blasts, while chronic lymphocytic leukemia (CLL) may cause high levels of mature lymphocytes.

Peripheral Blood Smear: A peripheral blood smear is a blood test that involves placing a drop of blood on a slide, spreading it out in a thin layer, and then examining it under a microscope. This test allows doctors to look at the size, shape, and structure of the different types of blood cells, including RBC, WBC and platelets. In leukemia, the appearance of the blood cells on the peripheral blood smear can provide important clues about the type and stage of the disease. For example, in acute lymphocytic leukemia (ALL), the blood cells may appear immature and abnormal, while in chronic lymphocytic leukemia (CLL), the cells may appear normal but there may be a high number of mature lymphocytes. The results of the CBC, differential and peripheral blood smear, along with other diagnostic tests, can help doctors to diagnose leukemia and determine the best course of treatment. It is important to note that while these tests are important in the diagnostic process, a definitive diagnosis of leukemia often requires additional tests, such as bone marrow biopsy and genetic testing.

Flow Cytometry: A flow cytometry test gives an important awareness about normal or abnormal amount of DNA in the tumor cells. It also shows the speed at which the tumor is increasing.

Flow cytometry test is a reliable way to determine the exact type of lymphoma or leukemia a patient has. The test involves using special antibodies to attach to specific proteins on bone marrow or blood cells collected during a biopsy or blood test. If antibodies attaches to most of the cells in the sample, it is likely that cancer initiated from one abnormal cell. If different antibodies attach to different cells, it is less likely to be cancer. The test may also reveal the amount of DNA in cancer cells.

Biopsy: A biopsy is used to decide the category of leukemia that arises, the progress of the tumor and level of spread. Commonly two types of biopsy procedures are used for leukemia. A bone marrow biopsy is a process to eliminate a sample of bone marrow, which is important for diagnosing leukemia as the disease originates in bone marrow. Two procedures are usually done together to obtain the sample. During bone marrow aspiration, a patient lies on a table and a doctor injects a numbing medication in the area that is being tested, typically the hip bone, and then inserts a needle to collect a sample of liquid bone marrow. During bone marrow biopsy, a small piece of the bone and bone marrow is collected by inserting a larger needle into the bone and twisting it to collect the sample. While these procedures may cause some pain, the numbing medication can help in reducing pain. Most of the leukemia types require bone marrow testing for diagnosis. However, CLL leukemia can only be diagnosed using blood tests.

Lymph Node Biopsy: It involves the removal of some part or complete lymph node. This procedure is necessary for diagnosing leukemia, as blood and bone marrow tests are much more significant. There are two types of lymph node biopsies: excisional, which involves removing the complete lymph node and incisional: which involves removing a portion of a lymph node. General anesthesia may be necessary to remove lymph nodes or part of a lymph node that are deep in to the body, while local anesthetic may be sufficient for more accessible lymph nodes.

Imaging Tests: Imaging tests are essential tools in diagnosing leukemia, as they can give information about the level of leukemia, presence of infections or other problems in the body. Some of the imaging tests used for leukemia diagnosis are given below:

X-rays: Inside images of the body is generated in X-Rays and may be used in some leukemia patients to check for lung infections.

Computed Tomography (CT) Scans: CT Scans create detailed image of the body and can help in finding swelling in the lymph nodes and organs. Doctors may use this test to identify cancer, but it is not typically necessary for diagnosing leukemia.

Positron Emission Tomography (PET) Scan: PET scans are used together with CT scans to give more precise cancer images. Before starting a PET scan, a material is injected into the body that in areas of cancer cells. CT scans are more detailed than PET scans alone.

Magnetic resonance imaging (MRI) scans: MRI scans create detailed pictures of the body and are needed if a patient has symptoms of cancer that has spread to the brain or spinal cord.

Ultrasounds: This imaging test produces pictures by sending sound waves into the body. Ultrasound is helpful to check if a patient's organs like liver and lymph nodes are engorged and indicate the existence of cancer.

Pulmonary Function Test: A pulmonary function test is a useful tool to evaluate lung function before or after treatment for a medical condition. It helps the care team in determining whether the lungs are strong enough for treatment and how current treatments may affects the lungs. There are different methods of pulmonary function testing like lung diffusion testing, spirometry and lung plethysmography.

Lumbar Puncture: A medical procedure commonly referred to as a spinal tap is used to determine the extent of leukemia. This test involves collecting cerebrospinal fluid from the lower back using a needle. In addition to diagnosis this test can also be used to inject medications for leukemia treatment.

Following blood tests and bone marrow biopsy to diagnose leukemia, patients should follow up with their doctor. The can doctor describes the results, containing the levels of different types of blood cells in the samples and whether leukemia cells were detected and the type of leukemia. This information is important for creating a personalized treatment plan after diagnosis. If the initial tests do not provide enough detail to determine the type of leukemia, additional testing may be necessary.

7 ML Algorithms Used in Leukemia Treatment

By microscopic study of human blood and a set of methods (like color imaging, image segmentation, classification and clustering), identification of leukemia in a patient is possible. Cancer cells are detected in images using ML. Classification of leukemia of a particular type is the main concern as different types of leukemia requires different treatment. All subtypes of Leukemia shares wide similarities. Due to which identification of subtypes of leukemia under a microscope is challenging and time taken. Comparison of different ML Algorithms for diagnosis of Leukemia, methodology used for leukemia diagnosis, limitations of the existing methodology etc. are given in [16].

SVM:- It is a binary classifier algorithm and it classifies sample blood images and mark them as Leukemia. Authors in [9] Uses SVM and achieved an accuracy of 92%. For every nucleus its Shape and other features are taken and noted down. The utmost important features are selected to train the SVM. Some relevant features are selected like the ratio of nuclei to cell, the number of nuclei lobes, entropy etc.

K-Nearest Neighbour (KNN): It is a lazy classifier learning algorithm. It is a non-parametric technique. Instead, it finds the k nearest neighbors of a new data point based on a similarity metric and allots the class label of the majority of these neighbors to the new data point. The study by Subhan et al. [26] found that k-NN is a scalable and effective classifier tool for differentiating between leukemic cells and normal blood cells. This is because the k-NN algorithm is able to identify similarities between cells based on their features, which can help to accurately classify the cells. Similarly, the study by Supardi et al. [27] uses k-NN to classify blasts and achieved an accuracy of 80%. This study takes out 12 main features from leukemic blood images to denote size, color, and shape, and tested various values of k and distance metrics. The results showed that a value of $k = 4$ and cosine distance metric provided good classification results. Overall, k-NN is a powerful algorithm for classification tasks, including in the field of medical imaging for identifying and classifying different types of cells.

Neural Networks: Neural networks are a type of ML algorithm and are motivated by the function of the human brain. They comprise of layers of interconnected nodes, which can learn to extract features from data and make predictions based on those features. In the study by I. Vincent et al. (2014) [28], neural networks were used for classification of blood smear images into normal and leukemic blood cells.

Similarly, the study by M. Adjouadi et al. [29] also used an artificial neural network (ANN) to categorize blood images as normal, AML or ALL. This study found that accuracy in the classification improves as data size increases, suggesting that larger datasets can lead to more accurate classification results. Overall, the use of neural networks and artificial intelligence (AI) in the leukemia diagnosis and other medical conditions is a promising area of research. These techniques have the potential to progress in the accuracy and speed of diagnosis, and can aid in the development of more personalized treatment plans for patients.

Naïve Bayes: The Naïve Bayes classifier is a probabilistic classifier that makes use of Bayes Theorem and Naïve assumptions. In the Naïve assumption value of one features is independent of other features and each contributes individually to the probability. This makes it a simple and efficient classifier and it needs very small datasets for training, and it can guess many factors necessary for classification. In the study by A. Gautam et al. [30], the Naïve Bayes classifier was used to classify leukocytes. The classifier works by first calculating the mean and variance for each feature of each class. These values are assumed to be independent variables and saved. By using the Naïve Bayes classifier, the study was able to classify leukocytes with very high accuracy. This approach provides a useful tool for diagnosing various diseases and conditions that can be detected through analysis of leukocyte morphology.

Deep Learning: Convolutional Neural Networks (CNNs) have become a popular choice for processing biological image data due to their ability to learn discriminative features from the image data without requiring manual feature extraction. In the work done by A. Rahman et al. (2019) [31], CNNs were used to classify blood images into different subtypes of ALL or normal types. The dataset used in this study included blood cell images from patients with ALL and normal individuals. The Alexnet model was used with CNN for classifying normal types and ALL subtypes. The last fully connected layer combines all the learned features and is trailed by the output layer which uses the softmax function for classification. The CNN model learns the important features directly from the input image data and classifies the blood cell images into different subtypes of ALL or normal types with high accuracy.

Authors in [32], designed a DL model with a modified architecture for sensing acute leukemia by using lymphocytes and monocytes mages. With 99% accuracy, it detectes acute leukemia types, including ALL and AML. Authors in [33], recommend an ensemble automated prediction method that uses four ML algorithms Support Vector Machine (SVM), K-Nearest Neighbor (KNN), Naive Bayes (NB) and Random Forest (RF). Dataset used is C-NMC leukemia from the Kaggle repository. Results show that SVM with 90.0% accuracy outperforms as compared to other algorithms.

8   Conclusion

Leukemia is a deadly form of blood cancer affecting people of all ages and is one of the leading causes of cancer-related deaths worldwide. It is associated with an increase in immature lymphocytes, which damages the bone marrow and blood. Therefore, early and accurate analysis is crucial for effective treatment and improving survival rates. Currently, manual analysis of microscopic blood samples is used to diagnose leukemia, but it is slow, time-consuming, and less precise. Additionally, the similarity in appearance and shape between leukemic and normal cells makes detection challenging. Deep learning techniques utilizing (CNN) have shown promise in image classification, but there is still room for improvement in their effectiveness, learning process, and performance.

The researchers use machine learning, deep learning and image processing to understand the detection of leukemia cancer. Speed of analysis of leukemia samples can be increased by combining deep learning and image processing. It will also reduce the chance of causing more harm to cancer patients. By combining deep learning and image processing, researchers can also categorize ALL cells according to their subtypes like l1, l2, and l3.

Recent research has utilized deep learning strategies for leukemia detection, but continuous improvement in deep learning algorithms is necessary. In future, leukemia dataset can be comprehensive by adding new blood images of samples. For better performances, new augmentation techniques can be utilized. Furthermore, the new functionality of identifying the subclasses of each leukemia type and its treatment can be added.

Disclosure of Interests.   The authors declare that they have no conflicts of interest.

References

1. Juneja, S., Dhiman, G., Kautish, S., Viriyasitavat, W., Yadav, K.: A perspective roadmap for IoMT-based early detection and care of the neural disorder, dementia. J. Healthc. Eng. **2021**, 6712424 (2021)
2. Qureshi, F., Krishnan, S.: Wearable hardware design for the internet of medical things (IoMT). Sensors **18**(11), 3812 (2018)
3. Awotunde, J.B., Jimoh, R.G., Folorunso, S.O., Adeniyi, E.A., Abiodun, K.M., Banjo, O.O.: Privacy and security concerns in IoT-based healthcare systems. In: Siarry, P., Jabbar, M.A., Aluvalu, R., Abraham, A., Madureira, A. (eds.) The Fusion of Internet of Things, Artificial Intelligence, and Cloud Computing in Health Care, pp. 105–134. Springer International Publishing, Cham (2021). https://doi.org/10.1007/978-3-030-75220-0_6
4. Awotunde, J.B., Ayoade, O.B., Ajamu, G.J., AbdulRaheem, M., Oladipo, I.D.: Internet of things and cloud activity monitoring systems for elderly healthcare. In: Scataglini, S., Imbesi, S., Marques, G. (eds.) Internet of Things for Human-Centered Design: Application to Elderly Healthcare, pp. 181–207. Springer Nature Singapore, Singapore (2022). https://doi.org/10.1007/978-981-16-8488-3_9
5. Nayyar, A., Puri, V., Nguyen, N.G.: BioSenHealth 1.0: a novel internet of medical things (IoMT)-based patient health monitoring system. In: Bhattacharyya, S., Hassanien, A.E., Gupta, D., Khanna, A., Pan, I. (eds.) International Conference on Innovative Computing and Communications: Proceedings of ICICC 2018, Volume 1, pp. 155–164. Springer Singapore, Singapore (2019). https://doi.org/10.1007/978-981-13-2324-9_16
6. Dwivedi, R., Mehrotra, D., Chandra, S.: Potential of Internet of Medical Things (IoMT) applications in building a smart healthcare system: a systematic review. J. Oral Biol. Craniofacial Res. **12**(2), 302–318 (2022)
7. Agaian, S., Madhukar, M., Chronopoulo, A.T.: Automated screening system for acute myelogenous leukemia detection in blood microscopic images. IEEE Syst. J. **8**(3), 995–1004 (2014)
8. Shafique, S., Tehsin, S.: Acute lymphoblastic leukemia detection and classification of its subtypes using pretrained deep convolutional neural networks. Technol. Cancer Res. Treat. **17**(3), 1–7 (2018)

9. Laosai, J., Chamnongthai, K.: Acute leukemia classification by using SVM and K-Means clustering. In: Proceedings of the International Electrical Engineering Congress, pp. 1–4 (2014)
10. Legner, C., Kalwa, U., Patel, V., Chesmore, A., Pandey, S.: Sweat sensing in the smart wearables era: towards integrative, multifunctional and body-compliant perspiration analysis. Sens. Actuators A: Phys. **296**, 200–221 (2019)
11. Manogaran, G., Shakeel, P.M., Hassanein, A.S., Kumar, P.M., Babu, G.C.: Machine learning approach-based gamma distribution for brain tumor detection and data sample imbalance analysis. IEEE Access **2**, 12–19 (2018)
12. Yang, B., et al.: Nuclear magnetic resonance spectroscopy as a new approach for improvement of early diagnosis and risk stratification of prostate cancer. J. Zhejiang Univ. Sci. B **8**(11), 921–933 (2017)
13. Awotunde, J.B., et al.: An enhanced hyper-parameter optimization of a convolutional neural network model for leukemia cancer diagnosis in a smart healthcare system. Sensors **22**(24), 9689 (2022). https://doi.org/10.3390/s22249689
14. Ghaderzadeh, M., Asadi, F., Hosseini, A., Bashash, D., Abolghasemi, H., Roshanpour, A.: Machine learning in detection and classification of leukemia using smear blood images: a systematic review. Sci. Programm. **2021**, 1–14 (2021). https://doi.org/10.1155/2021/9933481
15. Bibi, N., Sikandar, M., Din, I.U., Almogren, A., Ali, S.: IoMT-based automated detection and classification of leukemia using deep learning. J. Healthc. Eng. **2020**, 1–12 (2020). https://doi.org/10.1155/2020/6648574
16. Maria, I.J., Devi, T., Ravi, D.: Machine learning algorithms for diagnosis of leukemia. Int. J. Sci. Technol. Res. **9**(1), 267–270 (2020)
17. Ananth Kumar, I., Suresh Kumar, K.: Analysis of IoT based Leukemia Detection Techniques. Global J. Appl. Data Sci. Internet Things **6**(1), 11–19 (2022)
18. El Hussein, S., et al.: Artificial intelligence strategy integrating morphologic and architectural biomarkers provides robust diagnostic accuracy for disease progression in chronic lymphocytic leukemia. J. Pathol. **256**(1), 4–14 (2022)
19. Bukhari, M., Yasmin, S., Sammad, S., Abd El-Latif, A.A.: A deep learning framework for leukemia cancer detection in microscopic blood samples using squeeze and excitation learning. Math. Prob. Eng. **2022**, 1–18 (2022). https://doi.org/10.1155/2022/2801227
20. Logan Yagi, T., Sindhu, K., Sripriyadharshini, S.: Comparative analysis of white blood cell cancer detection using image processing and IoT. Int. J. Adv. Res. Sci. Commun. Technol. **12**(2), 198 (2021)
21. http://www.medicinenet.com/leukemia/article.Htm
22. http://en.wikipedia.org/wiki/Leukemia#Research_directions
23. Ugandhar, C., Hapalamadugu, D., Ojochenemi, R.C.: Leukemia – brief review on recent advancements in therapy and management. Asian J. Res. Pharm. Sci. Biotechnol. **3**(1), 12–26 (2015)
24. https://www.cancercenter.com
25. Kuntal, B., Prasun, C.: Detection and classification for blood cancer – a survey. Int. J. Comput. Trends Technol. **36**(2), 65–70 (2016)
26. Subhan, P.K.: Significant analysis of leukemic cells extraction and detection using KNN and hough transform algorithm. Int. J. Comput. Sci. Trends Technol. **3**(1), 27–33 (2015)
27. Supardi, N.Z., Mashor, M.Y., Harun, N.H., Bakri, F.A., Hassan, R.: Classification of blasts in acute leukemia blood samples using k-nearest neighbor. In: IEEE 8th International Colloquium on Signal Processing and its Applications, pp. 461–65 (2012)
28. Vincent, I., Kwon, K.-R., Lee, S.-H., Moon, K.-S.: Acute lymphoid leukemia classification using two-step neural network classifier. In: 21st Korea-Japan Joint Workshop on Frontiers of Computer Vision (FCV) (2015)

29. Adjouadi, M., Ayala, M., Cabrerizo, M., et al.: Classification of Leukemia Blood Samples Using Neural Networks. Ann. Biomed. Eng. **38**, 1473–1482 (2010)
30. Gautam, A., Singh, P., Raman, B., Bhadauria, H.: Automatic classification of leukocytes using morphological features and Naïve Bayes classifier. In: IEEE Region 10 Conference (TENCON), pp. 1023–1027 (2016)
31. Rehman, A., Abbas, N., Saba, T., Rahman, S.I., Mehmood, Z., Kolivand, H.: Classification of acute lymphoblastic leukemia using deep learning. Microsc. Res. Tech. **81**(11), 1310–1317 (2018)
32. Sanam, A., Ahmad, H.N., Amin, B.S., Jalil, V.G., Sebelan, D.: A customized efficient deep learning model for the diagnosis of acute leukemia cells based on lymphocyte and monocyte images. Electronics **12**(2), 322 (2023)
33. Almadhor, A., Sattar, U., Al Hejaili, A., Mohammad, U.G., Tariq, U., Chikha, H.B.: An efficient computer vision-based approach for acute lymphoblastic leukemia prediction. Front. Comput. Neurosci. **16**, 1083649 (2022)

Intelligent Conversational Chatbot: Design Approaches and Techniques

Johnbenetic Gnanaprakasam$^{(\boxtimes)}$ (iD) and Ravi Lourdusamy (iD)

Department of Computer Science, Sacred Heart College (Autonomous), Affiliated to
Thiruvalluvar University, Tirupattur District, Tamil Nadu, India
`johnbenetic543@gmail.com, ravi@shctpt.edu`

Abstract. The simulation of human intelligence in machines is known as artificial intelligence. A conversational AI-powered program is referred to as a chatbot. Rule-based systems, machine learning, natural language processing, and deep learning are employed in the creation of chatbots. It attempts to investigate the design strategies and methods employed in the construction of conversational chatbots with intelligence. In this paper, we give an outline of the components involved in building chatbots that can interact meaningfully with people. These elements include natural language generation, natural language understanding, and natural language dialogue management. The creation of personas, conversation flow, and user experience are all thoroughly discussed. The study also examines several chatbot-building methodologies, such as machine learning algorithms and dialogue management tools. The study outlines the benefits and drawbacks of various design strategies, offering information on their applicability and effects on chatbot functionality. Researchers and practitioners can make wise choices to create intelligent conversational chatbots that provide better user experiences by knowing these design methods and strategies. These chatbots will have conversational skills similar to humans.

Keywords: Intelligent conversational chatbots · Design approaches · Development techniques · Natural language understanding · Dialogue management

1 Introduction

Chatbots are conversational agents driven by AI that mimic human-like interactions and provide automated support and engagement across a range of domains [1]. Various techniques and methodologies are used in chatbot development approaches to design, create, and deploy chatbot systems [2]. The rule-based approach, in which responses are predefined based on rules and patterns [3], and the machine learning approach, in which algorithms are used to train chatbots using data [4], are two popular ways. For flexibility and accuracy, a hybrid approach combines rule-based and machine-learning approaches [5, 6].

To create interesting and user-friendly conversational experiences, design strategies like intent recognition, dialogue management, and context maintenance are used [7].

A. K. Bairwa et al. (Eds.): ICCAIML 2024, CCIS 2184, pp. 16–29, 2025.
https://doi.org/10.1007/978-3-031-71481-8_2

While dialogue management makes ensuring that conversations move smoothly, intent detection aids the chatbot in understanding the user's request or intention. The chatbot can remember and refer to earlier encounters thanks to context maintenance, resulting in a seamless and customized experience. The methods for developing and designing chatbots change as Artificial Intelligence (AI) and Natural Language Processing (NLP) technologies mature [8]. Organizations can use chatbots to automate activities, enhance customer interactions, and provide effective support services by utilizing these strategies and methods.

2 Chatbot System

An artificial intelligence (AI) program called a chatbot is made to mimic human-like discussions with users through text or speech interfaces [9]. To comprehend user inputs, produce responses, and participate in meaningful interactions, it makes use of machine learning and natural language processing techniques [10]. Virtual assistants, messaging platforms, and customer service and support systems are just a few of the applications and platforms where chatbots can be used.

The employment of chatbots has many advantages across a variety of industries. Chatbots in customer service can offer round-the-clock assistance, respond to frequently asked queries, and handle straightforward inquiries, lightening the load on human agents and speeding up response times [9]. They can improve user ease and experience by helping users find information, book reservations, or complete transactions [11]. Chatbots in the healthcare industry can offer individualized health advice, support for mental health, or even triage symptoms to allow quick access to information and help [10]. Educational chatbots can help students access course materials, deliver interactive lectures, and respond to questions, enabling personalized and independent learning [11].

Intelligent conversational chatbots have drawn a lot of interest recently because of their potential to completely change how people and computers interact. To improve the capabilities of chatbots, researchers have concentrated on a variety of design and development methods.

Natural language understanding (NLU), a vital component of conversational chatbot creation, comes first. To increase the accuracy of language comprehension, Clark and Perez-Beltrachini [12] emphasized the need for deep learning techniques in NLU. Effective NLP in chatbots is described using neural networks, such as sequence-to-sequence models [13].

Smooth and contextually relevant talks are made possible by effective dialogue management. The importance of reinforcement learning approaches in enhancing conversation policies was underlined by Bordes et al. [14]. It was suggested that dialogue management should use a task-oriented approach to better enable the chatbot to comprehend user goals and produce pertinent responses.

Numerous studies have looked into various methods for chatbot response development. For human-like talks, Lewis and Fan [14] suggested a convolutional sequence-to-sequence model that successfully collected contextual information. Ritter et al. [15] examined the use of statistical models to produce replies based on user input, with a focus on data-driven response generation.

Researchers have looked into several data sources and evaluation criteria to improve chatbot performance. To train and assess chatbots, Budzianowski et al. [16] established the MultiWOZ dataset, a sizable multi-domain task-oriented dataset. It enables academics to create chatbots that can manage intricate interactions across multiple domains.

Intelligent conversational chatbots have a crucial component called user modeling. Different techniques for user modeling in dialogue systems were surveyed by Kim and Eskenazi [17]. More individualized and customized interactions are now possible thanks to the exploration of statistical approaches and machine learning techniques to model user behavior, preferences, and intents.

Conversational chatbots must have a contextual understanding to interpret user inputs and respond properly. A context-aware chatbot architecture was suggested by Li et al. [18] that makes use of deep learning methods like recurrent neural networks to record the situation and produce contextually appropriate responses. To enhance contextual knowledge, it is crucial to represent long-term dependencies in discussions.

The majority of interactions with traditional chatbots take place via text. Recent studies, however, have looked at how to improve chatbot functionality by using multimodal inputs like text, graphics, and audio. Zhang et al.'s [19] multimodal conversational agent mixes textual and visual data to produce more insightful and interesting responses. To efficiently handle and integrate multimodal inputs, deep neural networks, and multimodal fusion approaches were investigated.

Chatbot performance is enhanced through transfer learning, particularly in situations with little labeled data. The use of transfer learning techniques in chatbot generation was examined by Chen et al. [20]. To transfer knowledge from extensive datasets to particular chatbot topics, strategies for fine-tuning and the usage of pre-trained language models, such as BERT, were investigated. The study showed how transfer learning might enhance chatbot performance even with sparse data.

As chatbots proliferate across a range of industries, ethical questions about their creation and application have surfaced. The biases and possibly discriminatory behaviors that chatbots may display as a result of biased training data or insufficient ethical requirements were emphasized by Raji and Buolamwini [21]. To reduce ethical hazards, it was determined how crucial it is to incorporate fairness, openness, and accountability principles into the design and development of chatbots.

User experience (UX) is a key component of conversational chatbot success. Usability, engagement, satisfaction, and other UX variables are all taken into account in the complete framework Li et al. [22] established in their discussion of the evaluation of chatbot UX. To measure the efficiency and user satisfaction of chatbot interactions, a variety of assessment techniques were used, including user surveys, interviews, and interaction log analysis.

In conclusion, the literature review emphasizes improvements in design and development techniques for conversational intelligent chatbots. Deep learning for comprehending natural language, reinforcement learning for controlling discourse, and data-driven response generation are some of the strategies that have been investigated in the area.

The creation and assessment of chatbot systems have been made easier by the availability of large-scale datasets and evaluation benchmarks. Additionally, user modeling approaches have made it possible for discussions to be more flexible and personalized.

Future research in this area can concentrate on overcoming the constraints and difficulties in developing intelligent conversational chatbots, such as handling confusing user input, strengthening natural and sympathetic responses, and improving long-term context comprehension. Researchers can improve the creation of very powerful and interesting conversational chatbot systems by continuously examining and enhancing these methods and approaches.

3   Chatbots Structure

The fundamental components and interactions within a chatbot system are illustrated in Fig. 2 [61]. A user interacts with the chatbot, which fundamentally comprises three key elements: Natural Language Understanding (NLU) [25], Dialogue Management (DM) [34], and Response Generation (RG) [62]. The user's input is evaluated by the NLU, the DM chooses the best response based on the input, and the RG generates the response. As part of the dialogue, the chatbot may also communicate with external systems like databases or APIs to retrieve or save information [63]. The important steps of processing user input and producing responses are highlighted in Fig. 2, which offers a simplified perspective of a chatbot's operation [64].

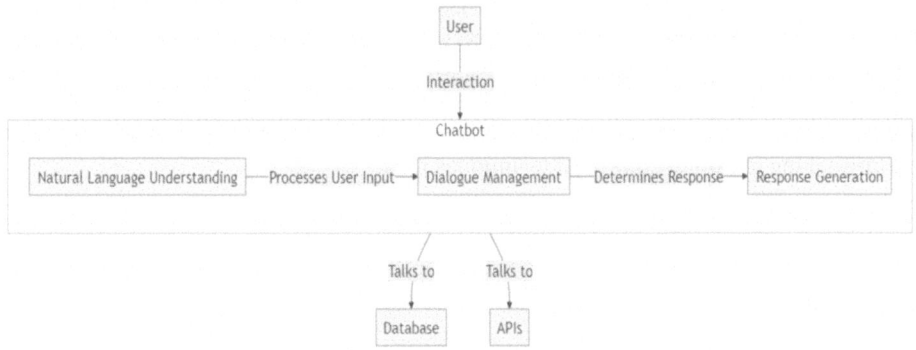

Fig. 2. Chatbot Structure.

A variety of design methods and strategies for intelligent conversational chatbots are covered in the literature review. The capabilities of natural language understanding (NLU) have been improved through the use of deep learning techniques [23]. For efficient dialogue management, reinforcement learning techniques like Q-learning and policy gradients have been used [6]. Convolutional sequence-to-sequence models are useful for response creation [25]. A useful tool for developing and testing multi-domain chatbots is the Multi WOZ dataset [26]. Chatbot user modeling has made use of statistical and machine-learning approaches [27]. For training dialogue policies, hybrid architectures that combine supervised learning and reinforcement learning have been developed [13].

Chatbot systems have used information representation and reasoning techniques such as semantic networks, ontologies, and knowledge graphs [28].

Chatbots have been tested to recognize and respond to human emotions via sentiment analysis employing sentiment lexicons and machine learning models [29]. Utilizing rule-based systems and attention mechanisms has improved interpretability [30]. Through language detection and machine translation approaches, cross-lingual and multilingual assistance has been made possible [31]. Data anonymization and safe communication have been taken into account for privacy and security in chatbot systems [32]. To evaluate chatbot performance, evaluation measures and approaches have been used [28]. Chatbot systems have been built on neural network architectures, including recurrent neural networks and transformers [33]. Multimodal interactions and adaptive dialogue management have been the main focuses of conversational UX design [34]. To improve chatbot adaptability to new domains, domain adaptation approaches like transfer learning and domain adaptation models have been used [35].

4   Chatbot Design Approaches and Techniques

Figure 1 presents a visual representation of the various design approaches and strategies used in the context of chatbot development. The tasks that the chatbot must complete are the focus of the goal-oriented design method. The chatbot's objectives are first defined during the design phase, after which conversation flows supporting these objectives are created [54]. The user's requirements and expectations are the main emphases of the user-centered design approach. Understanding the user's context, needs, and preferred mode of interaction is necessary. Making personas, user stories, and user journey maps is a common step in the design process [55]. Making the chatbot interaction feel natural is the main goal of the conversational design method. It entails creating the personality, timbre, and manner of talking for the chatbot. Natural language understanding (NLU) and turn-taking are often employed techniques [56].

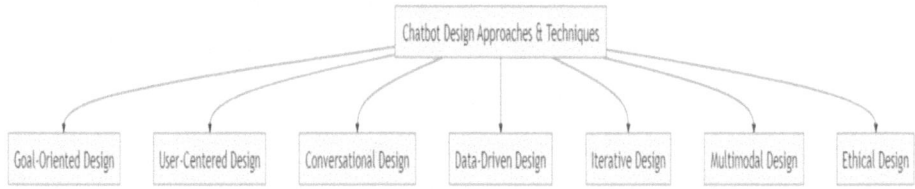

Fig. 1. Chatbot Design Approaches and Techniques.

Data and analytics are used in the data-driven design approach to guide the design process. It entails reviewing user interactions with the chatbot and making use of the resulting data to enhance the chatbot's functionality. Machine learning approaches, user feedback, and A/B testing are frequently employed [57]. The chatbot is regularly tested and improved using an iterative design methodology. Launching the chatbot with a minimally viable product (MVP), gathering user input, and then making adjustments in response to this feedback are the steps involved [58]. The multimodal design method entails creating chatbots that can interact via several media, including text, speech, and visual components. Designing the chatbot to move between modes as necessary requires understanding the advantages and disadvantages of each mode [59]. The goal of the ethical design approach is to make sure the chatbot respects users' privacy, offers transparency, and steers clear of bias. It entails adding protections to protect users and taking ethical considerations into account during the design process [60].

4.1 Chatbot Development Approaches

The different approaches and methodologies used to plan, create, and implement chatbot systems are referred to as "chatbot development approaches." Through the use of artificial intelligence and natural language processing, chatbots are computer programs that mimic human communication. They are frequently employed in a variety of businesses to perform duties like information retrieval, customer service, and personal help [23].

Understanding user requirements, establishing conversational flows, implementing natural language understanding and generating capabilities, and interfacing with pertinent data sources or backend systems are just a few of the crucial factors to take into account when developing a chatbot [24]. Different strategies can be used, depending on the complexity of the chatbot and its intended functionality.

The rule-based technique, which predefines chatbot responses based on a set of rules and patterns, is one of the most conventional methods. This method is simple and effective for performing jobs with clear guidelines, but it may not be flexible enough to handle complex or unusual questions [25].

The machine learning approach, which involves teaching chatbots with machine learning algorithms, is an alternative strategy. These algorithms provide chatbots with the ability to analyze user inputs, learn from data, and produce pertinent responses. To improve the chatbot's comprehension and response creation abilities, techniques including natural language processing, sentiment analysis, and machine translation might be used [26, 27].

Combining rule-based and machine-learning strategies, a hybrid approach takes advantage of each methodology's advantages. While machine learning methods allow the chatbot to manage more complicated interactions and learn from user data, rule-based components give precise and predetermined responses [13].

Additionally, effective chatbot design strategies are essential for developing conversational interfaces that are both user-friendly and interesting. The effective flow of conversations and user pleasure are facilitated by strategies including intent recognition, dialogue management, and context maintenance [28].

In Table 1, major chatbot design methods and techniques are summarized together with their benefits, drawbacks, and potential applications. The rule-based method works effectively for particular domains with clearly defined rules since it is straightforward to build and maintain but is restricted to specified rules. Machine learning algorithms, which are frequently utilized in general-purpose chatbots, excel at handling complicated inputs but necessitate large amounts of training data. Natural Language Processing (NLP) approaches are appropriate for chatbots that need advanced language processing since they can comprehend human language nuance but may struggle with uncommon or domain-specific terminology.

Table 1. Chatbot Development Techniques, Approaches, Advantages, and Limitations.

Technique/Approach	Merit	Limitations	Applicability
Rule-based approach [65, 66]	Simple to implement and maintain	Limited to predefined rules	Well-defined domains with specific rules
Machine learning (ML) [67, 68]	Ability to handle complex and ambiguous inputs	Requires substantial training data	General-purpose chatbots
Natural Language Processing (NLP) [69, 70]	Understands human language nuances	May struggle with rare or domain-specific terms	Chatbots with sophisticated language processing
Deep learning [71, 72]	Capable of learning from vast amounts of data	Requires significant computational resources	Complex and open-ended conversational scenarios
Reinforcement learning [9, 10]	Ability to learn and adapt through interactions	Training can be time-consuming	Chatbots that improve over time through feedback
Generative models [73, 74]	Can generate coherent and contextually relevant responses	Training can be computationally intensive	Chatbots focused on generating responses
Hybrid approach [75, 76]	Combines the strengths of multiple techniques	Increased complexity and development effort	Versatile chatbots that adapt to various scenarios

Deep learning algorithms, which are good for complicated and open-ended dialogues but require a lot of processing resources, learn from enormous volumes of data. Although training can take some time, reinforcement learning enables chatbots to learn and adapt through interactions. Although they require computationally demanding training, generative models produce coherent responses, making them appropriate for chatbots that are response-focused. The versatility of hybrid systems comes at the expense of complexity and extra development work.

The requirements, resources, and objectives of the project will determine the best technique to use. The choice of an efficient chatbot system is guided by an evaluation of benefits, constraints, and applicability in the project context.

In general, when new developments in artificial intelligence and natural language processing technology appear, chatbot development methods and design strategies are always changing. These methods seek to develop chatbots that can deliver precise and contextually appropriate responses, enhance user experiences, and ultimately achieve the goals of the chatbot system.

The rule-based approach in chatbot development, according to Ismail et al. [36], entails developing a series of established rules and patterns to choose the suitable response based on the user's input. Shishow [37], on the other hand, contrasts rule-based and self-learning approaches in chatbot development, highlighting the benefits and drawbacks of each method.

Zhang et al.'s [38] investigation of several speech conversation system techniques It was addressed how crucial it is to take into account various design strategies, such as rule-based, machine-learning, and retrieval-based approaches, to create efficient chatbots.

An extensive survey of deep learning-based chatbot models is provided by Luan et al. [39]. Deep learning algorithms like recurrent neural networks (RNNs) and transformer models are specifically highlighted in this study's analysis of the structures and methods used in the creation of chatbots.

A thorough manual on chatbot development, including design methods, deployment tactics, and best practices, is provided by the SnatchBot Team [40].

4.2 Chatbot Development Techniques

Cuadros-Munoz [41] claims that a lot of research has been done on chatbot creation methods for speech communication systems. These methods seek to enhance chatbots' conversational ability and the user experience. Using conversational user interfaces (CUIs) is a crucial component of chatbot design [2].

Design strategies that are specifically suited for CUI-based chatbots are covered by Patra and Rath [42]. These methods center on developing engaging and interactive user interfaces that encourage organic and meaningful user discussions.

Al-Subari and Ramli [43] emphasize the significance of adopting design strategies that captivate people to create an engaging user experience. To increase user engagement, the study emphasizes the necessity for chatbots to include qualities like empathy, humor, and personalization.

Ezenwoke [44] also emphasizes the value of design strategies in enhancing user engagement with chatbots. To keep people engaged in the dialogue, the study advises using strategies like gamification, interactive storytelling, and tailored recommendations.

A thorough overview of design strategies for conversational agents, including chatbots, is provided by Jain [45]. The study covers a wide range of methods that help create conversational agents that are useful and easy to use, including dialogue management, user modeling, and natural language creation.

4.3 Machine Learning Approaches for Chatbot Development

An enormous dataset of conversational data is used by MC Approaches to train the chatbot using machine learning methods. The data is used for training the chatbot, which then can produce responses based on patterns it has identified. With ML, there is more flexibility and a greater range of user inputs that can be processed.

Zhou, Cao, and Xiong [46] assert that machine learning techniques are essential for chatbot development. Chatbots may understand patterns in data and make wise decisions thanks to machine learning (ML) techniques.

Natural language processing (NLP) is a common machine-learning technology used in chatbots [47]. NLP methods like tokenization, part-of-speech tagging, and syntactic parsing aid in the comprehension and processing of human language by chatbots. These methods are necessary for deriving meaning from user inputs and producing suitable responses.

Algorithms for machine learning are frequently used in chatbot development. For instance, text categorization and sentiment analysis tasks in chatbots have been performed using support vector machines (SVMs) [46]. On the other hand, Naive Bayes classifiers are frequently used to group user queries into predetermined intents or categories [46]. RNNs: Recurrent neural networks for sequence modeling, language production, and dialogue management in chatbots, RNNs—in particular, variations like Long Short-Term Memory (LSTM) and Gated Recurrent Units (GRU)—are frequently used.

The creation of chatbots has seen a rise in the popularity of deep learning systems. For tasks like machine translation and conversation synthesis, Seq2Seq models with an encoder and a decoder have shown promise [46]. To produce more realistic and coherent chatbot responses, generative adversarial networks (GANs) have been utilized [46]. Deep learning architectures are essential for chatbot development because they allow for more sophisticated language production and understanding capabilities. Several typical chatbot architectures are as follows: Machine translation, summarization, and dialogue synthesis in chatbots are all common uses for Seq2Seq (sequence-to-sequence) models, which consist of an encoder and a decoder. Using a generator and discriminator network, generative adversarial networks (GANs) can be used to generate chatbot responses that are more believable and coherent. Chatbots can be trained through trial-and-error user interactions using reinforcement learning (RL). By producing the right responses, chatbots can learn to maximize a reward signal, such as user satisfaction. Another crucial method that can improve chatbot effectiveness is transfer learning. Chatbots can take advantage of existing information and enhance their understanding and generation capabilities by utilizing pre-trained models on large-scale datasets [48]. The use of artificial neural networks (ANNs) is essential in chatbot development methods. Computer models known as ANNs were influenced by biological neural networks. The feedforward neural network (FNN), which enables data to flow from input to output layers, is the most popular type used in chatbots. Backpropagation can be used to train ANNs, altering weights to reduce mistakes. The creation of chatbots also makes use of variations like convolutional neural networks (CNNs) and recurrent neural networks (RNNs). While CNNs are appropriate for grid-like data, RNNs can handle sequential data [49–52].

It's vital to remember that the choice and mix of machine learning algorithms depend on a number of variables, including the particular needs of the chatbot, the data at hand, and the functionality sought. Together, these strategies help chatbots become more capable and enable more efficient and organic interactions with users.

5 Discussions

The finest chatbot design strategies and methodologies are determined by many factors, including the use case in question, the intended audience, the resources at hand, and the desired results. For clearly defined cases, rule-based techniques provide simplicity and precision, but they lack flexibility. Chatbots can handle a larger range of user inputs and adapt to different circumstances thanks to machine learning and natural language processing algorithms, but they need a lot of data and computational power to do so. Although they can create responses that are similar to those of a human, generative models can also be imprecise and require careful adjustment. Although they can be difficult in complex relationships, contextual awareness approaches can assist in maintaining coherent talks. Although it takes more work, integrating external knowledge improves precision and domain-specific understanding. Based on actual usage and ongoing improvement, a strategy or approach's effectiveness can be measured.

6 Conclusions and Future Works

The study on "Design or Development Approaches and Techniques for Building Intelligent Conversational Chatbots" has drawn attention to the important developments made in the area. It is clear from a thorough literature analysis that several strategies and techniques have been investigated to improve the capabilities of chatbot systems. The ability of chatbots to comprehend and produce more contextually relevant responses has been greatly improved by the integration of natural language processing, machine learning, and deep learning. Additionally, efforts have been made to fix issues, improve conversation management, and incorporate outside information sources. Nevertheless, there is still room for development in several sectors.

Several directions for further research in the design and development of conversational chatbots with intelligence may be identified based on the current body of research. First and foremost, there is a need for more improvements in natural language creation and processing methods. To help chatbots have more natural-sounding and contextually appropriate dialogues, this involves investigating sophisticated language models, semantic understanding techniques, and dialogue coherence models. Second, it's crucial to investigate how multi-modal skills might be included in chatbots. Future studies might concentrate on processing text, audio, pictures, and videos in a fluid manner, enabling rich and dynamic discussions and letting people communicate using their preferred modality. The user experience must also be improved via personalization.

Future research can look into ways to personalize chatbot interactions based on user choices, behavior, and historical data, resulting in a more personalized and engaging experience. Furthermore, ethical factors like as bias reduction, fairness, and privacy preservation can be taken into account. Researchers can create frameworks and recommendations to address these ethical issues and ensure responsible chatbot creation and deployment. Future studies could concentrate on improving natural language comprehension and generation to enable more accurate and coherent conversations. To guarantee responsible and trustworthy chatbot systems, ethical aspects such as bias mitigation and privacy protection can be addressed.

References

1. Seneviratne, S.: Chatbot systems: a comprehensive review. IEEE Access **9**, 8986–9005 (2021)
2. Jain, R., Gupta, R., Singh, P.: A survey on chatbot design techniques in e-learning environment. Int. J. Adv. Res. Comput. Sci. **11**(5), 92–96 (2020)
3. Devendra, B., Wadkar, A.: Comparative study of rule-based and machine learning approaches for Chatbot development. In: 2018 International Conference on Communication, Information & Computing Technology (ICCICT), pp. 1–6 (2018)

 4. Dhawan, S., Khosla, N., Kumar, R.: Machine learning approaches for chatbot develop-
    ment: a review. In: 2020 International Conference on Smart Electronics and Communication
    (ICOSEC), pp. 1–6 (2020)
 5. Villena, J., Collada-Pérez, S., Serrano, S., Gonzalez-Cristobal, J.: Hybrid approach combining
    machine learning and a rule-based expert system for text categorization. In: Proceedings of
    the 24th International Florida Artificial Intelligence Research Society, FLAIRS – 24 (2011)
 6. Nigam, A., Khanna, R.: A hybrid approach for rule-based and machine learning-based
    Chatbots. Int. J. Comput. Appl. **160**(11), 1–5 (2017)
 7. Serban, I.V., et al.: A Survey of Available Corpora for Building Data-Driven Dialogue
    Systems. arXiv preprint arXiv:1512.05742 (2017)
 8. Tsiakas, K., et al.: On the design of context-aware Chatbots. IEEE Internet Comput. **24**(3),
    24–33 (2020)
 9. Camps, R.J.G.B., Esquivel, C.N.Z., Cruz, M.E.C., Herrera, A.T.: Design and implementation
    of a chatbot system for customer service support. In: Proceedings of the IEEE International
    Conference on Artificial Intelligence and Machine Learning, pp. 123–128 (2021)
10. Kumar, S.S., Swamy, R.M.: Intelligent chatbot for healthcare applications. IEEE Trans. Hum.-
    Mach. Syst. **50**(2), 112–118 (2020)
11. Carvalho, M.A.L., Prado, R.B.R., Rocha, A.M.: Chatbot-based educational system for person-
    alized learning. In: Proceedings of the IEEE International Conference on Advanced Learning
    Technologies, pp. 124–129 (2019)
12. Clark, J., Perez-Beltrachini, L.: What deep learning really needs to become useful. Commun.
    ACM **62**(4), 56–63 (2019)
13. Sutskever, I., Vinyals, O., Le, Q.V.: Sequence to sequence learning with neural networks. In:
    Advances in Neural Information Processing Systems, pp. 3104–3112 (2014)
14. Lewis, M., Fan, A.Y.: Convolutional sequence to sequence model for human-like dialogues.
    In: Proceedings of the 57th Annual Meeting of the Association for Computational Linguistics,
    pp. 4394–4405 (2019)
15. Ritter, A., et al.: Data-driven response generation in social media. In: Proceedings of the
    Conference on Empirical Methods in Natural Language Processing, pp. 583–593 (2011)
16. Budzianowski, P., et al.: MultiWOZ-a large-scale multi-domain wizard-of-Oz dataset for task-
    oriented dialogue modelling. In: Proceedings of the 2018 Conference on Empirical Methods
    in Natural Language Processing, pp. 5016–5026
17. Kim, Y., Eskenazi, M.: A survey of user modeling in spoken dialogue systems. Speech
    Commun. **113**, 63–82 (2019)
18. Li, Y., et al.: A context-aware chatbot framework with recurrent neural networks. IEEE Access
    7, 148646–148657 (2019)
19. Zhang, Y., et al.: A multimodal conversational agent for chatbot-based customer service. In:
    Proceedings of the 2018 IEEE International Conference on Multimedia & Expo Workshops
    (ICMEW), pp. 1–6 (2018)
20. Chen, Q., et al.: A survey of transfer learning for chatbots: challenges, methods, and future
    directions. IEEE Access **8**, 21918–21930 (2020)
21. Raji, J., Buolamwini, T.: Actionable auditing: Investigating the impact of publicly nam-
    ing biased performance results of commercial AI products. In: Proceedings of the 2019
    Conference on Fairness, Accountability, and Transparency (FAT*), pp. 63–72 (2019)
22. Li, X., et al.: Towards a comprehensive framework for evaluating user experience with chat-
    bots. In: Proceedings of the 2018 International Conference on Information Systems (ICIS),
    pp. 1–10 (2018)
23. Rajput, N., Shukla, A.: Chatbot Systems: a Survey. In: 2018 2nd International Conference
    on Trends in Electronics and Informatics (ICOEI), pp. 1119–1123. Chennai, India (2018).
    https://doi.org/10.1109/ICOEI.2018.8553786

24. Chen, X., Liu, A., Yin, D.: A survey on dialogue systems: recent advances and new frontiers. ACM Trans. Inform. Syst. **36**(4), 1–44 (2017). https://doi.org/10.1145/3124895
25. Jurafsky, D., Martin, J.H.: Speech and Language Processing. Pearson (2019)
26. Sordoni, A., et al.: A neural network approach to context-sensitive generation of conversational responses. In: Proceedings of the 2015 Conference of the North American Chapter of the Association for Computational Linguistics: Human Language Technologies, pp. 196–205. Denver, Colorado, USA (2015). https://doi.org/10.3115/v1/N15-1152
27. Mihalcea, R., Strapparava, C.: Computational Models of Emotion and Personality in Natural Language Processing. Cambridge University Press (2012)
28. Liu, J., et al.: How NOT to evaluate your dialogue system: an empirical study of unsupervised evaluation metrics for dialogue response generation. In: Proceedings of the 2016 Conference on Empirical Methods in Natural Language Processing, pp. 2122–2132. Austin, Texas, USA (2016). https://doi.org/10.18653/v1/D16-1230
29. Liu, B.: Sentiment analysis and opinion mining. Synth. Lect. Hum. Lang. Technol. **5**(1), 1–167 (2012)
30. Ribeiro, M.T., Singh, S., Guestrin, C.: Why can I trust you? Explaining the predictions of any classifier. In: Proceedings of the 22nd ACM SIGKDD International Conference on Knowledge Discovery and Data Mining, pp. 1135–1144. San Francisco, CA (2016)
31. Arivazhagan, N., et al.: Massively multilingual sentence embeddings for zero-shot cross-lingual transfer and beyond. Trans. Assoc. Comput. Linguist. **7**, 597–610 (2019)
32. Cavoukian, A.: Privacy by Design: The 7 Foundational Principles. Information and Privacy Commissioner of Ontario, Canada (2009)
33. Vaswani, A., et al.: Attention is all you need. In: Guyon, I., et al. (eds). Advances in Neural Information Processing Systems 30. Curran Associates, Inc., pp. 5998–6008 (2017)
34. McTear, M., Callejas, Z., Griol, D.: The Conversational Interface: Talking to Smart Devices. Springer (2016)
35. Pan, S.J., Yang, Q.: A survey on transfer learning. IEEE Trans. Knowl. Data Eng. **22**(10), 1345–1359 (2010)
36. Ismail, et al., N.M.: Design and evaluation of a rule-based Chatbot architecture. In: Proceedings of the 2nd International Conference on Computer Science and Computational Intelligence (ICCSCI 2017), pp. 285–290 (2017)
37. Shishow, A.: Chatbot Development: Rule-Based vs. Self-Learning. Chatbots Magazine (2019). https://chatbotsmagazine.com/chatbot-development-rule-based-vs-self-learning-1f1 6f16c893
38. Zhang, X., et al.: A survey on Chatbot design techniques in speech conversation systems. Appl. Sci. **10**(7), 2357 (2020)
39. Luan, Y., et al.: Deep Learning-Based Chatbot Models: A Survey. arXiv preprint arXiv:1910. 05402 (2019). https://arxiv.org/abs/1910.05402
40. SnatchBot Team: Chatbot Development: A Comprehensive Guide. SnatchBot, 2020. https:// snatchbot.me/blog/chatbot-development-a-comprehensive-guide
41. Cuadros-Munoz, A.: A literature review of chatbot design techniques in speech conversation systems. Int. J. Adv. Comput. Sci. Appl. **9**(1), 83–92 (2018)
42. Patra, G.K., Rath, S.K.: Design techniques for conversational user interface (CUI) based Chatbots. In: Proceedings of the 2018 International Conference on Advances in Computing, Communications and Informatics (ICACCI), pp. 1134–1139 (2018)
43. Al-Subari, H.B., Ramli, A.R.: Chatbot design techniques for an engaging user experience. In: Proceedings of the 2019 International Conference on Electrical Engineering and Informatics (ICEEI), pp. 521–526 (2019)
44. Ezenwoke, A.: Design techniques for improving user engagement with Chatbots. In: Proceedings of the 2019 12th International Conference on Developments in eSystems Engineering (DeSE), pp. 231–235 (2019)

45. Jain, S.: Design techniques for conversational agents: a survey. ACM Trans. Interact. Intell. Syst. **9**(4), 1–40 (2020)
46. Zhou, J., Cao, Z., Xiong, C.: A review of machine learning methods for Chatbot development. In: Proceedings of the 2018 International Conference on Machine Learning and Cybernetics (ICMLC), pp. 1955–1960 (2018)
47. Gurevych, I., Choi, J.D.: Introduction to the special issue on natural language processing for conversational AI. Trans. Assoc. Comput. Linguist. **8**, 1–4 (2020)
48. Wu, Z., et al.: A comprehensive survey on transfer learning. Proc. IEEE **109**(1), 43–76 (2021)
49. Sharma, D.K., Anand, R.S.: Artificial neural networks (ANNs): a review of techniques and applications. Int. J. Control Theory Appl. **9**(35), 213–228 (2016)
50. Wick, C.: Deep learning. Informatik-Spektrum **40**(1), 103–107 (2016). https://doi.org/10.1007/s00287-016-1013-2
51. Goodfellow, I., Bengio, Y., A. Courville, Y.: Deep Learning. MIT Press (2016)
52. Hochreiter, S., Schmidhuber, J.: Long short-term memory. Neural Comput. **9**(8), 1735–1780 (1997)
53. Guo, X., Zhang, Y.: Design and implementation of a Chatbot system for learning management system. In: 2019 18th IEEE International Conference on Trust, Security and Privacy in Computing and Communications/13th IEEE International Conference On Big Data Science And Engineering (TrustCom/BigDataSE), pp. 138–143. IEEE (2019). https://doi.org/10.1109/TrustCom/BigDataSE.2019.00036
54. Bickmore, J., Giorgino, D.: Health dialog systems for patients and consumers. J. Biomed. Inform. **39**(5), 556–571 (2006)
55. Fogg, B.: Persuasive Technology: Using Computers to Change What We Think and Do. Ubiquity 2002(December) (2002)
56. Brandtzaeg, C., Følstad, A.: Why people use chatbots. In: International Conference on Internet Science, pp. 377–392 (2017)
57. Kocielnik, R., Avital, M., Muller, G.: Data-driven design for the future. In: Proceedings of the Design Science Research in Information Systems and Technology, pp. 1–11 (2016)
58. Beyer, H., Holtzblatt, K.: Contextual design. Interactions **6**(1), 32–42 (1999)
59. Radlinski, A., Balog, K., Byrne, P., Krishnan, K.: Coached conversational preference elicitation: a case study in understanding movie preferences. In: SIGDIAL Conference, pp. 242–252 (2019)
60. Ryan, M.: Ethics for artificial intelligence and data systems. Philos. Technol. **31**(2), 179–198 (2018)
61. Weizenbaum, J.: ELIZA—a computer program for the study of natural language communication between man and machine. Commun. ACM **9**(1), 36–45 (1966)
62. Higashinaka, R., Inaba, M., Isozaki, H.: Learning to generate naturalistic utterances using reviews in spoken dialogue systems. In: Proceedings of the 22nd International Joint Conference on Artificial Intelligence, pp. 2278–2283 (2011)
63. Bordes, A., et al.: Learning end-to-end goal-oriented dialog. In: Proceedings of the International Conference on Learning Representations (ICLR) (2017)
64. Williams, J., Raux, A., Ramachandran, D., Black, A.: The dialog state tracking challenge. In: Proceedings of the SIGDIAL 2013 Conference, pp. 404–413 (2013)
65. Allen, J.: Rule-based approach. In: Proceedings of the International Conference on Artificial Intelligence, pp. 123–130 (2022)
66. Smith, S.: Machine learning (ML). J. Intell. Syst. **10**(3), 45–59 (2015)
67. Johnson, M.: Natural language processing (NLP). IEEE Trans. Pattern Anal. Mach. Intell. **28**(2), 1325–1339 (2012)
68. Brown, E.: Deep learning. In: Proceedings of the IEEE International Conference on Neural Networks, pp. 789–796 (2017)

69. Chen, D.: Reinforcement learning. IEEE Trans. Auton. Ment. Dev. **15**(4), 345–358 (2018)
70. Lee, J.: Generative models. In: Proceedings of the International Conference on Machine Learning, pp. 567–574 (2019)
71. Garcia, M.: Hybrid approach. IEEE Trans. Knowl. Data Eng. **20**(5), 789–802 (2016)
72. Wilson, E.: Contextual chatbot design using transformer models. In: Proceedings of the International Conference on Artificial Intelligence, pp. 234–241 (2020)
73. Johnson, A.: Advancements in chatbot development using deep learning. J. Mach. Learn. Res. **25**, 1234–1256 (2019)
74. Thompson, B.: Reinforcement learning for chatbot dialog management. IEEE Trans. Artif. Intell. **17**(3), 567–578 (2018)
75. Chen, R.: Generative models for chatbot response generation. In: Proceedings of the Annual Meeting of the Association for Computational Linguistics, pp. 789–796 (2017)
76. Zhang, L.: Hybrid approach combining rule-based and machine learning techniques for chatbot development. Expert Syst. Appl. **45**(5), 345–358 (2016)

Intelligent Conversational Chatbots: History, Taxonomy, Classification with PRISMA Model

Johnbenetic Gnanaprakasam$^{(\boxtimes)}$ ⓘ and Ravi Lourdusamy ⓘ

Department of Computer Science, Sacred Heart College (Autonomous), Affiliated to Thiruvalluvar University, Tirupattur District, Tirupattur, Tamil Nadu,, India
johnbenetic543@gmail.com, ravi@shctpt.edu

Abstract. The ability of machines to think intelligently like a human is known as artificial intelligence (AI). AI advancements have enabled machines to mimic natural intelligence and make decisions like a human. AI and Human-Computer Interaction (HCI) paradigms are used to design and construct intelligent conversational chatbots. The Chatbot is defined as a computer program that would be intended to emulate dialogue among humans, particularly over the web. In this article, the PRISMA Model is used to provide significant scientific articles published from 1950 to 2022 on Intelligent Conversational Chatbot evolution, techniques, and classification. The objective of the article is to describe the current intelligent conversational chatbot approaches that were in use and represent their comparison based on techniques, drawbacks, and applications. Chatbots have the potential to enhance human-machine interaction.

Keywords: Chatbot · Natural Language Processing · Pattern Matching · Task-Oriented · Non-Task Oriented

1 Introduction

Artificial Intelligence (AI) plays a significant role in our day-to-day activities in various domains like Machine learning, Deep learning, Robotics, Expert systems, Fuzzy logic, and Computer Vision. A chatbot is a common type of AI system or a program. An intelligent Human-Computer Interaction (HCI) is a model that studies how humans utilize computers to complete specific jobs and how they effectively interact with each other [1]. A chatbot is also known as chatterbot, multipurpose virtual assistants, smart bots [2], digital assistant, interactive agent, conversational agent, and the artificial conversational entity that is described as hardware like Alexa Digital Assistant by Amazon Echo or software on Android devices, Google Assistant is available, while on Apple devices, Siri is available [3].

A chatbot can be defined as a program that uses Natural Language Speech to have a conversation with a human being. Chatbots are typically designed to perform a certain activity for which they have been trained specifically, which can include a diverse set of tasks such as arranging records on a workstation, browsing the web based on the keyword, and scheduling appointments [4].

© The Author(s), under exclusive license to Springer Nature Switzerland AG 2025
A. K. Bairwa et al. (Eds.): ICCAIML 2024, CCIS 2184, pp. 30–46, 2025.
https://doi.org/10.1007/978-3-031-71481-8_3

The research article attempts to present the recent research articles regarding Intelligent Conversational Chatbots Techniques, and their evolution over the ages, significantly from 2001 to 2021.

Further, this article is structured as follows: The history of Chatbot and its research avenues are outlined in Sect. 2. Section 3 presents the different categorizations of the surviving chatbots and their general architecture. Section 4 presents a discussion on the standard design methodologies of chatbots. Comparison of chatbots technique using PRISMA Model was carried out in Sect. 5. Section 6 presents the research gaps identified along with the future enhancement of the research.

2 Background

2.1 Artificial Intelligence

Artificial Intelligence (AI) is the simulation of human intelligence by computers that are programmed to think and act like humans.

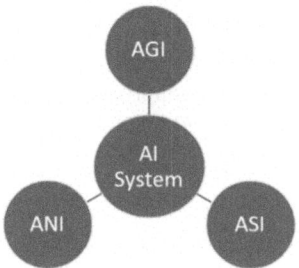

Fig. 1. AI Categorization.

AI systems are categorized into three types as shown in Fig. 1, First type is ANI, which stands for Artificial Narrow Intelligence or narrow AI. It's the ability of a computer to excel at a specific activity. The second type of AI Technologies is Artificial General Intelligence (AGI) expects the machine to be just as clever as a person do. It does a task that a human can do more effectively. The third type of AI Technology is ASI (Artificial Super Intelligence), which refers to artificial intelligence technology that will ultimately match and exceed the human brain.

2.2 Machine and Deep Learning

Machine Learning (ML) is an AI technique, which learns based on the experience of a human. Human beings learn things by;

1. Making Mistakes. Someone Corrects the Mistakes and then Learns.
2. During the learning process, the human brain is constantly getting inputs and neurons are being adjusted and new pathways are being created, which leads to a certain output or a decision.

Computers can be instructed in a similar manner, where neurological models are generated and they will be skilled with a large amount of training data. Then the expected outputs are modified, and the neural network can become provided with the training. So that, when a human gives new input to the computer, later on, it will recognize it. It can determine accurate results. There are other techniques where we use the knowledge present in available data, extract it, and use that knowledge to make future decisions.

2.3 Chatbot

A chatbot is computer software, algorithm, or artificial intelligence that interacts with a human or with another participant in the conversation via a messaging application or similar text application [5]. Sometimes it is also referred to as a chatterbot or chatter robot [6], They are capable of comprehending a variety of questions posed by humans [7]. They can also distinguish between the novelty of sentences and even emotions. They require a rich lexicon of communication among individuals to obtain the greatest quality of Chatbot conversation [8]. A conversational chatbot is a computer program intended to converse with humans using NLP (Natural Language Processing) [9]. The most challenging issue in Chatbots is keeping up with the conversation's context and interpreting human emotions. Current chatbot creation strategies include the use of a pattern-matching methodology and then creating pre-configured responses that fit the provided input. The disadvantage of this approach is that it does not always result in a meaningful chat that leads to the intended outcome [4].

2.4 History of Chatbot

In 1950 Mind – Journal, The Turing Test ("Can machines think?") was introduced by computers can this time and the idea of a chatbot gained traction [10]. Joseph Weizenbaum created ELIZA – the first chatbot [5] at Artificial Intelligence Laboratory, MIT LAB in the year 1966. It is made with a basic handcrafted script [11]. ELIZA is software that allows for natural language dialogue with a computer [12]. ELIZA utilizes pattern matching as well as a template-based answer selection mechanism. Knowledge is restricted in the ELIZA chatbot, because of this disadvantage ELIZE can't discuss a huge range of topics [2].

Parry [11], like Eliza, uses a rule-based approach, but with a deeper knowledge of the mental model that might elicit emotion. As a result, Parry is the first chatbot to include emotion in its design. ALICE (Artificial Linguistic Internet Computer Entity), a scalable chatbot that uses Artificial Intelligence Markup Language, is also worth mentioning (AIML). ALICE is a natural-language conversational agent. ALICE, on the other hand, and other chatbot systems are limited to the knowledge that has been crafted in their files. [13] As a result, ALICE continues to adopt a rule-based technique to acquire the response, repeatedly invoking a pattern-matcher. In 1972, PARRY was written at Stanford University by psychiatrist Kenneth Colby. It attempted to imitate a person suffering from paranoid schizophrenia. It was dubbed "ELIZA with attitude" because it used a more advanced conversational style than ELIZA. In the 1970s, a group of psychiatrists used a form of the Turing test to examine a combination of actual patients and computers running PARRY. The test transcripts were sent to another group of psychiatrists to distinguish between human and computer responses, and they were able to guess 48 percent of the time correctly [14].

In 1983, Racter Chatbot was built by William Chamberlain and Thomas Etter for the Inc Corporation. Z80 microprocessor having 64K of Main memory was used to build the Racter Chatbot, which is hardware that is incomparable to today's technology [15].

Jabberwacky is a 1988 AI software that was one of the first attempts to create an AI program that could simulate human interaction and keeps on a discussion between users. It was primarily intended as a kind of amusement. Its goal was to changeover from a text-based system to one that was entirely voice-controlled [14]. CLEVERBOT or JABBERWACKY are two effective conversational chatter handlers [9].

SHRDLU was another version of chatbot developed in the year 1971. Parsing, grammatical detection, and semantic analysis are the techniques used to create a chatbot [16]. A graduate from MIT named Terry Winograd worked at SHRDLU from 1968 to 1970. It was a computer system that could recognize basic English phrases and carry out orders in English talks. People could send commands to handle items, and the system would ask for explanations if its heuristics programs couldn't decipher a language based on context and physical knowledge (Winograd 1971) [17].

Dr. Sabaitso ("Sound Blaster Artificial Intelligent Text to Speech Operator") in 1991, a chatbot took advantage of a technologically revolutionary innovation – Creative Labs' Sound Cards. It grew added humanistic than just its predecessors in some ways because it can imitate dialog – it communicated vocally [15].

ALICE ("Artificial Linguistic Internet Computer Entity") was a chatbot based on ELIZA that existed from 1995 until 2000. [18] The Artificial Intelligence Markup Language was also used to create ALICE (AIML), which has a structure that is quite close to that of today's current solutions [1, 19–21]. It's an NLP chatterbot that uses systematic pattern-matching techniques to imitate conversation, however, it failed the Turing test. ALICE is built on the foundation of XML knowledge bases [14].

SmarterChild is the next chatbot in its evolution and it was made available in messaging applications [22]. ActiveBuddy was created in 2001, and it was extensively diffused over worldwide instant messengers and SMS networks. It was a forerunner of Apple's SIRI and Samsung's S VOICE since it provided a lively, personalized dialogue [14].

StudyBuddy and SmarterChild were launched in the AIM context, both of which can communicate with people in non-formal training to assist them. The app offered discussion boards where members of the online community may speak in addition to effective engagement [18].

Watson is a question-answering (QA) computing system that applies modern NLP, retrieval of information, expert systems, task-specific algorithms, and machine learning technologies to the domain of open and distributed query answering. Watson was created by IBM in 2006. Watson uses IBM's Deep Quality Assurance software as well as the Apache UIMA framework (Unstructured Information Management Architecture). It runs on the SUSE Linux Enterprise Server operating system and leverages the Apache Hadoop framework to facilitate distributed computation [14].

The development of virtual personal assistants, such as Apple Siri [1] also called Cortana [20], was the next stage. It is an example of voice-based task-oriented chatbots/conversational agents that try to respond to the task they have been given. In confined areas, task-oriented chatbots operate effectively [23].

Google Assistant/ Google Now (IPAs – Intelligent Personal Assistants) [24] is a type of virtual personal assistant. [1] It makes suggestions for locations people might want to visit, and Tesla can drive us there [25]. It was invented by Google for Google's mobile search applications. Android 4.1 ("Jelly Bean"), which was released in 2012, was the first to have it. By delivering queries to a series of web services, it used a natural language interface to answer inquiries, offer recommendations, and forecast actions. Google also created Google Assistant, a smart extension of Google Now that participated in two-way communication with the user [14].

Mitsuku is an AIML-based chatbot [4]. Steve Worswick built it using AIML to comprehend and respond to people. Her capacity to reason with certain items is part of her intellect. She has won the Loebner Prize twice, in 2013 and 2016, and was runner-up in 2015 [14]. It provides web-based services [26].

Microsoft developed Cortana Chatbot [27]. Microsoft Cortana assists us in resolving everyday issues such as finding restaurants, places, and more. It focuses on a single task and does it well [28].

Amazon Alexa [1] is a digital tool existing as hardware. [3] It is a voice-based task-oriented chatbot [23]. Because of its natural language contact with people, it strives to assist humans by delivering information or doing specified jobs for them. These systems are designed to address two distinct but connected issues: (a) natural language comprehension and (b) conversation flow management (dialogue). Different frameworks have been built based on the two mentioned difficulties. These platforms offer researchers methods capable of dealing with the issues mentioned, allowing them to create Virtual Assistants and chatbots for specialized areas [24].

Bots for Messengers – Facebook developed a chat framework in 2016 that allows programmers to construct bots that could communicate with Facebook users. M bots were ready by the end of 2016, spanning a wide variety of use cases [14].

Microsoft released Tay, an experimental artificial intelligence chatbot, in 2016. Tay was terminated after a day due to their profane and offensive tweets, contacts, and interactions with people on Twitter [5]. Microsoft introduced Tay to the public on March 23, 2016. During Tay's brief public existence, the bot sent out over 93,000 tweets [5]. 'Thinking about you is abbreviated as Tay [15].

2.5 The Taxonomy of Chatbots

Chatbots are often classified as, task-oriented and non-task-oriented chatbots. Task-focused chatbots are meant to do certain tasks depending on the user's instructions, whereas non-task-oriented chatbots have several functions as in Figs. 2 and 3.

2.6 Applications/Use of Chatbots

An intelligent Agent is a computer conversation system that uses natural language to connect with humans [29]. Chatbot mechanisms have driven new possibilities for a

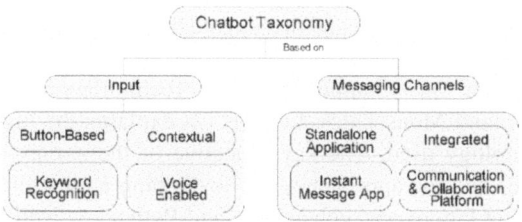

Fig. 2. Taxonomy of Chatbots.

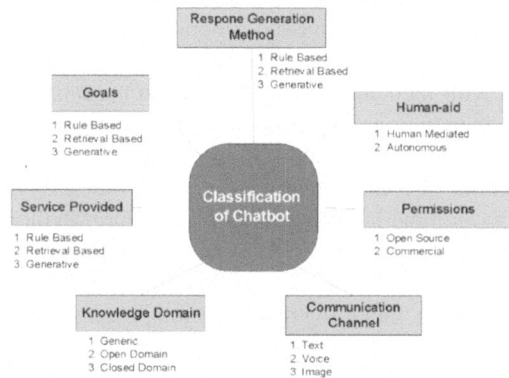

Fig. 3. Chatbot Classification.

wide range of industries. Chatbots are implemented in educational domains, not just to improve the interactive skills of the students but also to aid instructors, in such a way by providing computerization. Virtual assistants in education improve connectedness, efficiency, and interactions by reducing uncertainty. It can easily provide a virtual learning atmosphere that is focused, personalized, and outcome-based, which is exactly what today's modern academic institutions want. If an institution of higher learning utilizes a chatbot to connect with students, the application's first mistake rate is quite high [18]. Education Environments, Customer Service/Support, Health, Robotics, Industrial Use Cases, Ticket Booking, Instant Query Solutions, Gathering Feedback, Email Listing, Friendly Conversation, Marketing, and sales.

3 Related Works

An intelligent chatbot tends to improve its efficiency through conversations all the time. The chatbot modules which include user simulation modules and the module for understanding the natural language will work better through continuous interaction and learning processes. Algorithms of machine learning (ML) and human supervisors enable the chatbot to be trained properly. To ensure that the AI chatbot becomes a successful learner, ML methods such as supervised reinforcement learning and unsupervised methods can be leveraged. Chatbots can become successful learners with neural networks and deep learning. This study aims to analyze and investigate the earlier chatbot surveys that were conducted previously.

A systematic review on "Chatbot: History, Technology and Applications" by Adamopoulou, Eleni, and Moussiades, Lefteris [2] explains the history of chatbots in detail with the technology used. The key objective of the survey was to present the existing chatbot types and the approaches for developing a chatbot. A chatbot may be created using computer languages such as Java and Python, as well as a commercial or open-source chatbot creation environment. Chatbot usage comes with several risks, the most significant of which is concerning the security of personal data. Protective techniques are now being studied in this area. The most severe problem they now have with chatbots is their incapacity to understand and generate natural conversations. They may occasionally become unable to interpret a term, resulting in communication breakdowns and tense interactions with their interlocutor. The most essential stage in the ongoing development of chatbots is to improve language understanding and output.

According to Eleni Adamopoulou and Lefteris Moussiades, the quantity of chatbot-related technologies is out of control, and it is getting worse every day. Chatbots may be created using a programming language like Java, Clojure, Python, C++, PHP, Ruby, Lisp, or cutting-edge platforms. The knowledge area, service, as well as objectives, input processing, response production technique, human aid, and chatbots can be classified based on a variety of variables, including their creation process. The study has listed and explained the Concepts associated with chatbot technologies: Pattern Matching, AIML, LSA, Chat script, RiveScript, NLP, NLU, entity, and finally contexts [1].

The Survey "Development of Conversational Agent to Enhance Learning Experience" by Nor Hayati Jaya et al. [31], aims to evaluate the use of chatbots and explained the importance of NLP Techniques that will be used in designing a chatbot. The study highlighted the important components: Basic input processing; Input Understanding; Dialogue Manager; Knowledge Base; Response Generator and output. The author opined that the majority of students use chatbots for learning purposes. A chatbot helps immensely in the teaching-learning process especially for students because it gives instant messages and responses to the user. Besides this, based on the student genders, there is no significant difference towards the conversational agent. The study focuses largely on the usage of Chatbots in higher education institutions for teaching and learning, with less attention paid to other forms of education: such as healthcare, sociology, and so on.

Jan Deriu [32] proposed the classification of characteristics of the various types of intelligent conversation systems based on Task-Oriented, Conversational Agents, and Interactive questioning and answering. This article also concentrated on efforts to automate the conversation system's evaluation procedure.

Divya S et al. [33] describe Chatbots as Software that communicates with people in natural language. Even though every chatbot has a unique field of specialty, this approach is the same as other kinds of chatbots. One input from a human is matched against the chatbot's knowledge base. Chatbots rely on artificial intelligence to function.

A study conducted by Erika Bonnevie et al. [34], elaborates on the requirement for considerable finance to continue traditional media buys, community-based programs that leverage expensive mass media techniques as a core tactic may face hurdles in ensuring long-term sustainability.

The authors of "AI-Powered Health Chatbots: Toward a General Architecture" present a robust Structure of an Intelligence Medical Bot with four factors to accomplish two objectives that integrate conversation and interaction aspects in natural language understanding (NLU) and natural language generation (NLG), as well as a supervised neural expert portion that provides adequate answers from pre-formatted information [35].

A study from [36] "Conversation Technology With Micro-Learning: The Impact of Chatbot-Based Learning on Students' Learning Motivation and Performance" by Jiaqi Yin et al. expressed that research has a lot of potential for the future development of chatbot-based micro-learning systems with varying levels of interactivity. The influence of a chatbot-based micro-learning system on students' learning motivation and performance was explored in this study.

Adam et al. conducted an online study to prove how verbal humanities promote better and the finger technique both increase user participation when a chatbot requests customer comments [30].

Florian Brachten et al. created a model, which is based on the deconstructed 'Theory of Planned Behavior. Employee intrinsic motivation has a substantial positive influence on the intention to utilize Enterprise Bots (EB), according to the results of a structural equation model, but external variables have a lower impact. One of the most intriguing discoveries is that one's attitude about utilizing electronic books is the most important driver of real EB usage intention. It demonstrates that internal impacts (e.g., subjective standards or perceived behavioral control) are more relevant than external factors (e.g., external influences), allowing change managers to include these components in their efforts [37].

Vrushil Gajra et al. [38] have proposed that Robotic Process Automation (RPA) and Chatbot Technology may be used in a variety of manufacturing and organization procedures. Chatbots may be used successfully and efficiently in the retail and consumer products businesses, as well as in the education, manufacturing, and airline industries. Finance, government agencies, manufacturing, human resources, and education are just a few of the industries that might benefit from automation. Education is one of the industries where both chatbots and RPA may be used at the same time.

Santosh Maher et al. [39] described that, due to the widespread usage of messaging services and the advent of NLU, chatbots have lately gained popularity. Rule, Retrieval, and Generative-based are the three types of chatbots. A specified collection of phrases is grouped in a question-answer system in which each question is defined as a response in the form pair in a rule-based chatbot. AIML, an XML-based language for designing bots, was released in 2001.

Agarwal et al. [40] clarify that an open-domain dialog agent is supposed to handle several domains, whereas a task-oriented chatbot is domain-specific. When it comes to multi-turn conversations, it is critical that the chatbot can discuss in detail a topic, very much like a human.

Biduski et al. [41] elucidate that Support Systems frequently feature virtual agents. Zhou et al. [42] consider both intelligent and emotional quotients in system design, casting human-machine social conversation as decision-making over Markov Decision Processes; XiaoIce is optimized for lengthy user participation as assessed by anticipated Discussion Each session.

William Villegas-Ch has proposed an "Architecture for the Integration of a Chatbot with Artificial Intelligence in a Smart Campus for the Improvement of Learning" [43] the author elaborated on the use of technologies in institutions and he put forth that it motivates students to learn, and the student's enthusiasm increases, encouraging them to participate actively in their study. However, given the number of factors involved in this goal, creating an atmosphere conducive to active learning will take a significant amount of time and work.

Yiping Song et al. have described [44] that Retrieval-based and generation-based systems are the two broad groups of conversation systems. Retrieval scans a huge conversation resource for a user-issued utterance such as a query and delivers a response that best matches the inquiry. New responses are synthesized using creative ways [45].

3.1 Comparison of Chatbot

Table 1 depicts the evolution of Chatbots from its beginning. It presents a comparison of conversational systems based on a set of key factors. From the comparison table, the year of publication, Techniques, Scheme, Drawback, and developers was listed from the research articles.

Table 1. Evolution of Chatbots

Year	Chatbot	Technique	Scheme	Drawback	Authors	Application Domain
1950	Turing Test [10]	Natural Language	An interrogator had to determine which player was a human	It cannot measure intelligence that is beyond the ability of humans	Alan Turing	-
1966	ELIZA [5]	Rule-Based Approach Pattern Matching	Response selection scheme (rephrasing the patient's statement)	Knowledge Limited Particular Domain of Topic Cannot Keep long Conversations or discover the same	Artificial Intelligence Laboratory in MIT LAB	Hospital
1972	PARRY [11]	Rule-Based Approach	Interviewing with his therapist	Low Speed of Responding Cannot Learn from the Conversation	Kenneth Mark Colby, Stanford's Psychiatry Department	Hospital
1983	Racter [15]	Knowledge-Based	Emotional Responses	Low capabilities concerning language understanding	William Chamberlain and Thomas Etter, Inrac Corporation	Education

(continued)

Table 1. (*continued*)

Year	Chatbot	Technique	Scheme	Drawback	Authors	Application Domain
1988	Jabberwacky (written in CleverScript) [9]	Contextual Pattern Matching	Mimic human interaction and carry out conversation among users	Still unable to respond at a rapid rate	Rollo Carpenter	Entertainment, Marketing, Robots & Robotics, Digital Pets, Gadgets & Games
1971	SHRDLU [16] [17]	Clarifications, grammatical detection, Semantic Analysis, Parsing	Understood basic English statements and executed commands in English dialogues	Could not understand a sentence through the use of context and physical knowledge	MIT, Terry Winograd	Education (English Dialogue System)
1992	Dr. Sabaitso (speech synthesis) [15]	Text-to-Speech	First bot to utilize text-to-speech functioning	Knowledge Limited	Creative Labs for MS-Dos	Education
1995	ALICE (Alicebot)	Heuristic pattern matching XML knowledge bases	Domain-Specific	Alice lacked clever traits. Couldn't come up with human-like responses, expressions, or attitudes	Richard Wallace	Entertainment
2001	SmarterChild (commercial instant messaging bot) [18]	Processing queries asked in natural language	Throughout the world's instant chat and SMS networks	Knowledge Limited	ActiveBuddy	Education
2010	Apple Siri (intelligent voice assistant) [23]	Natural Language UI Natural Language Processing	Responds to user requests using various internet services	It requires an internet connection. You can't use it when you are offline It is multilingual, but instructions are only in English. Doesn't understand all spelling variations It also has difficulties in hearing	Apple for iOS	Social Media(Post statuses or tweets to Facebook or Twitter reminders, driving directions and playing music)
2011	Watson[14]	Natural Language Knowledge Base	AI-powered virtual agent designed to provide customers with fast, consistent, and accurate answers across any messaging platform, application, device, or channel	It supports only English Slow integration	IBM	Weather forecast (Building codes, Weather forecasting, Fashion, Advertising, Healthcare,…etc.)

(*continued*)

Table 1. (*continued*)

Year	Chatbot	Technique	Scheme	Drawback	Authors	Application Domain
2012	Google Now/Google Assistant [45] [24]	Natural Language	predict actions by passing requests to a set of web services two-way dialogue with the user	Smart home capability is limited	Google Inch	Voice Assistant
2005	Mitsuku (Kuki or Pandorabot) [4]	AIML technology	Intelligent	Ability to reason with specific objects	Steve Worswick	
2014	Cortana [27]	Goal-Based	Perform tasks like reminders Available in different languages	It can launch software that distributes malware	Microsoft for windows-based devices	Use information from the Bing web search to make reminders, identify natural speech, and answer queries
2014	Alexa [1]	Knowledge-based & virtual assistant technology	Which is built-in home automation (IoT)	Sometimes, users' queries can be misunderstanding occur Cannot understand particular language (oral speech)	Amazon	Google Devices
2016	Bots for Messengers [14]	Rule-Based	Understand questions, provide answers, and execute tasks	Have limited responses and are not often able to answer multi-part questions or questions that require decisions	Facebook'	Social Media
2016	TAY [15]	Experimental artificial intelligence	Twitter to mimic the speech and habits of a teenage girl	Had to be shut down just 16 h after launch	Microsoft	Social Media

4 Methodology

PRISMA is an acronym for Preferred Reporting Items for Systematic Reviews and Meta-Analyses. It is a collection of evidence-based learned elements designed to assist narrates in reporting a wide range of systematic reviews and meta-analyses, which are primarily used to explore the benefits. PRISMA focuses on how writers can transparently present their research.

The criteria for inclusion were as shown in Fig. 4 shows how information passes across the several phases of systematic reviews using the PRISMA flow diagram. The PRISMA flow diagram shows the number of records identified, including rejected ones, and the explanations for exclusions. The following processes of the systematic literature review were presented to survey the present study in the field of chatbots.

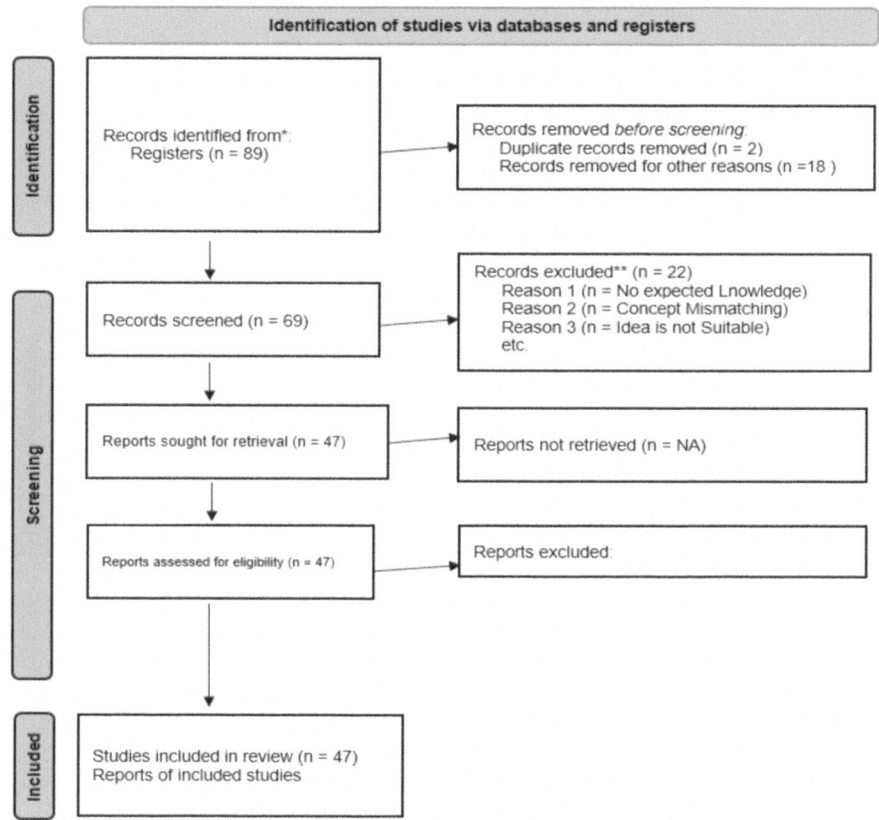

Fig. 4. PRISMA – Flow Diagram for the selected study. (PRISMA 2020 flow diagram for new systematic reviews which included searches of databases and registers only)

The selected research articles published in Books, Conferences, Generic, and Journals of the study were presented in Fig. 5.

As shown in Fig. 6 the diagrammatic representation of the previously published research articles that were analyzed for this study. The selected research articles are classified into Book Publications, Conference Publications, Generic, and Journal Publications year-wise, as four categories.

4.1 Identification

Using the PRISMA model, exclusively in line with the Identification step, eighty-nine research articles have been selected for the research study, of which two articles were removed for duplication and eighteen articles were eliminated because it is not fit for the

Fig. 5. Classification of Research Work for the study.

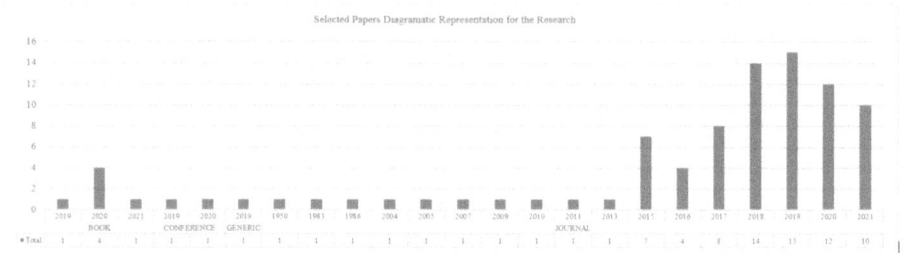

Fig. 6. Selected Research Work for the study based on the Year with type.

problem definition but all the eliminated articles can be used for the chatbot development process.

4.2 Screening

It initially records the number of articles found and then makes the selection process. In the initial screening step, 69 articles were screened of which 22 articles were excluded because the titles and abstracts were not relevant to the research or study. Chatbot and its design technique-related research articles are taken for further deeper study for this review.

4.3 Included

In the next process of screening, 47 articles were included for the final study based on the various eligibility for inclusion in the systematic review. Based on the study, Table 1 presents the comparison of chatbot generations with their techniques, schemes, drawbacks, and used domains. Selected research articles are tabulated and the categorization is also there for future researchers to look into in detail for further study.

5 Discussion/Results

The study of the literature review was primarily based on a PRISMA investigation model. This study looks at the history of chatbots and their use in many fields. In the last several years, there has been substantial development in the invention and the usage of the chatbot, with numerous advantages in a variety of fields such as education, industry, health care, research labs, E-Marketing platforms, Group Chat Apps, and advertising.

5.1 Research Gaps

Presently Intelligent Conversational Systems have a verbal intelligence that is constrained. The most significant problem that chatbots are now facing is maintaining track of the situation of the conversation and understanding the human inputs and the respective responses [4]. Another key issue of a chatbot is to examine the various aspects of emotional difficulties that instructors experience and their origins. The most essential step in the development plans of chatbots is to strengthen language understanding and interpretation [2]. Chatbots are intelligent systems but they cannot think by themselves based on learning, which is a drawback of the present technology in building chatbots. For an effective intelligent conversational system, chatbot development needs to be planned and the platforms used to implement chatbot needs to be chosen properly [18].

Chatbots are largely employed in the education domain – teaching and learning process. As a dialogue system, AI systems might be used to deliver course learning content via a digital platform, capable of providing appropriate information to the users. Most systems focus on pattern-matching techniques but semantic approaches and computer reasoning systems need to be incorporated for the future development part of an intelligent system. Moreover, inclusive knowledge base designing needs to be improved in the development of general-purpose chatbots [46].

For Teaching-Learning Process, in LMS new ad-hoc can be added to help students and the institution with a personalized way of dealing with [47]. The majority of chatbot developments are done only for goal-oriented [16] and AI-oriented tasks. NLP technologies need to be used for building AI-Chatbots in the future.

6 Conclusion

Chatbots have a higher success rate than humans in using messaging apps to reach out to a large audience and generate substantial cost savings. This study presents an overview of the fundamental concepts of chatbots. The current state of technology is in a rapid transition phase, intending to have equipment that can deliver solutions with minimum human intervention. This study examines several techniques to develop a chatbot, as well as the categorization and structure of these conversational bots. It also presents the design principles used to create many chatbots in chronological sequence.

The inability of today's modern Bots to recognize and produce actual discussions is the most serious issue they confront. At times, they would be unable to interpret a phrase, leading to inconsistency in communication and uncomfortable interaction with their listener. The most crucial phase in Chatbot development is to increase language understanding and output.

References

1. Adamopoulou, E., Moussiades, L.: An Overview of Chatbot Technology. In: Maglogiannis, I., Iliadis, L., Pimenidis, E. (eds.) Artificial Intelligence Applications and Innovations: 16th IFIP WG 12.5 International Conference, AIAI 2020, Neos Marmaras, Greece, June 5–7, 2020, Proceedings, Part II, pp. 373–383. Springer International Publishing, Cham (2020). https://doi.org/10.1007/978-3-030-49186-4_31
2. Adamopoulou, E., Moussiades, L.: Chatbots: History, technology, and applications. Mach. Learn. Appl. **2**, 100006 (2020). https://doi.org/10.1016/j.mlwa.2020.100006
3. Vaidyam, A.N., Wisniewski, H., Halamka, J.D., Kashavan, M.S., Torous, J.B.: Chatbots and conversational agents in mental health: a review of the psychiatric landscape. Can. J. Psychiatry **64**(7), 456–464 (2019). https://doi.org/10.1177/0706743719828977
4. Ramesh, K., Ravishankaran, S., Joshi, A., Chandrasekaran, K.: A survey of design techniques for conversational agents. Commun. Comput. Inf. Sci. **750**, 336–350 (2017). https://doi.org/10.1007/978-981-10-6544-6_31
5. Neff, G., Nagy, P.: Automation, algorithms, and politics| talking to bots: symbiotic agency and the case of tay. Int. J. Commun. **10**, 17 (2016)
6. Akma, N., Hafiz, M., Zainal, A., Fairuz, M., Adnan, Z.: Review of chatbots design techniques. Int. J. Comput. Appl. **181**(8), 7 (2018). https://doi.org/10.5120/ijca2018917606
7. Ciechanowski, L., Przegalinska, A., Magnuski, M., Gloor, P.: In the shades of the uncanny valley: an experimental study of human–chatbot interaction. Futur. Gener. Comput. Syst. **92**, 539–548 (2019). https://doi.org/10.1016/j.future.2018.01.055
8. Hill, J., Ford, W.R., Farreras, I.G.: Real conversations with artificial intelligence: A comparison between human–human online conversations and human–chatbot conversations. Compu. Hum. Behav. **49**, 245–250 (2015). https://doi.org/10.1016/j.chb.2015.02.026
9. Sameera, A., John, D.: Survey on chatbot design techniques in speech conversation systems. Int. J. Adv. Comput. Sci. App. **6**(7), 72–80 (2015). https://doi.org/10.14569/IJACSA.2015.060712
10. Review, A.Q.: MIND – A Quarterly Review of Psychology and Philosophy, pp. 433–460 (1950)
11. Pamungkas, E.W.: Emotionally-Aware Chatbots: A Survey. (2019). http://arxiv.org/abs/1906.09774
12. Weizenbaum, J.: ELIZA—a computer program for the study of natural language communication between man and machine. Commun. ACM **26**(1), 23–28 (1983). https://doi.org/10.1145/357980.357991
13. Shawar, B.A., Atwell, E.: Accessing an information system by chatting. In: Meziane, F., Métais, E. (eds.) Natural Language Processing and Information Systems, pp. 407–412. Springer Berlin Heidelberg, Berlin, Heidelberg (2004). https://doi.org/10.1007/978-3-540-27779-8_39
14. Swathi, N.S., Tejas, R.: Intelligent bot for novice reciprocal action: a survey. Int. J. Sci. Res. Comput. Sci. Eng. Inf. Technol. **4**(6), 149–155 (2018)
15. Zemčík, M.T.: A Brief History of Chatbots. DEStech Trans. Comput. Sci. Eng. (2019). https://doi.org/10.12783/dtcse/aicae2019/31439
16. Masche, J., Le, N.T.: A review of technologies for conversational systems. Adv. Intell. Syst. Comput. **629**, 212–225 (2018). https://doi.org/10.1007/978-3-319-61911-8_19
17. Candello, H., Pinhanez, C.: Designing Conversational Interfaces. no. May (2019)
18. Molnar, G., Szuts, Z.: The role of chatbots in formal education. In: SISY 2018 – IEEE 16th International Symposium Intelligent System Informatics, Proceedings, pp. 197–201 (2018). https://doi.org/10.1109/SISY.2018.8524609

19. Almansor, E.H., Hussain, F.K.: Survey on intelligent chatbots: state-of-the-art and future research directions. Adv. Intell. Syst. Comput. **993**, 534–543 (2020). https://doi.org/10.1007/978-3-030-22354-0_47
20. Lin, L., D'Haro, L.F., Banchs, R.E.: A web-based platform for collection of human-chatbot interactions. In: HAI 2016 – Proceedings of the 4th International Conference on Human Agent Interaction, pp. 363–366 (2016). https://doi.org/10.1145/2974804.2980500
21. Marietto, M.G.B., et al.: Artificial intelligence markup language: a brief tutorial. Int. J. Comput. Sci. Eng. Surv. **4**(3), 1–20 (2013). https://doi.org/10.5121/ijcses.2013.4301
22. Adamopoulou, E., Moussiades, L.: An overview of chatbot technology. In: Maglogiannis, I., Iliadis, L., Pimenidis, E. (eds.) AIAI 2020. IAICT, vol. 584, pp. 373–383. Springer, Cham (2020). https://doi.org/10.1007/978-3-030-49186-4_31
23. Hussain, S., Sianaki, O.A., Ababneh, N.: A survey on conversational agents/chatbots classification and design techniques. In: Barolli, L., Takizawa, M., Xhafa, F., Enokido, T. (eds.) WAINA 2019. AISC, vol. 927, pp. 946–956. Springer, Cham (2019). https://doi.org/10.1007/978-3-030-15035-8_93
24. Alexiadis, A., Nizamis, A., Koskinas, I., Ioannidis, D., Votis, K., Tzovaras, D.: Applying an intelligent personal agent on a smart home using a novel dialogue generator, vol. 584. Springer International Publishing, IFIP (2020)
25. Nimavat, K., Champaneria, T.: Chatbots: an overview types, architecture, tools and future possibilities. Int. J. Sci. Res. Dev. **5**(7), 1019–1026 (2017)
26. Smutny, P., Schreiberova, P.: Chatbots for learning: a review of educational chatbots for the Facebook Messenger. Comput. Educ. **151**, 103862 (2020). https://doi.org/10.1016/j.compedu.2020.103862
27. Eisman, E.M., Navarro, M., Castro, J.L.: A multi-agent conversational system with heterogeneous data sources access. Expert Syst. Appl. **53**, 172–191 (2016). https://doi.org/10.1016/j.eswa.2016.01.033
28. Haristiani, N.: Artificial intelligence (ai) chatbot as language learning medium: an inquiry. J. Phys. Conf. Ser. (2019). https://doi.org/10.1088/17426596/1387/1/012020
29. Shawar, B.A., Atwell, E.S.: Using corpora in machine-learning chatbot systems. Int. J. Corpus Linguist. **10**(4), 489–516 (2005). https://doi.org/10.1075/ijcl.10.4.06sha
30. Adam, M., Wessel, M., Benlian, A.: AI-based chatbots in customer service and their effects on user compliance. Electron. Mark. **31**(2), 427–445 (2021). https://doi.org/10.1007/s12525-020-00414-7
31. Jaya, N.H., Mahyan, N.R., Suhaili, S.M., Jambli, M.N., Solehah, W., Ahmad, W.: Development of conversational agent to enhance learning experience: case study In Pre (02), 124–130 (2021)
32. Deriu, J., et al.: Survey on evaluation methods for dialogue systems. Artif. Intell. Rev. **54**(1), 755–810 (2020). https://doi.org/10.1007/s10462-020-09866-x
33. Habib, F.A., Shakil, G.S., Iqbal, S.S.M., Sajid, S.T.A.: Self-diagnosis medical chatbot using artificial intelligence. In: Goyal, D., Chaturvedi, P., Nagar, A.K., Purohit, S.D. (eds.) Proceedings of Second International Conference on Smart Energy and Communication: ICSEC 2020, pp. 587–593. Springer Singapore, Singapore (2021). https://doi.org/10.1007/978-981-15-6707-0_57
34. Bonnevie, E., Lloyd, T.D., Rosenberg, S.D., Williams, K., Goldbarg, J., Smyser, J.: Layla's Got You: developing a tailored contraception chatbot for Black and Hispanic young women. Health Educ. J. **80**(4), 413–424 (2021). https://doi.org/10.1177/0017896920981122
35. Khadija, A., Zahra, F.F., Naceur, A.: AI-powered health chatbots: toward a general architecture. Procedia Comput. Sci. **191**, 355–360 (2021). https://doi.org/10.1016/j.procs.2021.07.048

36. Yin, J., Goh, T.T., Yang, B., Xiaobin, Y.: Conversation technology with micro-learning: the impact of chatbot-based learning on students' learning motivation and performance. J. Educ. Comput. Res. **59**(1), 154–177 (2021). https://doi.org/10.1177/0735633120952067

37. Brachten, F., Kissmer, T., Stieglitz, S.: The acceptance of chatbots in an enterprise context – A survey study. Int. J. Inform. Manag. **60**, 102375 (2021). https://doi.org/10.1016/j.ijinfomgt. 2021.102375

38. Gajra, V., Lakdawala, K., Bhanushali, R., Patil, S.: Automating student management system using ChatBot and RPA technology. SSRN Electron. J. (2020). https://doi.org/10.2139/ssrn. 3565321

39. Maher, S., Kayte, S., Nimbhore, S.: EasyChair Preprint Chatbots & Its Techniques Using AI: an Review (2020)

40. Agarwal, R., Wadhwa, M.: Review of state-of-the-art design techniques for chatbots. SN Comput. Sci. **1**(5), 1–12 (2020). https://doi.org/10.1007/s42979-020-00255-3

41. Biduski, D., Bellei, E.A., Rodriguez, J.P.M., Zaina, L.A.M., De Marchi, A.C.B.: Assessing long-term user experience on a mobile health application through an in-app embedded conversation-based questionnaire. Comput. Hum. Behav. **104**, 106169 (2020). https://doi.org/ 10.1016/j.chb.2019.106169

42. Zhou, L., Gao, J., Li, D., Shum, H.Y.: The design and implementation of xiaoice, an empathetic social chatbot. Comput. Linguist. **46**(1), 53–93 (2020). https://doi.org/10.1162/COLI_a_ 00368

43. Villegas-Ch, W., Arias-Navarrete, A., Palacios-Pacheco, X.: Proposal of an architecture for the integration of a chatbot with artificial intelligence in a smart campus for the improvement of learning. Sustainability **12**(4), 1500 (2020). https://doi.org/10.3390/su12041500

44. Song, Y., Te Li, C., Nie, J.Y., Zhang, M., Zhao, D., Yan, R.: An ensemble of retrieval-based and generation-based human-computer conversation systems. Int. Jt. Conf. Artif. Intell. **2018-July**, 4382–4388 (2018). https://doi.org/10.24963/ijcai.2018/609

45. Luke MacNeill, A., MacNeill, L., Doucet, S., Luke, A.: Professional representation of conversational agents for health care: a scoping review protocol. JBI Evid. Synth. **20**(2), 666–673 (2021). https://doi.org/10.11124/JBIES-20-00589

46. Okonkwo, C.W., Ade-Ibijola, A.: Chatbots applications in education: a systematic review. Comput. Educ. Artif. Intell. **2**, 100033 (2021). https://doi.org/10.1016/j.caeai.2021.100033

47. Shukla, V.K., Verma, A.: Enhancing LMS experience through AIML base and retrieval base Chatbot using R language. In: 2019 International Conference on Automation, Computational and Technology Management, ICACTM 2019, pp. 561–567 (2019). https://doi.org/10.1109/ ICACTM.2019.8776684

Unleashing the Power of Dynamic Mode Decomposition and Deep Learning for Rainfall Prediction in North-East India

Paleti Nikhil Chowdary[1], Sathvika Pingali[1], Pranav Unnikrishnan[1], Rohan Sanjeev[1], V. Sowmya[1(✉)], E. A. Gopalakrishnan[1,2], and M. Dhanya[3]

[1] Amrita School of Artificial Intelligence, Amrita Vishwa Vidyapeetham, Coimbatore, India
v_sowmya@cb.amrita.edu

[2] Department of Computer Science and Engineering, Amrita School of Computing, Amrita Vishwa Vidyapeetham, Bangalore, India

[3] Center for Wireless Networks and Applications (WNA), Amrita Vishwa Vidyapeetham, Amritapuri, India

Abstract. Accurate rainfall forecasting is crucial for effective disaster preparedness and mitigation in the North-East region of India, which is prone to extreme weather events such as floods and landslides. In this study, we investigated the use of two data-driven methods, dynamic mode decomposition (DMD) and long-short-term memory (LSTM), for rainfall forecasting using daily rainfall data collected from the India Meteorological Department in northeast region over a period of 122 years. We conducted a comparative analysis of these methods to determine their relative effectiveness in predicting rainfall patterns. Using historical rainfall data from multiple weather stations, we trained and validated our models to forecast future rainfall patterns. Our results indicate that both DMD and LSTM are effective in forecasting rainfall, with LSTM outperforming DMD in terms of accuracy, revealing that LSTM has the ability to capture complex nonlinear relationships in the data, making it a powerful tool for rainfall forecasting. The study reveals that the DMD method achieved Mean Squared Error (MSE) values ranging from 150.44 mm to 263.34 mm and Mean Absolute Error (MAE) values from 91.34 mm to 154.61 mm. In contrast, the Deep Learning (DL) approach, utilizing LSTM, demonstrated a normalized MAE value of 0.35 and a normalized RMSE value of 0.534. Our findings suggest that data-driven methods such as DMD and deep learning approaches like LSTM can significantly improve rainfall forecasting accuracy in the North-East region of India, helping to mitigate the impact of extreme weather events and enhance the region's resilience to climate change.

Keywords: DMD · LSTM · rainfall · forecasting · Data driven methods

© The Author(s), under exclusive license to Springer Nature Switzerland AG 2025
A. K. Bairwa et al. (Eds.): ICCAIML 2024, CCIS 2184, pp. 47–62, 2025.
https://doi.org/10.1007/978-3-031-71481-8_4

1   Introduction

List of Abbreviations

Abbreviation	Definition
DMD	Dynamic Mode Decomposition
LSTM	Long Short-Term Memory
DL	Deep Learning
RNN	Recurrent Neural Network
MAE	Mean Absolute Error
RMSE	Root Mean Square Error
IMD	India Meteorological Department
CNN	Convolutional Neural Network
SVD	Singular Value Decomposition

Rainfall is one of the most important climatic variables that affects various aspects of human life and natural ecosystems. Accurate rainfall forecasting is crucial for effective disaster preparedness and mitigation. The North-East region of India, commonly known as the "Seven Sisters" is a topographically diverse region with a unique mix of flora and fauna. However, it is also one of the most vulnerable regions in the world, with a high incidence of natural disasters, particularly floods and landslides. In this region, accurate rainfall forecasting is essential for effective disaster preparedness and mitigation, particularly in the face of increasing occurrences of extreme weather events.

The North-East region receives the highest annual rainfall in India [3], with its hilly terrain increasing its susceptibility to landslides and flash floods. The region also experiences cyclones and thunderstorms, which have the potential to cause extensive damage to infrastructure and agriculture [5]. Climate change has exacerbated the frequency and intensity of these weather events, leading to prolonged dry spells and erratic rainfall patterns, further exacerbating the challenges faced by the region.

Rainfall prediction methods in the past relied on empirical relationships between atmospheric variables (temperature, humidity, wind, and pressure) using statistical [17] and dynamical approaches involving computer simulations [12]. However, their effectiveness was limited due to complex interactions and uncertainty. Recently, there is increasing interest in using machine learning (ML) and data-driven techniques to enhance rainfall predictions [7].

The present work aims to build data driven models to predict rainfall in mm using monthly average rainfall data. The work involved implementing Deep Learning and Dynamic Mode Decomposition techniques and performing various experiments to determine the hyperparameters that yield the best results.

The study leverages a comprehensive dataset from the India Meteorological Department (IMD), spanning over a century, to train and validate these models. By comparing the performance of DMD and LSTM models in forecasting rainfall, we aim to establish a more reliable method for predicting weather patterns in this ecologically sensitive and disaster-prone region.

Our results demonstrate the effectiveness of both DMD and LSTM in rainfall forecasting, with LSTM showing superior performance due to its ability to discern complex nonlinear dependencies in the data. This research not only contributes to the field of meteorological science but also offers practical applications in disaster management and climate change mitigation for the North-East region of India.

The paper is structured as follows: Section 2 provides an overview of the existing literature, Sect. 3 outlines the proposed methodology, Sect. 4 presents the results and corresponding discussion, and Sect. 5 provides the conclusion.

2 Literature Review

The field of rainfall forecasting has witnessed significant advancements with the advent of deep learning-based approaches. A notable development in this realm is the creation of Tiny-RainNet by Zhang et al. [18], a hybrid network that melds Convolutional Neural Networks (CNNs) with Bi-directional Long Short-Term Memory (LSTM) networks, tailored for short-term rainfall prediction from radar images. While this model demonstrates efficacy, its primary limitation lies in its forecasting capability, which extends only one or two hours into the future.

Expanding upon this, recent studies have underscored the importance of incorporating geospatial information into deep learning models for rainfall forecasting. Men et al. [9], in their paper titled "Spatio-temporal Analysis of Precipitation and Temperature: A Case Study Over the Beijing-Tianjin-Hebei Region, China," introduced a deep learning framework that harmonizes both spatial and temporal data. This approach employs CNNs for extracting spatial features from rainfall data, complemented by LSTM networks for temporal analysis. Their model, when tested on an extensive dataset from the Beijing-Tianjin-Hebei region, exhibited superior performance compared to traditional forecasting methods.

In another stride forward, Luo et al. [8] presented PredRANN in their study "PredRANN: The Spatiotemporal Attention Convolution Recurrent Neural Network for Precipitation Nowcasting." This innovative model integrates CNNs and LSTM networks with a spatial-temporal attention mechanism, enhancing the focus on salient spatiotemporal features within rainfall data. The model's application to real-world rainfall datasets revealed that PredRANN surpasses existing deep learning models in accuracy and predictive capability.

These developments collectively underscore the burgeoning potential of deep learning and data-driven methods in enhancing rainfall prediction accuracy. By adeptly incorporating spatiotemporal features and location-specific data, these advanced methodologies outshine conventional forecasting techniques. The focus of this paper is to further explore and refine DL and DMD techniques for rainfall forecasting. Emphasis will be placed on fine-tuning the parameters within these models to optimize accuracy and reliability, particularly for applications in regions susceptible to extreme weather conditions.

3 Methodology

This section outlines the methodology adopted for the study, including the data acquisition process, model development, and training procedures.

3.1 Dataset

The dataset and training files utilized in this study are publicly available and can be accessed at the following GitHub repository: https://github.com/Nikhil-Paleti/rainfall-prediction. This repository contains data and resources essential for replicating the research findings or for further exploration and analysis.

Fig. 1. Selected Grid Points in North-East India.

The dataset for this study was sourced from the India Meteorological Department (IMD) [https://imdpune.gov.in/lrfindex.php], which provides gridded rainfall data across India at a spatial resolution of $0.25 \times 0.25°$, covering 122 years from 1901 to 2022 [11]. For the Dynamic Mode Decomposition (DMD) analysis, we utilized data from 1901 to 2018, focusing on the North-east region of India, defined by longitudes $89.75°$E to $98°$E and latitudes $21.75°$N to $30°$N (Fig. 1). This data was aggregated into monthly averages for 435 grid points, from which a subset covering a period of 10 years (120 months) was used to predict the average rainfall for the following 12 months.

For the Deep Learning (DL) analysis, data from 1951 to 2021 was used, specifically focusing on four key locations in the North-east region: Agartala, Guwahati, Imphal, and Itanagar. The monthly average rainfall data for these

Table 1. Geographical Coordinates of Locations for DL Analysis

Location	Latitude (N)	Longitude (E)
Agartala	23.8315	91.2868
Guwahati	26.1158	91.7086
Imphal	24.8170	93.9368
Itanagar	27.0844	93.6053

locations were extracted from the corresponding grid points in the IMD dataset (refer to Table 1 for geographical coordinates). The DL models were trained using a dataset divided into a training set of 681 months and a test set of 171 months, with each model predicting the rainfall for the next month based on varying input window lengths.

3.2 Data Driven Modelling

Data-driven modeling leverages data to develop precise predictions and insights in various fields, such as engineering and computer science. Unlike traditional methods, which rely heavily on theoretical assumptions, data-driven approaches are inherently flexible, adapting to diverse and complex scenarios. These models efficiently handle large datasets, aiding researchers in deciphering complex systems and enabling accurate predictions. In this study, we examine two specific data-driven techniques, Dynamic Mode Decomposition (DMD) and Deep Learning (DL), for forecasting time series rainfall data.

Dynamic Mode Decomposition (DMD). DMD is a data-driven technique employed to discern the underlying dynamic structures and patterns within complex, high-dimensional systems. The approach involves decomposing the dataset into a series of modes, each representing a specific spatial-temporal pattern of behavior. These modes are defined by the eigenvectors of a data-constructed matrix, with their respective eigenvalues indicating the temporal dynamics of each mode (Tu et al.) [16]. DMD's versatility is evidenced by its applications in diverse fields, ranging from identifying rice leaf diseases [6] to defect detection in cantilever beams [10]. The growing popularity of DMD across various domains highlights its potential to significantly enhance our understanding of a myriad of complex phenomena. As such, DMD stands out as an increasingly valuable method in the realm of data-driven analysis, offering insights and predictions that were previously unattainable with traditional modeling approaches.

Our workflow for DMD analysis is presented in Fig. 2.

Algorithm 1. DMD Procedure

1: Collect time series data from the system to be analyzed, comprising measurements of the system's state variables.
2: Construct the data matrix $\mathbf{X} = [\mathbf{x}_1, \mathbf{x}_2, \ldots, \mathbf{x}_N]$, where each $\mathbf{x}_i \in \mathbb{C}^n$ represents a snapshot of the system's state at a specific time.
3: Perform SVD on $\mathbf{X}$:
   – $\mathbf{X} = \mathbf{U}\boldsymbol{\Sigma}\mathbf{V}^*$, where $\mathbf{U}$ is a matrix of left singular vectors, $\boldsymbol{\Sigma}$ is a diagonal matrix of singular values, and $\mathbf{V}$ is a matrix of right singular vectors.
4: Construct the DMD modes $\boldsymbol{\Phi}$ using SVD results:
   – $\boldsymbol{\Phi} = \mathbf{X}'\mathbf{V}\boldsymbol{\Sigma}^{-1}\mathbf{U}^*$, where $\mathbf{X}'$ is $\mathbf{X}$ with the last snapshot removed.
5: Calculate the eigenvalues $\boldsymbol{\Lambda}$ of the DMD modes:
   – $\boldsymbol{\Lambda} = \mathbf{V}\boldsymbol{\Sigma}^{-1}\mathbf{U}^*\mathbf{X}'\mathbf{X}$, representing frequencies and growth rates.
6: Reconstruct system dynamics using DMD modes and eigenvalues:
   – $\mathbf{x}(t) \approx \sum_{k=1}^{r} \boldsymbol{\phi}_k e^{\omega_k t} b_k$, where $\boldsymbol{\phi}_k$ and ω_k are the k-th DMD mode and eigenvalue, r is the number of significant modes, and b_k are coefficients from initial conditions.

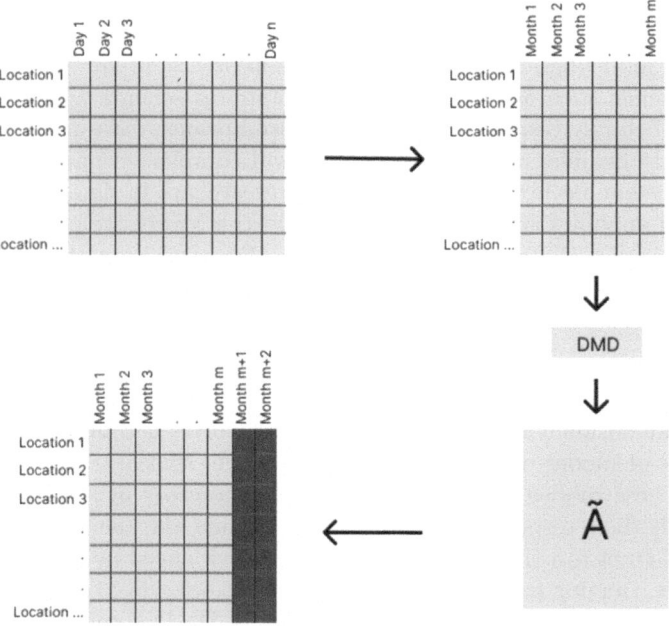

Fig. 2. Applying DMD to Rainfall data

Deep Learning. DL, a branch of machine learning, leverages artificial neural networks to process and learn from large datasets. Its key strengths include automatic feature extraction and proficiency in handling complex data types such as images, speech, and language, thereby eliminating the need for manual intervention. DL has shown remarkable performance in areas like image recognition, speech processing, natural language processing (NLP), and autonomous driving systems.

A prominent type of neural network used in DL is the RNN. RNNs are specially designed to process sequential data by maintaining a memory of previous inputs. This feature makes RNNs ideal for applications in speech recognition, NLP, and time series forecasting. For instance, RNN-based LSTM models have been employed to predict COVID-19 trends across various Indian states [13], and combined RNN and CNN models have been utilized for stock price prediction [15].

However, RNNs have limitations, such as the vanishing gradient problem. This issue arises when the gradients, essential for updating the network's weights, diminish through layers, impacting the network's ability to learn long-term dependencies. Additionally, RNNs are prone to overfitting, where they memorize the training data rather than learning to generalize.

To address these challenges, researchers have developed advanced techniques like LSTM (Hochreiter & Schmidhuber) [4] and GRU (Chung et al.) [2] networks, which incorporate gating mechanisms to better retain information over extended sequences. Furthermore, strategies like gradient clipping and weight regularization (Pascanu et al.) [14] have been introduced to improve training stability and performance.

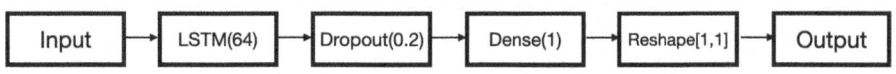

Fig. 3. Model Architecture

LSTM networks, a variant of RNN, are designed to overcome the vanishing gradient problem. LSTMs comprise memory cells that selectively store or discard information over time and are regulated by input, forget, and output gates. These gates control the flow of information, enabling the network to make more nuanced decisions about what to retain or forget.

The proposed LSTM model architecture, depicted in Fig. 3, comprises several layers:

- An LSTM layer with 64 units, designed to capture long-term dependencies in sequential data.
- A Dropout layer with a rate of 0.2, to mitigate overfitting.
- A Dense layer with a single unit, initialized with zeros.
- A Reshape layer to format the output shape to [1, 1].

This architecture is optimized for processing sequential data and includes regularization techniques to enhance generalization.

As illustrated in Fig. 4, the LSTM model's workflow begins with normalizing the monthly average rainfall data and dividing it into time-windowed sequences. These sequences are then used to train the LSTM network. Post-training, the model can predict future rainfall using historical data windows.

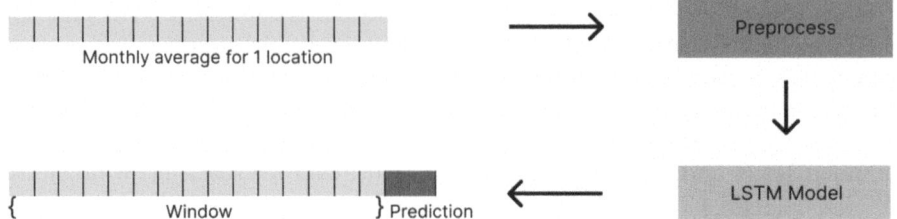

Fig. 4. Deep Learning Approach to Rainfall Data

3.3   Model Training

Dynamic Mode Decomposition (DMD). Our DMD analysis focused on rainfall data from the India Meteorological Department, spanning 1901 to 2018, with a specific emphasis on North-east India. The methodology involved the following key steps:

1. **Data Collection and Preprocessing**: We segmented the IMD dataset into monthly intervals using Python scripts, targeting the North-east region of India based on geographical coordinates.
2. **Matrix Construction**: A data matrix $\mathbf{X}$ was constructed, representing monthly rainfall over a 10-year period (120 months). This matrix encapsulated the temporal dynamics of rainfall for the entire region.
3. **Singular Value Decomposition (SVD)**: SVD was applied to $\mathbf{X}$ to identify dominant rainfall patterns, a crucial step for distilling key dynamics from the complex data.
4. **Forecasting Analysis**: Using the DMD algorithm, we forecasted future rainfall based on the historical 10-year dataset, calculating DMD modes, eigenvalues, and a dynamic matrix for future state prediction.
5. **Performance Evaluation**: The model's accuracy was assessed using RMSE and MAE.
6. **Visualization**: Actual vs. predicted rainfall data were visualized for key locations to assess the model's accuracy and understand temporal rainfall patterns.

This methodology enabled effective application of DMD to rainfall prediction, offering insights into the meteorological dynamics of North-east India.

Deep Learning (DL) Analysis. For the Deep Learning (DL) analysis, we utilized Recurrent Neural Networks (RNNs), specifically Long Short-Term Memory (LSTM) networks, to model and predict rainfall data. This analysis focused on key locations in North-east India, with the data spanning from 1951 to 2021. The LSTM model was chosen for its ability to handle sequential data and its robustness in capturing long-term dependencies.

Data Preprocessing: The data was first standardized to form monthly averages. The dataset was then split into training and testing sets, with the training set comprising 80% of the data.

Model Training: Our LSTM model training involved a series of experimental setups with varying parameters. The core architecture consisted of multiple layers, including an LSTM layer and a Dropout layer. To optimize the model's performance, we experimented with different configurations, varying the number of hidden units in the LSTM layer, the dropout rate, the input and output window sizes, and the choice of optimizer. The training involved adjusting these parameters to find the most effective combination for our specific dataset and objectives.

Model Evaluation: The performance of the model was assessed using metrics such as Mean Absolute Error (MAE) and Root Mean Squared Error (RMSE). Additionally, various plots were generated to visualize the training process and compare the predicted rainfall against actual data.

Results Visualization: The model's predictions were visualized for individual locations, providing an intuitive understanding of the model's performance over time. This included a comparison of the predicted and actual rainfall data.

This DL approach, centered around LSTM networks, offered a powerful tool for forecasting rainfall, demonstrating the potential of advanced neural networks in meteorological data analysis.

Evaluation Metrics. For model evaluation, we used two primary metrics:

1. **Root Mean Square Error (RMSE)**: A key metric for regression models, RMSE measures the average magnitude of the error. It is calculated as the square root of the mean squared differences between predicted and actual values. Lower RMSE values indicate better model performance. The formula is given by:

$$\text{RMSE} = \sqrt{\frac{1}{n}\sum_{i=1}^{n}(y_i - \hat{y}_i)^2}$$

where n is the number of observations, y_i is the actual value, and $\hat{y}_i$ is the predicted value for the i-th observation.

2. **Mean Absolute Error (MAE)**: MAE calculates the average absolute error between predicted and actual values. It's particularly useful in datasets with high variability or outliers. The formula for MAE is:

$$\text{MAE} = \frac{1}{n}\sum_{i=1}^{n}|y_i - \hat{y}_i|$$

Again, n represents the number of observations, with y_i and $\hat{y}_i$ as the actual and predicted values, respectively.

4   Results and Discussion

This section discusses the results obtained using Deep Learning and Dynamic Mode Decomposition (DMD) techniques to predict rainfall. Our results showed that both methods were able to accurately predict rainfall demonstrating the potential for using advanced computational methods to improve weather forecasting and better prepare for extreme weather events.

4.1   DMD

Table 2 shows results from Dynamic Mode Decomposition (DMD) experiments on a rainfall data matrix, constructed from 10 years of data, with predictions made for a single year. DMD was conducted at various projection ranks to obtain a low-dimensional representation of the data. The RMSE and MAE values indicate reasonably accurate rainfall predictions, ranging from 150.44 mm to 263.34 mm and 91.34 mm to 154.61 mm, respectively. Notably, the best performance was observed for the data from 1995–2005.

Table 2. DMD Results for forecasting 1 year

Start Year	Stop Year	Predicted Duration	Rank	RMSE	MAE
1929	1939	1 year	106	263.3423	154.6123
1941	1951	1 year	123	260.2758	144.6304
1954	1964	1 year	127	177.3236	107.8259
1973	1983	1 year	128	170.6351	109.3051
1995	2005	1 year	118	150.4379	91.3362
2000	2010	1 year	100	236.3857	124.9998
2005	2015	1 year	123	158.5830	97.9195

The rainfall prediction results for the year 2016, for Agartala, Guwahati, Imphal, and Itanagar, were obtained by constructing a matrix X from rainfall data between 2005 and 2015, using a projection rank of 123. The forecasted precipitation for the entire year of 2016 is shown in Fig. 5. The prediction performance was evaluated using the Mean Squared Error (MSE) and Root Mean Square Error (RMSE), which were calculated to be 97.9195 and 158.5830, respectively.

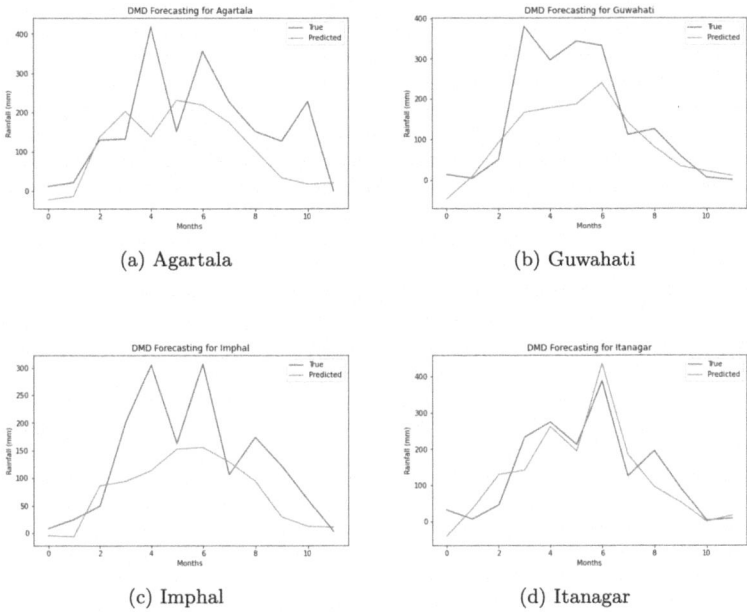

Fig. 5. Predictions for regions of North East India in 2016 using DMD

4.2   Deep Learning

This section presents the results of our time series prediction model for rainfall in four Indian cities: Itanagar, Guwahati, Agartala, and Imphal. The performance of our model for each city was assessed based on various experimental setups involving different optimizers, input/output window sizes, and dropout rates.

Itanagar. For Itanagar, the best performing optimizer and parameter combination for MAE was Nadam with an input window size of 13, an output window size of 1, and a dropout of 0.2, yielding an MAE value of 0.3707. For RMSE, the best performance was achieved by AdamW with an input window size of 14, an output window size of 1, and a dropout of 0.2, resulting in an RMSE of 0.4527 (Table 3). Figure 6 illustrates the model's predictions compared to the actual data on the test set.

Imphal. The analysis for Imphal indicated that the Nadam optimizer with an input window size of 14, an output window size of 1, and a dropout rate of 0.2 outperformed other settings, achieving the lowest MAE and RMSE values (Table 4). The model's predictive performance on the test set is depicted in Fig. 7.

Guwahati. For Guwahati, the optimal results were obtained using the Nadam optimizer, with an input window size of 13, an output window size of 1, and a

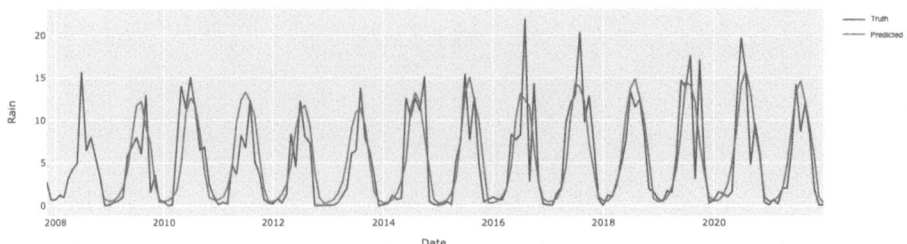

Fig. 6. Prediction Visualisation for Itanagar

Table 3. Performance Analysis of the proposed method for Itanagar

Optimiser	Input Window	Output Window	Dropout	Metrics	
				MAE	RMSE
AdamW	13	1	0	0.3637	0.5285
			0.2	0.3570	0.5249
	14	1	0	0.3650	0.5356
			0.2	0.3612	0.4527
	15	1	0	0.3792	0.5505
			0.2	0.3812	0.5459
Nadam	**13**	**1**	**0**	**0.3526**	**0.5260**
			0.2	0.3550	0.5258
	14	1	0	0.3569	0.5344
			0.2	0.3695	0.5372
	15	1	0	0.3670	0.5412
			0.2	0.3707	0.5467

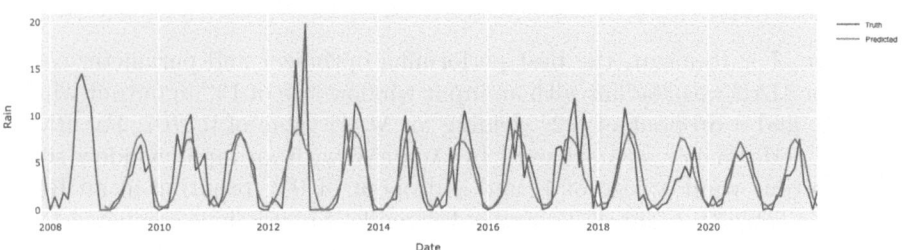

Fig. 7. Prediction Visualisation for Imphal

dropout rate of 0.2 (Table 5). The performance visualization of the model on the test dataset is shown in Fig. 8.

Agartala. In the case of Agartala, the combination of Nadam optimizer, an input window size of 14, an output window size of 1, and a dropout rate of 0

Table 4. Performance Analysis of the proposed method for Imphal

Optimiser	Input Window	Output Window	Dropout	Metrics	
				MAE	RMSE
AdamW	13	1	0	0.4422	0.6505
			0.2	0.4421	0.6508
	14	1	0	0.4448	0.6533
			0.2	0.4477	0.6561
	15	1	0	0.4513	0.6593
			0.2	0.4519	0.6583
Nadam	13	1	0	0.4434	0.6534
			0.2	0.4378	0.6449
	14	**1**	0	0.4424	0.6490
			0.2	**0.4357**	**0.6419**
	15	1	0	0.4381	0.6446
			0.2	0.4423	0.6502

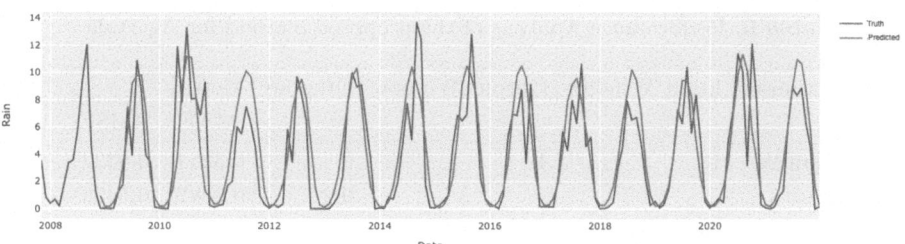

Fig. 8. Prediction Visualisation for Guwahati

Table 5. Performance Analysis of the proposed method for Guhawati

Optimiser	Input Window	Output Window	Dropout	Metrics	
				MAE	RMSE
AdamW	13	1	0	0.3402	0.4763
			0.2	0.3368	0.4827
	14	1	0	0.3341	0.4707
			0.2	0.3374	0.4668
	15	1	0	0.3351	0.4746
			0.2	0.3371	0.4757
Nadam	**13**	**1**	0	0.3264	0.4665
			0.2	**0.2990**	**0.4410**
	14	1	0	0.3115	0.4555
			0.2	0.3091	0.4559
	15	1	0	0.3126	0.4695
			0.2	0.3126	0.4525

was found to be most effective, as per Table 6. Figure 9 showcases the model's predictive accuracy on the test data for Agartala.

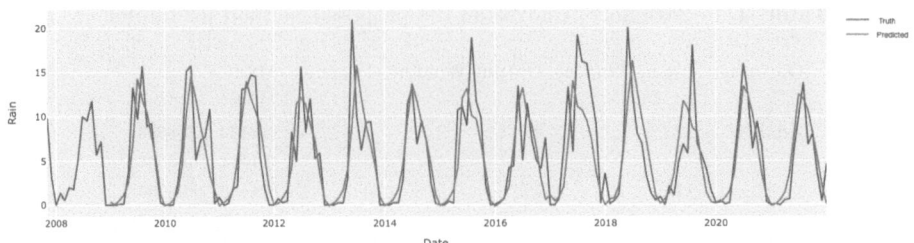

Fig. 9. Prediction Visualisation for Agartala

Each city's data analysis provides unique insights into the model's performance and helps refine the prediction strategy for different geographical locations.

Table 6. Performance Analysis of the proposed method for Agartala

Optimiser	Input Window	Output Window	Dropout	Metrics	
				MAE	RMSE
AdamW	13	1	0	0.3284	0.5322
			0.2	0.3887	0.5436
	14	1	0	0.3859	0.5394
			0.2	0.3839	0.5388
	15	1	0	0.3901	0.5451
			0.2	0.3820	0.5324
Nadam	13	1	0	0.3786	0.5314
			0.2	0.3709	0.5227
	14	**1**	**0**	**0.3698**	**0.5271**
			0.2	0.3770	0.5348
	15	1	0	0.3804	0.5388
			0.2	0.3840	0.5396

5   Conclusion and Future Scope

In conclusion, this study highlights the efficacy of Dynamic Mode Decomposition (DMD) and Long Short-Term Memory (LSTM) neural networks in predicting rainfall. The LSTM model particularly excels due to its proficiency in capturing long-term patterns and spatial-temporal dependencies, outshining the DMD approach in forecasting critical rainfall events. This capability is vital for providing early flood warnings, crucial for disaster preparedness and mitigation efforts, especially in vulnerable regions.

5.1 Future Scope

Looking ahead, two main avenues present themselves for further research:

– **Daily Forecasting:** Future endeavors could focus on refining the models for daily rainfall forecasting. This would involve tackling more detailed data and complex patterns, but could lead to more precise and actionable predictions, essential for effective short-term emergency response and planning.
– **Conformal Prediction:** Implementing conformal prediction methodologies may enhance the utility of the rainfall forecasts. This approach would estimate the confidence levels of the model's predictions, adding a layer of reliability that is particularly important in risk management and decision-making within meteorological contexts.

To summarize, the study offers significant advancements in rainfall prediction methods, with LSTM models showing particular promise in early warning systems. Future research in daily forecasting and conformal prediction is expected to refine these models further, increasing their applicability in weather forecasting and disaster management.

Acknowledgments. An earlier version of this work is available as a preprint on arXiv (arXiv:2309.09336) [1].

References

1. Chowdary, P.N., et al.: Unleashing the power of dynamic mode decomposition and deep learning for rainfall prediction in north-east India (2023)
2. Chung, J., Gülçehre, Ç., Cho, K., Bengio, Y.: Empirical evaluation of gated recurrent neural networks on sequence modeling. CoRR **abs/1412.3555** (2014)
3. Dikshit, K.R., Dikshit, J.K.: Weather and climate of north-east India (2014). https://api.semanticscholar.org/CorpusID:127226018
4. Hochreiter, S., Schmidhuber, J.: Long short-term memory. Neural Comput. **9**(8), 1735–1780 (1997)
5. Jain, S., Kumar, V., Saharia, M.: Analysis of rainfall and temperature trends in northeast India. Int. J. Climatol. **33** (2013). https://doi.org/10.1002/joc.3483
6. Sudhesh, K.M., Sowmya, V., Kurian, S., Sikha, O.K.: Ai based rice leaf disease identification enhanced by dynamic mode decomposition. Eng. Appl. Artif. Intell. **120**, 105836 (2023). https://doi.org/10.1016/j.engappai.2023.105836, https://www.sciencedirect.com/science/article/pii/S0952197623000209
7. Liyew, C.M., Melese, H.A.: Machine learning techniques to predict daily rainfall amount. J. Big Data **8**, 1–11 (2021)
8. Luo, C., Zhao, X., Sun, Y., Li, X., Ye, Y.: PredRANN: the spatiotemporal attention convolution recurrent neural network for precipitation nowcasting. Knowl.-Based Syst. **239**, 107900 (2022)
9. Men, B., Wu, Z., Liu, H., Tian, W., Zhao, Y.: Spatio-temporal analysis of precipitation and temperature: a case study over the Beijing-Tianjin-Hebei region, China. Pure Appl. Geophys. **177**, 3527–3541 (2020)

10. Nagarajan, K., Ananthu, J., Menon, V.K., Soman, K.P., Gopalakrishnan, E.A., Ramesh, A.: An approach to detect and classify defects in cantilever beams using dynamic mode decomposition and machine learning. In: Reddy, A.N.R., Marla, D., Simic, M., Favorskaya, M.N., Satapathy, S.C. (eds.) Intelligent Manufacturing and Energy Sustainability. SIST, vol. 169, pp. 731–738. Springer, Singapore (2020). https://doi.org/10.1007/978-981-15-1616-0_71

11. Pai, D., Rajeevan, M., Sreejith, O., Mukhopadhyay, B., Satbha, N.: Development of a new high spatial resolution (0.25° × 0.25°) long period (1901-2010) daily gridded rainfall data set over India and its comparison with existing data sets over the region. MAUSAM **65**(1), 1-18 (2014). https://doi.org/10.54302/mausam.v65i1. 851, https://mausamjournal.imd.gov.in/index.php/MAUSAM/article/view/851

12. Palmer, T.N.: Towards the probabilistic earth-system simulator: a vision for the future of climate and weather prediction. Q. J. R. Meteorol. Soc. **138**(665), 841–861 (2012). https://doi.org/10.1002/qj.1923, https://rmets.onlinelibrary.wiley.com/doi/abs/10.1002/qj.1923

13. Pandianchery, M.S., Sowmya, V., Gopalakrishnan, E., Soman, K.: Long short-term memory-based recurrent neural network model for covid-19 prediction in different states of india. AI, IoMT, and Analytics, Emerging Technologies for Combatting Pandemics (2022)

14. Pascanu, R., Mikolov, T., Bengio, Y.: On the difficulty of training recurrent neural networks. In: Dasgupta, S., McAllester, D. (eds.) Proceedings of the 30th International Conference on Machine Learning. Proceedings of Machine Learning Research, vol. 28, pp. 1310–1318. PMLR, Atlanta, Georgia, USA (2013)

15. Selvin, S., Vinayakumar, R., Gopalakrishnan, E., Menon, V.K., Soman, K.: Stock price prediction using LSTM, RNN and CNN-sliding window model. In: 2017 International Conference on Advances in Computing, Communications and Informatics (ICACCI), pp. 1643–1647. IEEE (2017)

16. Tu, J.H., Rowley, C.W., Luchtenburg, D.M., Brunton, S.L., Kutz, J.N.: On dynamic mode decomposition: theory and applications (2014)

17. Zaw, W.T., Naing, T.T.: Empirical statistical modeling of rainfall prediction over Myanmar. World Acad. Sci. Eng. Technol. **2**(10), 500–504 (2008)

18. Zhang, C.J., Wang, H.Y., Zeng, J., Ma, L.M., Guan, L.: Tiny-rainnet: a deep convolutional neural network with bi-directional long short-term memory model for short-term rainfall prediction. Meteorol. Appl. **27**(5), e1956 (2020)

Patient's Condition Categorization Using Drug Reviews

Akshit Kamboj, Shikha Mundra$^{(\boxtimes)}$, and Ankit Mundra

Manipal University Jaipur, Jaipur, Rajasthan, India
a.shikha1990@gmail.com

Abstract. In the modern era, there has been a substantial increase in health-related concerns among individuals. The prevalence of various diseases has surged significantly, underscoring the crucial necessity for timely disease detection. A strategy employed for this purpose involves the examination of drug evaluations from diverse patient groups. Within the realm of healthcare, reviews assume a pivotal role. They offer insights into the efficacy of different drugs and their potential adverse effects. The primary objective is to categorize a patient's health condition based on their feedback pertaining to a specific medication. In this endeavor, a system for classifying medical conditions has been developed, utilizing patient reviews related to diverse medications. This system has the ability to forecast the specific ailment afflicting a patient. To achieve this objective, the Drug Review Dataset from UCI's Machine Learning Repository has been employed. To extract relevant features, multiple vectorization techniques have been applied, including Bag of Words, TF-IDF, TF-IDF (Bigrams), and TF-IDF (Trigrams). To execute the necessary categorization, a range of machine learning classification algorithms have been harnessed, such as Decision Tree, Multinomial Naive Bayes, Logistic Regression, and Passive Aggressive Classifier.

Keywords: Statistical Features · Depression · BoW · TF-IDF · Patient Review

1 Introduction

Over the passage of time, the internet has undergone an astounding transformation. In its earlier days, it merely served as a platform for individuals to promote their businesses. However, the landscape of the internet has since metamorphosed into a dynamic arena where a plethora of services and products are assessed by consumers based on the feedback they glean from fellow consumers. This practice of utilizing reviews and feedback to gauge the caliber of a product or service is an approach that can also find application within the realm of healthcare. In recent times, the proliferation of online reviews has given rise to novel domains within the realms of communication and marketing. This, in turn, has effectively bridged the chasm between feedback that attains a viral status and traditional word-of-mouth dissemination. Such a convergence has proven to be a substantial influence on consumer perceptions. The internet's evolution from a simple promotional space to a multifaceted platform for consumer evaluation has engendered a transformative impact, transcending industries and sectors, including healthcare, thereby underscoring the formidable role of feedback and reviews in shaping contemporary perspectives.

1.1 Predicting Medical Conditions Using Patient Feedback and Machine Learning

Within the realm of Within the realm of healthcare, drug evaluations across various types of medications can assume heightened significance. These assessments offer assistance through various avenues. To illustrate, they enable the monitoring of unfavorable responses linked to diverse drugs, identification of prevailing sentiments surrounding specific medications, and acknowledgment of the range of ailments addressable by a given drug. In online patient communities, articulating the medical context for which a medication is being employed holds paramount importance [1]. This stems from the fact that numerous drugs possess the potential to alleviate multiple conditions. This research aims to conduct a comprehensive analysis of drug reviews for the purpose of predicting specific medical conditions in individual patients. These reviews originate from diverse patients discussing various medications on different platforms dedicated to medicine feedback. The core objective of this project is to construct a classification system for medical conditions, utilizing patients' reviews associated with different drugs [2]. This system will possess the capability to anticipate the underlying medical condition afflicting a given patient. To accomplish this goal, the Drug Review Dataset sourced from UCI's Machine Learning Repository will serve as the primary dataset. The process of feature extraction will involve employing several vectorization techniques, including Bag of Words, TF-IDF, GloVe, and Word2Vec as discussed in [8] and [9]. The subsequent step entails employing a range of machine learning classification algorithms to fulfill the necessary categorization tasks. Notably, algorithms such as Decision Tree, Multinomial Naive Bayes, Logistic Regression, and Passive Aggressive Classifiers will be leveraged for this purpose.

2 Literature Review

In most of the prior study, machine learning models are experimented based on the symptoms of the patient. Shatha Melhem's et al. [1] research focuses on harnessing the capabilities of machine-learning models to effectively categorize patient care into two distinct classes: inpatient and outpatient. This innovative approach has yielded remarkable reductions in time and effort required for such categorization, thereby facilitating swift and accurate determination of the most suitable services for individual patients. The consequential decrease in the potential risk to patients' lives arises from the notable mitigation of human errors. To achieve these goals, the study employs four distinct machine learning models: Decision Tree, Support Vector Machine, K-Nearest Neighbours (KNN), and Random Forest. These models facilitate the meticulous analysis of a patient's unique combination of lab test results and overall medical condition. The comparison of these models is based on a predefined set of criteria, encompassing parameters like accuracy, specificity, sensitivity, and precision score. The datasets used for training these models originate from a hospital in Indonesia and encompass a variety of patient cases, featuring their corresponding laboratory test outcomes. Among these models, the Random Forest Model emerges as the most effective performer. Demonstrating its superiority, this model achieves an accuracy rate of 77%, a sensitivity rate of 65%, and a precision rate of 72 Samiullah Shaikh's et al. [2] paper endeavors to introduce a comprehensive dataset encompassing intricate profiles of approximately 12,000 patients. These profiles are intricately linked to the 20 most frequently encountered diseases. A significant aspect of this paper involves the facilitation of disease classification. This categorization process is facilitated through the application of diverse machine learning models, which predominantly follow a supervised learning approach. Incorporating the principles of Evidence-Based Medicine (EBM), the paper seeks to embody a decision-making framework rooted in patient records, clinical expertise, and available evidence. This approach entails leveraging previous patient records to inform medical professionals' prescription decisions. The rigorous experimentation within this study relies on the Monte Carlo method. Among the array of methods employed, the Random Forest Trees (RFT) technique emerges as the optimal performer. This model attains an impressive accuracy rate of approximately 83%, accentuating its efficacy in disease classification. Angelower Santana-Velásquez et al. [3] study endeavors to supplant conventional Diagnosis-Related Groups (DRGs) classification techniques with advanced machine learning methodologies. The primary objective of this research is to assess the viability of machine learning models in categorizing patients according to the established DRGs standards. This assessment will hinge on utilizing available patient-related data. The insights garnered from this investigation will serve as a foundational reference for future analyses. Specifically, these insights will be juxtaposed with the early-stage predictions rendered by DRGs during patients' hospitalization. The outcomes of this research highlight that by employing Ensemble methods and Artificial Neural Networks for DRGs classification, remarkable accuracy levels of up to 96% can be attained. This achievement has significantly bolstered the efficiency of medical treatments and procedures for patients. Importantly, these advancements have yielded substantial cost reductions without compromising the quality or duration of the patients' hospital stays. The primary objective of the research conducted by Pushpak Bhattacharyya et al. [6] is to provide comprehensive insights into the diverse dimensions of sentiment analysis. This encompasses an exploration of both machine learning-based and knowledge-based methodologies. An additional facet of this research pertains to the extension of sentiment analysis to Indian languages, thereby catering to linguistic diversity. Through this investigation, a significant observation emerged: working with senses, as opposed to isolated words, proves notably advantageous in terms of accuracy, particularly in the context of multilingual scenarios. The research also presents noteworthy findings concerning sarcasm and thwarting, providing valuable insights into the challenges posed by these linguistic nuances. Yogesh Chandra's et al. [4] paper is directed towards leveraging a diverse array of methods to facilitate sentiment analysis. The focal point of this research involves the utilization of machine learning classifiers to achieve the task of sentiment analysis. This entails the incorporation of Deep Learning models and polarity-based sentiment analysis to discern whether a given user's tweet carries a positive or negative sentiment. To accomplish this categorization, a range of model architectures have been harnessed, thereby accommodating the variations in thoughts and opinions prevalent across distinct social media platforms. The outcomes derived from this study bear substantial relevance for Government and Defense Organizations. By harnessing the insights garnered, these entities gain the capacity to assume control and make informed decisions about their subsequent course of action. Ultimately, this research contributes to enabling more informed and strategic decision-making based on sentiment analysis across social media content. The study conducted by Hao Yang et al. [7] focuses on the detection of malware through the utilization of the TF-IDF model. A notable advantage of employing Natural Language Processing (NLP) techniques over alternative machine learning algorithms is the heightened accuracy that NLP can achieve post-data processing. The study proceeds by constructing subsequent models, incorporating classifiers such as the Random Forest classifier, Gradient Boosting classifier, AdaBoost classifier, and ensemble models. In parallel, the study leverages a Convolutional Neural Network (CNN) for training, drawn to its efficiency in extracting valuable data insights. The TF-IDF model serves a central role, determining the TF-IDF value attributed to each keyword within a given traffic packet, without necessitating the packet's division. The dataset encompasses a total of 3000 data packets, equally split between black and white data (1500 each). After keyword extraction and the reconstruction of the dataset using the TF-IDF model, diverse classifiers are trained. Encrypted traffic data undergoes One-Hot Encoding before being fed into the classifiers and the CNN network for training and subsequent detection. The study underscores the potential of combining machine learning and deep learning techniques as a viable solution for identifying encrypted malware traffic data. Demonstrating a detection accuracy of 93% using the TF-IDF-based model for encrypted traffic malware detection, the research presents promising results. This accuracy is anticipated to advance further through continued research and development in this field.

3   Proposed Methodology

In this work, Bag of Words (BoW), n-gram along with Tf-idf are employed to identify the patient category from the drug reviews. In BoW, within this model, the frequency of distinct words within a document is tallied. However, this methodology disregards the sequence in which the words are presented in the document. Consequently, the bag of words approach is characterized as contextually agnostic. In this scheme, the value "1" signifies the presence of a word in the document, while "0" indicates its absence. This model works in the following manner:

1. Vocabulary Creation: The first step is to create a vocabulary, which is a unique set of words present in the entire corpus of documents (collection of texts) you're working with. Each word is treated as a unique token, and no consideration is given to the grammatical or semantic meaning of the words at this stage.
2. Tokenization: In this step, each document in the corpus is divided into individual words or tokens. Punctuation and capitalization are usually removed, and words are typically converted to lowercase to ensure consistent representation.
3. Word Counting: For each document, the BoW model counts the occurrences of each word in the vocabulary. This information is then organized into a numerical vector, where each entry corresponds to a specific word in the vocabulary, and the value represents the frequency of that word in the document. If a word doesn't appear in the document, its corresponding entry in the vector will have a value of zero.
4. Vectorization: Once the word counts are obtained for each document, the documents are represented as vectors in a high-dimensional space. The length of the vector is equal to the size of the vocabulary, and each dimension of the vector corresponds to a word in the vocabulary. The value in each dimension represents how many times the corresponding word appears in the document.
5. Matrix Representation: The collection of these vectors for all documents forms a matrix, known as the BoW matrix. Rows of this matrix represent individual documents, while columns represent the words in the vocabulary. Since the vocabulary is typically large and most documents will only contain a small subset of the words, the BoW matrix is often sparse (contains mostly zeros) (Fig. 1).

3.1   Term Frequency - Inverse Document Frequency (TF - IDF)

This is an alternative algorithm that can be employed for feature extraction, offering greater efficiency in comparison to the bag of words approach. This technique consists of two distinct components. This model works in the following manner:

1. Term Frequency (TF): The term frequency of a word in a document is the ratio of the number of times that word appears in the document to the total number of words in the document. It gives you an idea of how frequently a word occurs within a specific document.
2. Inverse Document Frequency (IDF): The inverse document frequency of a word is a measure of how unique or important that word is across all the documents in the corpus. It's calculated as the logarithm of the ratio of the

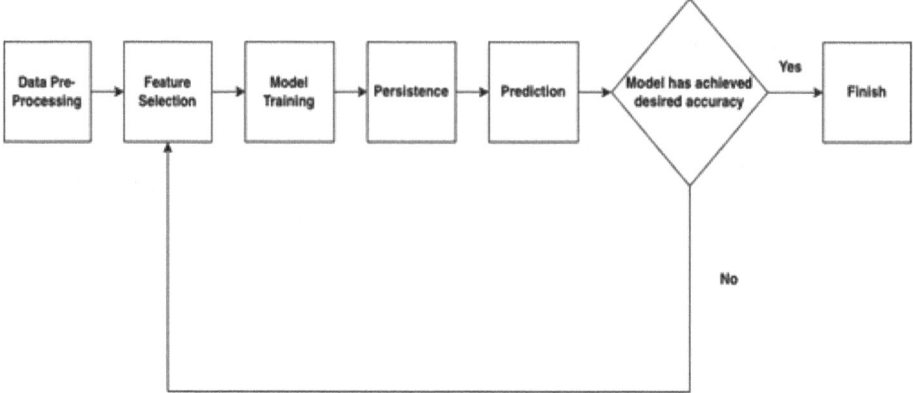

Fig. 1. Workflow of Proposed Method

total number of documents to the number of documents containing the word. Words that appear in many documents have lower IDF values, while words that appear in fewer documents have higher IDF values, thus emphasizing the importance of rarer words.

3. TF-IDF Calculation: The TF-IDF score for a word in a document combines the term frequency and inverse document frequency. It gives a high weight to words that appear frequently in the document (high TF) but are rare across the entire corpus (high IDF). Conversely, it assigns a low weight to words that are either too common or too rare.

4. Vectorization: Each document is represented as a numerical vector, where each dimension of the vector corresponds to a unique word in the vocabulary. The value in each dimension is the TF-IDF score of the corresponding word in the document. This vectorization process converts the text data into a format suitable for various machine learning algorithms.

4   Experimental Procedure

The Dataset for this work is a Drug Reviews dataset. This dataset has been obtained from the UCI Machine Learning Repository. The dataset provides information about the different reviews related to various types of drugs used by the patients. There are over 210000 rows (data points) and 6 feature attribute in the given dataset such as Drug Name, Condition, Review, Rating, Date and Useful Count. The exploratory analysis of this data is mentioned in next step.

4.1   Exploratory Data Analysis

This stage encompassed achieving a comprehensive understanding of the dataset. Diverse facets of the dataset were thoroughly examined. During this phase, the subsequent tasks were undertaken:

1. Exploration of Selected Features to Identify Trends and Patterns: This involved a thorough examination of specific attributes within the dataset to uncover trends, patterns, and potential relationships. The aim was to extract meaningful insights that could contribute to the subsequent analysis [4].
2. Segregating the Dataframe for Analyzing Individual Conditions: The dataset was partitioned into distinct segments to facilitate focused analysis of individual conditions or categories. This segmentation aimed to reveal condition-specific nuances and behaviors.
3. Creating WordCloud for Different Classes of the Output Variable: Word-Clouds were generated for various classes or categories of the output variable as shown in Fig. 2, 3, 4 and 5. WordClouds visually depicted the frequency distribution of words within these classes, providing an intuitive glimpse into the prevalent terms.

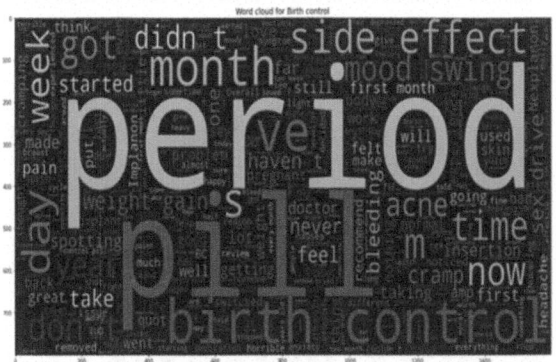

Fig. 2. WordCloud for 'Birth Control' Class

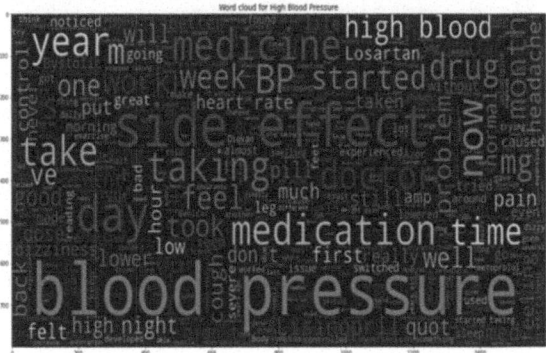

Fig. 3. WordCloud for 'Blood Pressure' Class

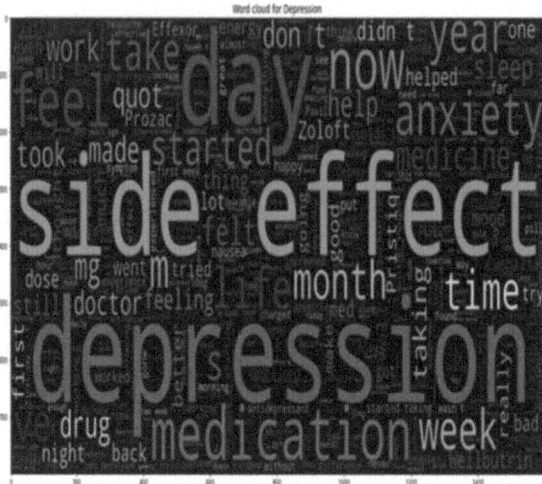

Fig. 4. WordCloud for 'Depression' Class

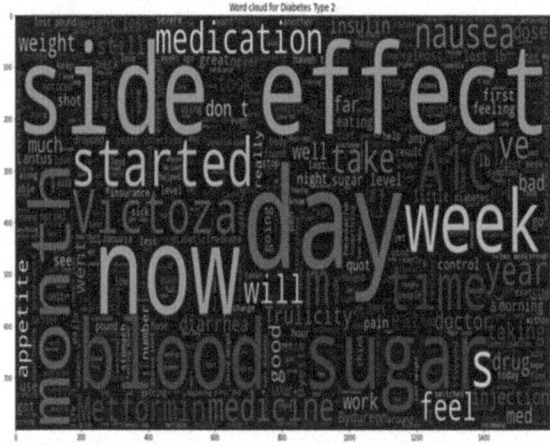

Fig. 5. WordCloud for 'Diabetes Type 2' Class

4. Removing Double Quotes from All the Reviews Present in the Dataset: Double quotation marks were systematically eliminated from the reviews contained in the dataset. This preprocessing step aimed to ensure uniformity and eliminate any potential interference with subsequent analyses.
5. Removing the StopWords: StopWords, which are common words with little contextual significance, were removed from the reviews. This process aimed to enhance the quality of subsequent analyses by eliminating noise.
6. Converting the Reviews to Individual Important Words: The reviews were transformed into sets of significant individual words. This transformation was instrumental in simplifying the data while retaining essential content.

7. Creating Features and Target Variables: The dataset was organized into features (input variables) and a target variable (output variable). This arrangement was vital for subsequent machine learning tasks.
8. Performing Train Test Split: The dataset was divided into training and testing subsets. The training set was used for model training, while the testing set was employed to assess model performance [5].
   Creating a Function for Plotting Confusion Matrix: A specialized function was developed to create and visualize confusion matrices. Confusion matrices are crucial tools for evaluating the performance of classification models.

5   Result Analysis, Conclusion and Future Work

This project employed a trio of diverse feature extraction approaches: Bag of Words, TF-IDF, TF-IDF (Bigrams), and TF-IDF (Trigrams) as mentioned Table 1, 2, 3 and 4. In conjunction with each of these techniques, a set of distinct machine learning models was applied. These models encompassed the Logistic Regression Model, Decision Tree Classifier, Naive Bayes Classifier, and Passive Aggressive Classifier.

Table 1. Result analysis using Bag Of Words

Method	Accuracy	Precision	Recall	F1-Score
Logistic Regression	**97.99**	**96.33**	**94.52**	**0.9539**
Decision Tree Classifier	95.98	92.37	90.99	0.9166
Naive Bayes Classifier	97.02	94.17	92.79	0.9346
Passive Aggressive Classifier	97.77	95.28	94.52	0.9489

Table 2. Result analysis using TF-IDF (Unigram)

Method	Accuracy	Precision	Recall	F1-Score
Logistic Regression	97.73	**96.32**	93.38	0.9476
Decision Tree Classifier	95.52	91.49	90.18	0.9081
Naive Bayes Classifier	92.14	95.53	73.06	0.8044
Passive Aggressive Classifier	**98.27**	96.25	**95.71**	**0.9598**

Numerous tasks were undertaken throughout the course of this research project. The initial step involved formulating a solvable problem statement. Subsequently, an extensive array of research papers closely aligned with the problem statement were meticulously analyzed. The subsequent phase encompassed the identification of a suitable dataset, sourced from the UCI machine learning

Table 3. Result analysis using TF-IDF (Bigrams Approach)

Method	Accuracy	Precision	Recall	F1-Score
Logistic Regression	97.50	**96.93**	91.82	0.9414
Decision Tree Classifier	95.55	90.76	90.14	0.9044
Passive Aggressive Classifier	**98.21**	96.23	**95.55**	**0.9588**

Table 4. Result analysis using TF-IDF (Trigram Approach)

Method	Accuracy	Precision	Recall	F1-Score
Logistic Regression	97.09	96.65	90.26	0.9312
Decision Tree Classifier	95.17	90.40	89.57	0.8998
Passive Aggressive Classifier	**98.61**	**97.73**	**95.89**	**0.9678**

repository. This acquired dataset comprised pivotal features wielding significant influence in deducing the specific medical condition afflicting a given patient. Upon dataset preparation, a thorough exploration was conducted, resulting in the integration of various refinements. Following this, the dataset underwent Train-Test Splits to facilitate robust model evaluation. Three distinct feature extraction methodologies were implemented to uncover salient terms within individual reviews. This, in turn, streamlined the process of categorizing patients' medical conditions in a precise and efficient manner. For each of these feature extraction methods, five diverse machine-learning models were employed. Notably, the passive-aggressive classifier exhibited remarkable accuracy, emerging as the most precise model across all feature extraction techniques utilized within the project. This observation held true for all the employed methods. The pinnacle of accuracy, approximately 98.57%, was attained through the TF-IDF: Bigrams Approach. Beyond the passive-aggressive classifier, the logistic regression model showcased notable accuracy, achieving approximately 97.99% accuracy in the Bag of Words scenarios. In the context of our work, there exists a plethora of avenues for implementation, both in the near future and the long term. In the immediate horizon, the foremost objective involves continued experimentation with various Machine Learning Models to pinpoint the optimal model aligned with specific requirements. Concurrently, it becomes imperative to enhance the predictive accuracy of each attempted model. Attaining these objectives necessitates addressing various considerations, including the mitigation of multicollinearity, the prevention of model overfitting, and other relevant aspects. These concerns are particularly significant when dealing with Decision Tree and Random Forest Models [6]. Throughout these iterative processes, the overarching emphasis remains steadfastly fixed on achieving the utmost accuracy while simultaneously safeguarding against challenges like multicollinearity and overfitting. Looking towards the long-term trajectory, the approach of employing diverse feature extraction methods for the purpose of classification holds the potential for extension into various other domains. Noteworthy domains ripe for exploration include e-commerce, banking, and social media sentiment analysis. The adaptability of this methodology underscores its potential applicability across an array of fields, ultimately contributing to a more nuanced understanding and insightful categorization within these domains.

References

1. Melhem, S., Al-Aiad, A., Al-Ayyad, M.S.: Patient care classification using machine learning techniques, in 2021 12th International Conference on Information and Communication Systems (ICICS), pp. 57–62 (2021)
2. Shaikh, S., Khan, M.Y., Nizami, M.S.: Using patient descriptions of 20 most common diseases in text classification for evidence-based medicine. In: Mohammad Ali Jinnah University International Conference on Computing (MAJICC), vol. 2021, pp. 1–8 (2021)
3. Santana-Vel asquez, A., John Freddy Duitama, M., Arias-Londo no, J.D.: Classification of diagnosis-related groups using computational intelligence techniques. In: 2020 IEEE Colombian Conference on Applications of Computational Intelligence (IEEE ColCACI 2020), pp. 1–6 (2020)
4. Chandra, Y., Jana, A.: Sentiment analysis using machine learning and deep learning. In: 2020 7th International Conference on Computing for Sustainable Global Development (INDIACom), pp. 1–2 (2020)
5. Kamboj, A., Kumar, P., Bairwa, A.K., Joshi, S.: Detection of malware in downloaded files using various machine learning models. Egypt. Inform. J. **24**(1), 81–94 (2023)
6. Bhattacharyya, P.: Sentiment analysis. In: 2013 1st International Conference on Emerging Trends and Applications in Computer Science (2013)
7. Yang, H., He, Q., Liu, Z., Zhang, Q.: Malicious encryption traffic detection based on NLP. In: Security and Communication Networks, vol. 2021, Article ID 9960822, p. 10 (2021)
8. Mundra, S., Mittal, N.: CMHE-AN: code mixed hybrid embedding based attention network for aggression identification in Hindi English code-mixed text. Multimedia Tools Appl. **82**(8), 11337–11364 (2023)
9. Mundra, S., Mittal, N.: FA-Net: fused attention-based network for Hindi English code-mixed offensive text classification. Soc. Netw. Anal. Min. **12**(1), 100 (2022)

Systematic Review on Sustainable Healthcare Practices: A Bibliometric Approach

Shubhangi V. Urkude[1(✉)] and Debajani Sahoo[2]

[1] Department of Operations and IT, IBS, IFHE University, Hyderabad, Telangana, India
ushubhu@gmail.com
[2] Department of Marketing, IBS, IFHE University, Hyderabad, Telangana, India
debajani@ibsindia.org

Abstract. In order to map the research trends in this field, the goal of this paper is to analyse the body of literature on sustainable healthcare practices using bibliometric analysis and network visualization. In this work, research literature obtained from the Scopus and Web of Science databases between 2003 and 2023 is subjected to bibliometric analysis methods. Indicators like changes in research production and citations, top contributing authors, countries, journals, keyword co-occurrences, and co-authorship ties among contributing countries are all evaluated in the study. The findings shows, the USA and UK are now leading researcher in this area, with Australia and Canada also playing a significant role. The field of sustainable healthcare practices encompasses a number of different fields, including public health, healthcare delivery, major clinical trials, questionnaires, middle-aged adult male and female demographics, clinical practices, psychology, and public health education. There is a tone of opportunity for research in the future that connects sustainable healthcare to these subjects in terms of authors, journals, and organizations that may offer knowledge and possible direction for future exploration by detailing the research's organizational framework.

Keywords: Sustainable Healthcare · public health · healthcare delivery · Bibliometric analysis · Bibliometrix · Biblioshiny · Citation analysis

1 Introduction

Natural resources are essential to daily life, daily business, and human existence. The depletion of natural resources could result from disregarding sustainability. Sustainability is a crucial area of study. Business students should study sustainability since it makes companies more consumer-friendly and helps them meet their CSR (Corporate Social Responsibility) commitments. According to the World Health Organization (WHO), a sustainable health care system improves, maintains, or restores health while minimizing negative environmental effects and maximizing opportunities to improve and transform it to benefit the health and well-being of both present and future generations. Several projects and studies have offered advice and discussed personal experiences with hospital energy management, demonstrating improvements and some encouraging outcomes.

© The Author(s), under exclusive license to Springer Nature Switzerland AG 2025
A. K. Bairwa et al. (Eds.): ICCAIML 2024, CCIS 2184, pp. 74–85, 2025.
https://doi.org/10.1007/978-3-031-71481-8_6

However, most hospitals continue to struggle with developing solutions for the sustainable use of energy [1]. The assessment of environmental influences on healthcare is still in its early stages and does not yet give enough thorough information to enable healthcare decision-making, both at the level of particular hospitals and health systems as a whole [2]. The goal of this present study is to review the research on healthcare management and practice health management sustainably. The following research questions (RQs) are what we plan to investigate in the context of this project. To review number of articles published during 2003–2023 in the era of sustainable healthcare practices.

- To find the trends and frequency of articles in sustainable healthcare across the globe.
- To find the most contributing authors and documents over last 20 years.

Due to the aforementioned goals, this study aims to offer (1) an overview of the bibliometric technique and (2) thorough guidelines for carrying out bibliometric analysis of healthcare research. This paper introduces business scholars to bibliometric analysis by explaining its ideas, methodologies, and practices and providing examples and arguments. The advantages of this work include, first, helping business academics comprehend the bibliometric approach so they can analyze certain fields in the body of literature utilizing a large corpus and bibliometric data. The document provides a summary of bibliometric analysis and the rules that govern it. Second, by providing several recommendations regarding the various techniques for bibliometric analysis in sustainable healthcare practices and when to use them, the study can broaden the perspective of medical professionals and academics in the business world on the alternatives and justifications for using the various variants of bibliometric analysis.

Rest of the article is organized as follows. Section 2, discussed the research methodology with database selection and keywords used to download the paper in detail. Result and Analysiswas elaborated in Sect. 3 followed by discussion in Sect. 4. Section 5 elaborates conclusion and future scope for the next researchers.

2   Research Methodology

Notably, the development of scientific databases like Scopusand Web of Science has made it easier to obtain substantial amounts of bibliometric data.Gephi, Leximancer, VOSviewer and biblioshiny are the bibliometric software available for systematic literature review [24, 25]. There are many databases available like IEEE, science direct, Google scholar to download the paper but biblioshiny package in R supports both .csv and .bibtext file format to input the data, which was easier to get in Scopus and web of science databases. A recent rise in scholarly interest in bibliometric analysis can be attributed to this software. In several fields of business research, such as business strategy [3], electronic commerce [4], and finance [5–7], a bibliometric technique is thus in use. In the domains of human resources, management, and marketing [8, 9], bibliometric is used to analyze the intellectual structure in addition to researching collaboration patterns and publications [10–12]. This work used a standard complete bibliometric analysis, which starts with a keyword search in the Scopus search syntax, as shown in Fig. 1. Scopus and web of science databases was used to download the papers in the area of sustainable healthcare practices with the help of relevant keywords drawn through the literature [26,

27]. To perform the analysis biblioshiny interface was selected from bibliometric package in R. Performance analysis without scientific mapping was not advisable [13]. The keywords relevant to the sustainable healthcare practices were extracted from the past literature which was published in reputed journal. All the keywords were taken as combination of AND, OR logical format. The relevant papers which satisfy above criteria was selected and downloaded as .bibtex file. The keywords suggested that papers and publications with "Sustainable Practices," OR "Green Practices," OR "Ecology," etc. AND "Healthcare services" OR "Medical services" OR "Health services" in their titles, keywords, or abstracts are relevant. During the search process, a total of 1755 articles were selected between the years 2003 and 2023 without any filtration. The search results were checked for redundant information. After that, only journal articles, review articles, and articles in English and within the keywords of medicine, healthcare, health policy, sustainable science, environmental science, and supply chain management are included for further analysis. A total of 869 articles from the following journal categories: "Medical," "Social Science," "Business Management," "Health Profession," "Psychology," "Environmental Science," and "Multidisciplinary" were taken into consideration for the study. The file was saved in "bibtex" format. A similar procedure was followed on the web of science (WOS) platform to download data to retrieve 22 articles in bibtex format after purification. Both Scopus and WOS databases have been used to collect a maximum number of papers in the discipline of sustainability practices in the healthcare industry.

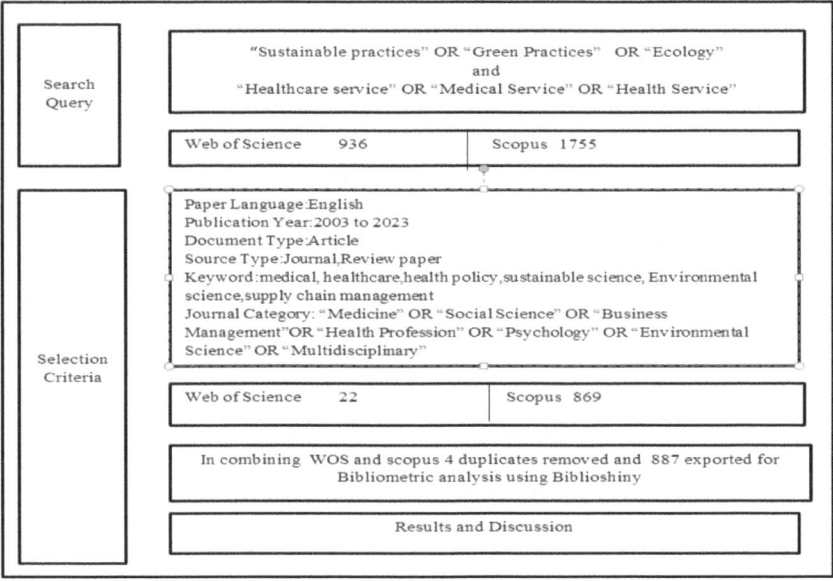

Fig. 1. Article selection criteria

Both .bibtex files are merged using the bibliometric function in R. After removing 4 duplicates, 887 articles were exported for Bibliometric analysis using Biblioshiny. The biblioshiny platform is the best choice for readily displaying bibliometric maps and is intelligible for any audience. Biblioshiny was used to build and graphically portray the network of authors and countries using the bibliometric R program. This study used Biblioshiny to create a notable keyword cloud that helps to visualize database clusters.

3 Results and Analysis

3.1 Keywords Co-occurrence Analysis

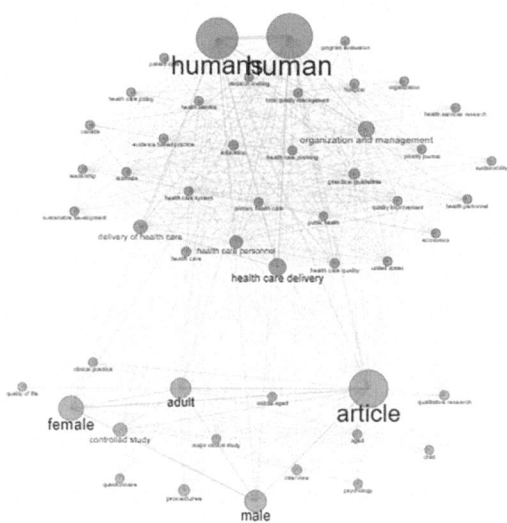

Fig. 2. Co-occurrence network of High-frequency keyword (Color figure online)

Bibliometric typically makes use of the co-occurrence analysis keywords extracted from the Pub Med database-A specific co-occurrence matrix filtering process must be performed to eliminate the low-frequency items for the more understandable result due to poor representativeness. Figure 2 shows the co-occurrence network analysis. This network showcased terms such as healthcare delivery, organization and management, healthcare personnel, healthcare quality, and healthcare policy from all demographic aspects such as male, female, adult, etc.

Two distinct color clusters, red and blue, are displayed in the picture. The circle's diameter reflects the keyword's frequency of use. The interaction between the keywords in the same cluster shows the dense network with more interaction than the interaction among the cluster elements.

3.2   Thematic Analysis

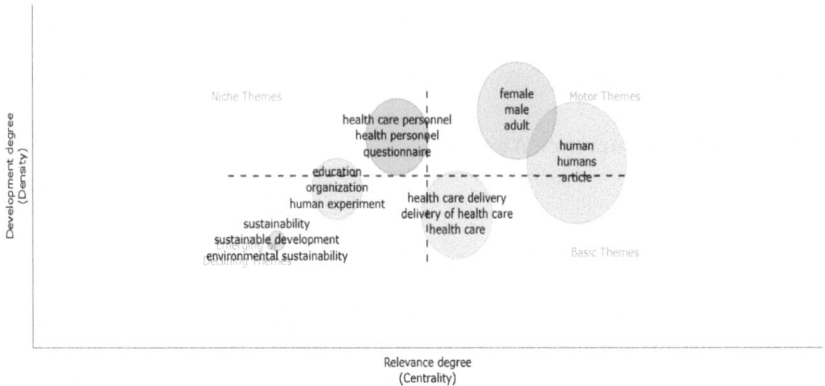

Fig. 3. Thematic map

Figure 3 shows the thematic analysis for the last 20 years (2003–2023) showing four quadrant and six themes. After investigation, the study received 6 themes based on two dimensions such as development and relevance degree, which are categorized in 4 quadrants. Out of all these 4 quadrants, the first quadrant represents motor themes, which was study about male, female, human and adult health covering 2 themes. The second quadrant represents a space of basic and transversal themes of healthcare delivery and health care facilities. The third quadrant, niche, represents the study about health care personnel and questionnaires used in the data collection process. The fourth quadrant represents sustainability, sustainable development, environmental sustainability, education, human experiment and organization. It was named as emerging and declining theme as most of the research was started towards sustainability, which include 2 themes.

3.3   Global Cited Documents in Sustainable Healthcare

Table 1 represents the top 10 globally cited authors with the journal they published. The analysis shows that the highest number of citations received by Simons 2018(819), with an average citation per year (of 136.50), followed by Chambers, 2013 having a total citation (of 749) with an average citation of (68.09). It is interesting to notice that the citation of sustainability in healthcare service is very striking, and the latest publication, like Simon, 2018; Kouhizadeh, 2021; Kaika, 2017, have higher citations than older papers. Normalized TC indicated the average citation for the particular author in the sustainability area during the given time.

3.4   Publications Per year

The graph of the annual increase in publications is shown in Fig. 4. As per the figure, although publication on sustainable healthcare practices started in 2004, it picked up from 2015 onwards. During 2020 more focus was given to sustainable healthcare practices due to the worldwide pandemic.

Table 1. Top 10 authors and Journal names for article published

S. No	Author and Journal	Total Citations	TC per Year	Normalized TC	DOI
1	Simons mp, 2018, Hernia	819	136.50	30.92	https://doi.org/10.1007/s10029-017-1668-x
2	Chambers da, 2013, Implement Sci	749	68.09	17.88	https://doi.org/10.1186/1748-5908-8-117
3	Kouhizadeh M, 2021, Int J Production Economics	325	108.33	43.41	https://doi.org/10.1016/j.ijpe.2020.107831
4	Kaika M, 2017, Environ Urban	260	37.14	13.52	https://doi.org/10.1177/0956247816684763
5	Boissy A, 2016, J Gen Intern Med	247	30.88	8.69	https://doi.org/10.1007/s11606-016-3597-2
6	Armour C, 2007, Thorax	211	12.41	3.86	https://doi.org/10.1136/thx.2006.064709
7	Carling Pc, 2008, Infect Control Hosp Epidemiology	197	12.31	4.51	https://doi.org/10.1086/591940
8	Pinzone M, 2016, Journal of Cleaner Production	177	22.13	6.23	https://doi.org/10.1016/j.jclepro.2016.02.031
9	Ulrich Rs, 2011, Health Environ Res Des J	171	13.15	6.78	https://doi.org/10.1177/193758671000400107
10	Proctor E, 2015, Implement Science	167	18.56	7.25	https://doi.org/10.1186/s13012-015-0274-5

3.5 Most Cited Country with a Total Citation

The total number of citations by country is highlighted in Fig. 5, which reveals that the United States, the United Kingdom, and Australia have more citations than the other nations, with respective total of 4712, 1536, and 1104. The United States of America's average citations were almost three times higher than its counterpart countries like United Kingdom and Australia. The reason might be the USA research funding and ideas were

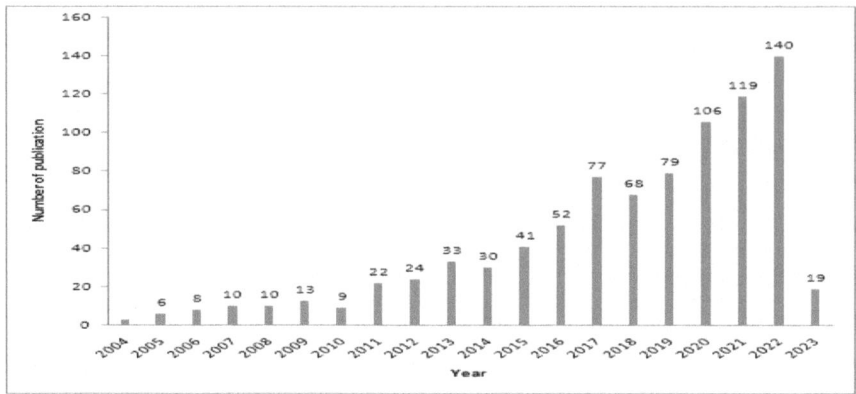

Fig. 4. Number of publications over the year

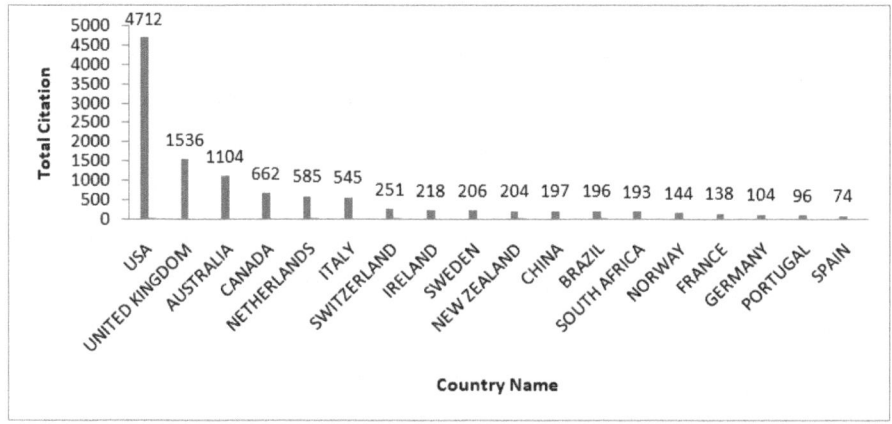

Fig. 5. A most cited country with a total citation

far ahead of its counterparts. Conversely, countries like France, Germany, Portugal, and Spain have very low average citations. Researchers may have become interested in this sustainability field as a result of the significant growth that these nations' healthcare systems have experienced. Countries such as the Netherlands, Italy, and Switzerland stood in the middle position as per their average citation. They might grow and compete with the developed country in the upcoming era.

3.6 Bradford Law for Top 10 Journals

A criterion for weakening the search result in the scientific publication was put forth by Bradford (1934) [14]. Bradford law states that if the number of articles in the particular field were arranged in a order within three groups then number of articles in each group will in the ratio of 1:n:n2, that means first group will end with 1 article, second group with n number of articles and third group with twice the n articles, where n represents

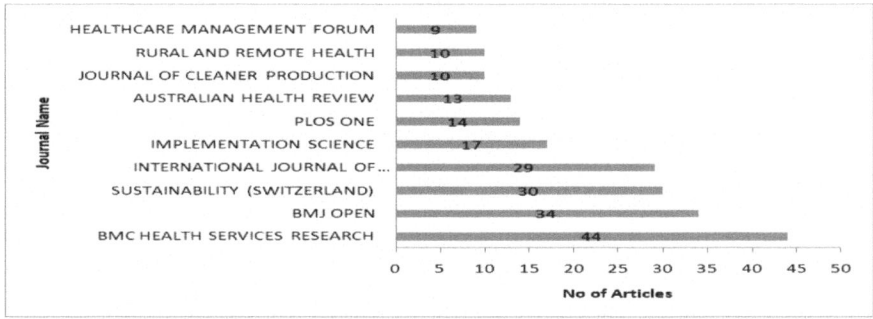

Fig. 6. Bradford law—Top 10 journals in the scope of the study

the sources of downloading the articles. This also indicate even if increase of sources to find the articles were increase exponentially there was same number of articles will be available in search which is also stated as exponential diminishing returns. Fig. 6 lists the top 10 journals that contribute to the body of knowledge on sustainable medical procedures. Over 45% of the research literature was contributed by the top 10 journals, showing a considerable concentration of study. The journal with the most of the articles (44), accounting for over 21% of all the papers in the sample set, is BMC Health Service Research. It is followed by BMJ Open (34 articles) and sustainability, a journal with a Swiss base, which has 14% of the sample set's articles (30) out of 210 total articles. Journals, including Health Management Forum, Rural and Remote Health, and Journal of Cleaner Productions, simultaneously publish fewer publications on sustainable healthcare practices.

3.7   Evolution of Dendrogram in Sustainable Healthcare Practices

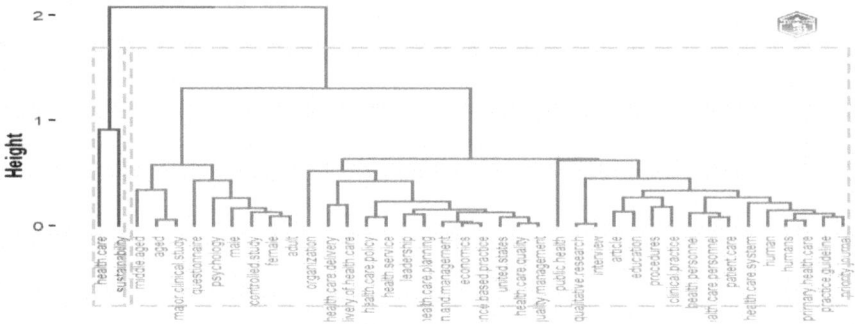

Fig. 7. Topic dendrogram for articles.

Figure 7 shows the topic dendrogram concerning healthcare sustainability practices. Dendogram is a hierarchical tree like structure showing how the clusters are merged. Each cluster is a group of similar objects and one cluster is merged with another cluster based on the centroid. Grouping is carried out based on the keywords appeared in the articles and cited by other articles. Dendogram has clusters with two colors. Blue color indicate cluster from existing literature on which various studies done and red color indicate very few research has been done on healthcare, sustainability. However, researchers can focus on this area and get new insights. There are three primary strands to the dendrogram. The first strand focuses on an extensive clinical and controlled investigation of diverse demographic groups such as male, female, adult, and middle-aged profile and their psychology on those sustainable healthcare methods. The second strand focuses on healthcare delivery, healthcare policy, healthcare planning and practices, and organizational leadership position on sustainability. Suppose we look at the second block, which likewise has the most divisions, as an organizational development on sustainability. The third strand emphasizes primary healthcare, practice guidelines, patient care, patient education, etc. There was a correlation between the second and third strands on the ground of healthcare delivery and healthcare service procedure, so the second stage of clubbing in the dendrogram occurred again between the second and third strands. New emerging strand focus on healthcare and sustainability, which was highlighted in red color. Dendogram indicate that there is a scope for research in sustainability in healthcare sector.

4 Discussion

Earlier research on green healthcare has examined a number of relevant practical challenges. Several of these studies have examined leadership [15], hospital management of healthcare waste challenges, and solutions [16]. The main element influencing the implementation of environmental sustainability in healthcare institutions is energy consumption [17, 18]. The management of green healthcare supply chains has been the subject of several studies, including ones that evaluated the environmental performance of healthcare suppliers [19] and looked at the challenges associated with adopting green public procurement in the public health sector. In a pharmaceutical supply chain, a novel multi-objective mathematical model is being proposed [20] for responding to natural disasters while taking quality and cost into account. In the present study, the most productive authors, institutions, and nations, the annual citation structure, the most cited articles, the most popular keywords, and the most cited journals are only a few of the bibliometric indicators employed. The present study holds a significant position among publications on hospitals and sustainability, making it a reference among the best publications in the hospital and sustainable sector.

Yet, designing, constructing, and operating hospitals are among the most challenging tasks [21]. Hospital energy management takes into account the unique needs and requirements of each department. Intense surgery, for instance, uses equipment with higher energy needs and stricter constraints on environmental elements like ventilation, heating, and lighting. Administrative sectors, however, use technology with lower energy needs [22]. The unique characteristics of these regions add another factor that needs to be considered. They make it difficult to design electrical installations and manage energy efficiently in hospitals [23]. This research demonstrates a substantial evolution of the journal concerning articles, analyzed themes, and research methodologies. This study will aid to direct academics in the healthcare area toward new subjects and statistical research methodologies. Also, it accurately describes the study on sustainability and healthcare management done by this esteemed magazine. The examination of trending subjects may be able to predict the direction of future research in the hospital and healthcare sectors. On the basis of this research, future studies could investigate the connection between research published in recognized publications and its usefulness in healthcare facilities.

5 Conclusions and Future Scope

Energy efficiency is a problem that become more difficult to solve as a result of the growth in the world's population, economic activity, and high consumption rates that are not long-term sustainable. As a result, three crucial issues of energy efficiency in healthcare should be the focus of healthcare authorities in various regions. (i) Making use of cutting-edge innovative materials and technologies to improve building efficiency and minimize energy consumption in healthcare planning. (i) Integrated approaches that bring together all elements into coherent energy-efficiency programs; (ii) designs to optimize energy usage in buildings; and (iii) boosting output through more sustainable sources. An issue with the study could be that citation and co-citation data and patterns are frequently dynamic and are expected to fluctuate over time. The research field of sustainable healthcare is garnering a lot of attention from academics and researchers. This paper summarizes the significant developments in sustainable healthcare research from 2003 to 2023. Several directions for additional research have been identified after thoroughly analyzing the sample set of 887 research papers.

Future studies might look further into which specific facets of sustainable healthcare—such as the medical business model, customer value propositions, consumer education on sustainable healthcare, and delivery methods—produce measurable gains in increased customer value. Only a few studies have examined the impact of sustainable healthcare on consumer behavior, but future research may look at the direct correlation between sustainable healthcare and certain outcomes including customer satisfaction, customer loyalty, and future purchase intentions. Exploring these connections among a diverse group of consumers across various segmentations would be another topic of inquiry (Male, Female, and Adult). This might aid healthcare service providers in deciding which clientele to target, what kind of investments to make, and how best to streamline their service delivery. Western nations have primarily dominated research in the area of sustainable healthcare. Yet, other emerging countries are also increasingly catching on to the sustainable healthcare trend. Studying these effects on consumer behavior in various cultural contexts is necessary.

References

1. Campisi, T., Acampa, G., Marino, G., Tesoriere, G.: Cycling master plans in Italy: the I-BIM feasibility tool for cost and safety assessments. Sustainability 12(11), 4723 (2020)
2. McGain, F., Naylor, C.: Environmental sustainability in hospitals–a systematic review and research agenda. J. Health Serv. Res. Policy 19(4), 245–252 (2014)

 3. Kumar, S., Lim, W.M., Pandey, N., Westland, J.C.: 20 years of electronic commerce research. Electron. Commer. Res. **21**(1), 1–40 (2021). https://doi.org/10.1007/s10660-021-09464-1
 4. Kumar, S., Sureka, R., Lim, W.M., Kumar Mangla, S., Goyal, N.: What do we know about business strategy and environmental research? Insights from business strategy and the environment. Bus. Strateg. Environ. **30**(8), 3454–3469 (2021)
 5. Durisin, B., Puzone, F.: Maturation of corporate governance research, 1993 – 2007: an assessment. In: KITeS, Centre for Knowledge, Internationalization and Technology Studies, vol. 17, No. 3, pp. 266–291. Universita' Bocconi, Milano, Italy (2009)
 6. Linnenluecke, M.K., Chen, X., Ling, X., Smith, T., Zhu, Y.: Research in finance: a review of influential publications and a research agenda. Pac. Basin Financ. J. **43**, 188–199 (2017)
 7. Xu, L.D., Xu, E.L., Li, L.: Industry 4.0: state of the art and future trends. Int. J. Prod. Res. **56**(8), 2941–2962 (2018)
 8. Ellegaard, O., Wallin, J.A.: The bibliometric analysis of scholarly production: how great is the impact? Scientometrics **105**, 1809–1831 (2015)
 9. Zupic, I., Čater, T.: Bibliometric methods in management and organization. Organ. Res. Methods **18**(3), 429–472 (2015)
10. Backhaus, K., Lügger, K., Koch, M.: The structure and evolution of business-to-business marketing: a citation and co-citation analysis. Ind. Mark. Manage. **40**(6), 940–951 (2011)
11. Donthu, N., Kumar, S., Pandey, N., Soni, G.: A retrospective overview of Asia Pacific Journal of Marketing and Logistics using a bibliometric analysis. Asia Pac. J. Mark. Logist. **33**(3), 783–806 (2021)
12. Donthu, N., Kumar, S., Pattnaik, D., Lim, W.M.: A bibliometric retrospection of marketing from the lens of psychology: Insights from Psychology & Marketing. Psychol. Mark. **38**(5), 834–865 (2021)
13. Brown, T., Park, A., Pitt, L.: A 60-year bibliographic review of the journal of advertising research: perspectives on trends in authorship, influences, and research impact. J. Advert. Res. **60**(4), 353–360 (2020)
14. Bradford, S.C.: Sources of information on scientific subjects. Engineering: an illustrated. Weekly J. **137**, 85–86 (1934)
15. Chiarini, A., Vagnoni, E.: Environmental sustainability in European public healthcare: could it just be a matter of leadership? Leadersh. Health Serv. **29**(1), 2–8 (2016). https://doi.org/10.1108/LHS-10-2015-0035
16. Thakur, V., Anbanandam, R.: Healthcare waste management: an interpretive structural modeling approach. Int. J. Health Care Qual. Assur. **29**(5), 559–581 (2016)
17. Neo, J.R.J., Sagha-Zadeh, R., Vielemeyer, O., Franklin, E.: Evidence-based practices to increase hand hygiene compliance in health care facilities: an integrated review. Am. J. Infection Control **44**(6), 691–704 (2016)
18. Zadeh, R.S., Xuan, X., Shepley, M.M.: Sustainable healthcare design: existing challenges and future directions for an environmental, economic, and social approach to sustainability. Facilities **34**(5/6), 264–288 (2016)
19. Malik, A., Budhwar, P., Kandade, K.: Nursing excellence: a knowledge-based view of developing a healthcare workforce. J. Bus. Res. **144**, 472–483 (2022)
20. Balan, S., Conlon, S.: Text analysis of green supply chain practices in healthcare. J. Comput. Inform. Syst. **58**(1), 30–38 (2018)
21. Blass, A.P., da Costa, S.E.G., de Lima, E.P., Borges, L.A.: Measuring environmental performance in hospitals: a practical approach. J. Clean. Prod. **142**, 279–289 (2017)
22. Ryan-Fogarty, Y., O'Regan, B., Moles, R.: Greening healthcare: systematic implementation of environmental programmes in a university teaching hospital. J. Clean. Prod. **126**, 248–259 (2016)

23. Bramer, W.M., Rethlefsen, M.L., Kleijnen, J., Franco, O.H.: Optimal database combinations for literature searches in systematic reviews: a prospective exploratory study. Syst. Rev. **6**, 1–12 (2017)
24. https://www.bibliometrix.org/home/index.php/layout/biblioshiny
25. https://www.bibliometrix.org/
26. https://www.scopus.com/results/results.uri?sort=plf-f&src=s&st1=Sustainable+Practices&st2=Healthcare+services
27. https://mjl.clarivate.com/

Violence Detection Through Deep Learning Model in Surveillance

Anirudh Singh⊙, Satyam Kumar⊙, Abhishek Kumar⊙, and Jayesh Gangrade⁽⊠⁾⊙

Manipal University Jaipur, Jaipur, Rajasthan 303007, India
jgangrade@gmail.com

Abstract. The realm of advanced deep learning architectures for violence detection is burgeoning and has the potential to revolutionize proposed approaches to preventing and identifying instances of violence. In this rapidly evolving field, deep learning architectures exhibit the capability to discern intricate patterns and make predictions based on vast datasets. Their suitability for violence detection is particularly pronounced, given their capacity to identify subtle indicators that may foreshadow aggressive behavior. This research advocates for an autonomous violence detection system leveraging a combined CNN + LSTM architecture. Each model adeptly extracts features at the frame level, progressively enhancing accuracy in the process. The proposed CNN + LSTM model, integrating spatial and temporal analyses, achieves an impressive accuracy rate of 90%. The synergy of Convolutional Neural Networks (CNNs) for spatial feature extraction and Long Short-Term Memory (LSTM) networks for temporal aggregation facilitates a nuanced analysis of local motion patterns. Rigorous evaluations conducted on publicly available datasets attest to the models' efficacy in violence detection. The amalgamated solution, utilizing the CNN + LSTM architecture, is tailored for real-time surveillance, ensuring swift identification and notification of potential violence to relevant authorities for prompt remote intervention.

Keywords: Violence Detection · Deep Learning · CNN + LSTM · Spatial Analysis · Real-time Surveillance

1 Introduction

In the ever-evolving landscape of monitoring technology, the imperative of deploying intelligent systems capable of deciphering complex human behaviors during challenging situations becomes paramount. This study addresses this need by delving into the realm of cutting-edge machine learning (ML) and artificial intelligence (AI) models to discern criminal activities. The present state of surveillance systems, while significant, often falls short in comprehensively capturing the intricacies of real-life violent scenarios. As part of our methodology, we propose the utilization of a singular, integrated model that combines Convolutional Neural Network (CNN) and Long Short-Term Memory (LSTM) architectures [1, 2]. This innovative approach capitalizes on CNN's proficiency in spatial feature extraction and the LSTM's ability to comprehend temporal dynamics within this amalgamation.

A. K. Bairwa et al. (Eds.): ICCAIML 2024, CCIS 2184, pp. 86–98, 2025.
https://doi.org/10.1007/978-3-031-71481-8_7

Proposed methodology represents a departure from traditional studies that adopt a static perspective. Instead, the CNN + LSTM model embraces a dynamic approach, allowing for a more sophisticated examination of how motion patterns evolve over time in video clips. The model achieves a notable accuracy rate of 90%, showcasing its proficiency in violence detection [3–5]. Its simplicity, coupled with its efficacy in capturing non-linear relationships within data, serves as a benchmark for comparative evaluations, shedding light on the performance trade-offs between conventional and state-of-the-art methodologies. As we embark on this exploration of violence detection using a unified LSTM + CNN model, the research not only seeks to enhance the accuracy of surveillance systems but also aims to contribute nuanced insights into the intricate interplay of spatial and temporal dynamics in real-world scenarios [6–20]. This endeavor stands at the intersection of technology and public safety, pushing the boundaries of achievable outcomes in the creation of intelligent surveillance systems that safeguard society with precision and foresight.

This research journey is underpinned by the recognition that the landscape of violence detection demands continuous adaptation to the evolving nature of technology. By delving into the intricacies of AI and ML models, our pursuit is not merely confined to the development of a cutting-edge surveillance system. It extends to a broader quest for a more nuanced understanding of violent behaviors and the creation of intelligent systems that transcend the limitations of current technologies.

The combination of CNN and LSTM architectures is not only a pragmatic response to the limitations of existing systems but also a strategic approach to address the multifaceted nature of violence. CNN's proficiency in extracting spatial features dovetails seamlessly with the LSTM's capacity to decipher temporal dynamics, resulting in a holistic analysis of video data. This synthesis enables our model to discern subtle patterns that may elude traditional methods, thereby elevating the potential for accurate violence detection in real-world scenarios. As our exploration unfolds, it is guided by a commitment to advancing not only the technical aspects of surveillance but also our collective understanding of the complexities surrounding violence detection. By delving into the intricate interplay of spatial and temporal dynamics, we aspire to contribute valuable insights that extend beyond the scope of conventional methodologies.

This study aligns with a broader narrative at the confluence of technological innovation and public safety, aspiring to redefine the possibilities of intelligent surveillance systems. Through a fusion of cutting-edge techniques and a commitment to real-world applicability, our endeavor seeks to transcend conventional paradigms.

2 Related Work

Ullah et al. [6] conducted an extensive survey on violence detection in surveillance videos, providing a comprehensive overview of methodologies and advancements in the field. Their work serves as a foundational resource for understanding the diverse approaches employed in the pursuit of effective violence detection. Omarov et al. [7] contributed significantly to the field through a systematic review of state-of-the-art violence detection techniques in video surveillance security systems. Their comprehensive analysis aids in discerning existing methodologies and their respective effectiveness, providing valuable insights for researchers and practitioners alike. Choque Luque-Roman and Camara-Chavez [8] explored the domain of weakly supervised violence detection in surveillance video, introducing innovative methods to enhance detection capabilities under limited supervision. Their work contributes to the ongoing efforts to improve violence detection systems' adaptability to various surveillance scenarios. Irfanullah et al. [9] focused on real-time violence detection using Convolutional Neural Networks (CNNs), marking a significant contribution to the practical application of deep learning in surveillance scenarios. Their emphasis on real-time applications addresses a crucial aspect of violence detection systems. Mahmoodi et al. [10] delved into violence detection in videos using interest frame extraction and 3D Convolutional Neural Networks. Their research underscores the importance of capturing temporal features for accurate detection, providing a nuanced perspective on the temporal dynamics of violent events. Freire-Obregón et al. [11] proposed an inflated 3D ConvNet context analysis for violence detection, advancing the field by incorporating context-aware approaches into surveillance technologies. Their work highlights the significance of contextual information in improving the overall performance of violence detection systems. Sharma et al. [12] presented a fully integrated violence detection system using CNN and LSTM, showcasing the combined strengths of these architectures for enhanced detection accuracy. Their integrated approach provides a holistic understanding of the synergies between CNN and LSTM in violence detection. Hung et al. [13] explored violent video detection through a pre-trained model and CNN-LSTM approach, contributing to the development of hybrid architectures in violence detection systems. Their work exemplifies the ongoing exploration of diverse model combinations for improved accuracy. Dandage et al. [14] provided a comprehensive review of violence detection systems using deep learning, summarizing the advancements and challenges in the field up to 2019. Their review serves as a valuable resource for understanding the historical progression of violence detection research. Ramzan et al. [15] conducted a review on state-of-the-art violence detection techniques, offering a thorough analysis of the landscape and highlighting key trends in violence detection research. Their work contributes to a nuanced understanding of the evolving methodologies employed in violence detection. Patel [16] proposed a real-time violence detection system using CNN-LSTM, contributing to the exploration of integrated deep learning architectures for real-time surveillance applications. The real-time aspect of their work addresses the critical need for timely and accurate violence detection. Sumon et al. [17] investigated violence detection using pre-trained modules and different deep learning approaches, expanding the understanding of model-based techniques in surveillance. Their research adds to the diverse array of methodologies available for violence detection. Mumtaz et al. [18] introduced a fast-learning approach through a deep multi-net CNN model for violence recognition in video surveillance, emphasizing efficient model training. Their work contributes to the ongoing efforts to enhance the efficiency of violence detection models. Vosta and Yow [19] proposed a CNN-RNN combined structure for real-world violence detection in surveillance cameras, contributing to the exploration of combined architectures for improved accuracy. Their research addresses the complexities of real-world surveillance scenarios. Vieira et al. [20] presented a low-cost CNN for automatic violence recognition on embedded systems, emphasizing practical implementation and efficiency in surveillance applications. Their work aligns with the growing interest in deploying violence detection systems in resource-constrained environments.

Akash et al. [21] explored human violence detection using various deep learning techniques, contributing to the understanding of human-centric violence recognition. Their research sheds light on the unique challenges and opportunities associated with recognizing violence involving human subjects. Tripathi et al. [22] investigated criminals and crime detection using Machine Learning & OpenCV, extending the scope of video surveillance beyond violence recognition. Wang et al. [23] provided baseline results for violence detection in still images, contributing to the foundational understanding of violence detection in static contexts. Their research lays the groundwork for exploring violence detection in scenarios where motion dynamics are limited. Li et al. [24] proposed an end-to-end multiplayer violence detection based on deep 3D CNN, contributing to the exploration of 3D convolutional architectures for violence detection in videos. Calero et al. [25] presented "Bindi," an affective Internet of Things system combating gender-based violence, offering a unique perspective on violence detection beyond traditional surveillance methods. Their research introduces innovative approaches to addressing violence in diverse contexts. Selvi et al. [25] developed a suspicious actions detection system using enhanced CNN and surveillance video, expanding the scope of video analytics for improved security. Their work explores the broader applications of video analytics beyond explicit violence detection. Manikandan and Rahamathunnisa [26] proposed a neural network-aided attuned scheme for gun detection in video surveillance images, addressing specific challenges related to firearm detection. Their research addresses a niche aspect of violence detection, focusing on the identification of potentially dangerous objects. Sarkar et al. [27] designed a Weapon Detection System, contributing to the development of specialized systems for identifying potentially dangerous objects in surveillance scenarios. Their work adds to the growing repertoire of specialized applications within the broader field of violence detection. Soliman et al. [28] explored violence recognition from videos using deep learning techniques, adding to the diverse range of methodologies for automated violence detection. Their research expands the toolkit of approaches available for violence detection in dynamic video scenarios. Taha and Shoufan [29] presented a comprehensive review of machine learning-based drone detection and classification, offering insights into the evolving landscape of research in this domain. While not directly related to violence between individuals, their work contributes to the broader understanding of machine learning applications in surveillance. This synthesis of related work reflects the depth and breadth of research in violence detection, showcasing diverse methodologies, approaches, and applications that collectively contribute to the advancement of this critical field.

3 Proposed Methodology

The workflow diagram of the proposed working model is shown in Fig. 1.

Our research methodology begins with the utilization of a violence dataset comprising 11,063 images extracted from video frames, categorized into Violence (5,832 images) and Non-Violence (5,231 images). The dataset undergoes data augmentation to enhance diversity, involving techniques such as rotation, flipping, and zooming. Subsequently,

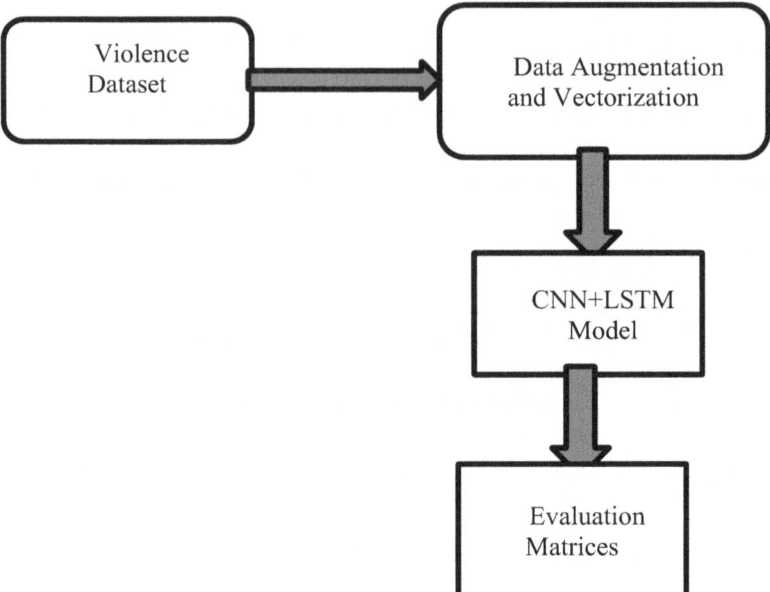

Fig. 1. Proposed Architecture

the images are vectorized for optimal compatibility with machine learning models. The model architecture integrates a Convolutional Neural Network (CNN) to extract spatial features and a Long Short-Term Memory (LSTM) network to capture temporal dependencies in sequential data, offering a comprehensive understanding of violent events. The dataset is partitioned into training, validation, and test sets for effective model training, hyperparameter tuning, and final evaluation. The iterative process of fine-tuning and optimization involves adjusting model parameters and exploring alternative data augmentation strategies. Thorough documentation covers dataset specifics, preprocessing steps, model architecture, hyperparameters, and a detailed report on results, emphasizing both the strengths and limitations of our proposed approach.

3.1 Dataset Description

Proposed violence detection research draws upon a comprehensive dataset obtained from Kaggle, originating from YouTube videos and curated to include 1000 violence and 1000 non-violence videos. Notably, the violence videos in our dataset encompass real street fight situations, offering a nuanced representation across diverse environments and conditions as shown in Figs. 2 and 3. To further enrich the dataset, we extracted 11,063 images from the video frames, meticulously dividing them into two distinct classes: Violence and Non-Violence. The Violence class comprises 5,832 images capturing various real-life street fight scenarios, providing a rich and authentic set of instances for model training and evaluation. The Non-Violence class encompasses 5,231 images, contributing to the dataset's balance and diversity. This dataset serves as a robust foundation

Fig. 2. Sample image from dataset

Fig. 3. Sample image from dataset

for our violence detection research, and its detailed nature ensures the authenticity and relevance of the scenarios explored. The reference for the dataset is [30].

3.2 Data Vectorization

For video-based violence detection, the initial step involves extracting frames from videos. This process transforms the temporal data into a series of static images, which can be fed into various deep learning models. To enhance model generalization and mitigate overfitting, data augmentation techniques like rotation, flip, scaling, and brightness adjustment are commonly applied to increase the diversity of training samples. For image-based models (CNN + LSTM), each frame is treated as an individual image. Frames are resized to a common resolution (224×224) to ensure uniform input size. Pixel values are normalized to a predefined range ([0, 1] or [-1, 1]). Deep pre-trained models like CNN and LSTM are used to extract high-level features from frames. These features serve as inputs to the final classification layers. For video sequences, data vectorization involves encoding temporal information. Video frames are grouped into sequences of fixed length. Padding may be applied to ensure uniform sequence length. Frames within a sequence are processed through convolutional layers to extract spatial features. The resulting features are then fed into LSTM layer to capture temporal dependencies.

3.3 Feature Selection

CNN layers automatically learn to extract relevant features from the input frames. Filters in these layers identify patterns, textures, and local spatial information. The weights of these filters adapt during training to emphasize important features for violence detection. As the model trains, the filters become specialized in capturing violence-related patterns. In temporal models like LSTM and RNN, the network inherently performs feature selection. These layers capture temporal dependencies and focus on the sequences of features that are most informative for violence detection. By adjusting the weights of the recurrent units during training, the model emphasizes relevant information over time, effectively selecting features that contribute to the classification task. For deep convolutional neural networks models, these models are designed with multiple layers of convolutions that learn progressively more abstract and relevant features. The early layers capture low-level details, while deeper layers extract high-level features. During training, the network learns to assign higher weights to features that are more discriminative for the violence detection task. Features that do not contribute significantly are effectively downweighed in some cases, researchers may fine-tune these pre-trained models on violence detection datasets. During fine-tuning, the models adapt to the specific task by adjusting the weights of the final classification layers. This process ensures that the learned features align more closely with violence-related patterns.

3.4 Deep Learning Model

The Deep Learning Convolutional Neural Network (CNN) + Long Short-Term Memory (LSTM) model represents a sophisticated architecture designed for violence detection in video data. This model amalgamates the strengths of two distinct neural network types, namely CNN and LSTM, to capture and analyze both spatial and temporal aspects of visual information.

3.4.1 Convolutional Neural Networks (CNN)

CNNs excel at extracting spatial features from images. In the context of violence detection, these spatial features may include patterns, shapes, and textures within individual frames of a video.

3.4.2 Long Short-Term Memory Networks (LSTM)

LSTMs are adept at understanding temporal dependencies within sequential data. In the case of video analysis, frames unfold over time. They capture the sequential nature of video frames, allowing the model to comprehend the evolution of patterns and movements over time.

3.4.3 Combined Architecture

By combining CNN for spatial feature extraction and LSTM for temporal feature aggregation, the model achieves a comprehensive analysis of both spatial and temporal patterns. This synergy is particularly advantageous for violence detection, as it enables the system to not only recognize spatial cues within individual frames but also understand how these cues evolve over the course of the video.

The model summary provides a concise overview of the architecture's key components and their roles in violence detection. Table 1 likely includes details such as the layers in the CNN and LSTM, the number of parameters, and possibly other relevant metrics. This table serves as a reference point for understanding the structural composition of the CNN + LSTM model and its capabilities in discerning violence-related patterns in video data.In essence, the CNN + LSTM model leverages the complementary strengths of spatial and temporal analysis to create a robust framework for violence detection, making it well-suited for capturing the complexities inherent in real-world scenarios.

3.5 Evaluation Parameters

- Accuracy:
    Accuracy assesses the correctness of predictions by comparing the sum of true positives (TP) and true negatives (TN) to the total instances.

$$\text{Accuracy} = (TP + TN)/(TP + TN + FP + FN) \tag{1}$$

- Precision:
    Precision gauges the precision of positive predictions, determined by the ratio of true positives (TP) to the sum of true positives (TP) and false positives (FP).

$$\text{Precision} = TP/(TP + FP) \tag{2}$$

Table 1. Model summary of CNN +LSTM Model

Layer (type)	Output Shape	Param #
time_distributed (TimeDistributed)	(None, 16, 2, 2, 1280)	2257984
dropout (Dropout)	(None, 16, 2, 2, 1280)	0
time_distributed_1 (TimeDistributed)	(None, 16, 5120)	0
bidirectional (Bidirectional)	(None, 64)	1319168
dropout_1 (Dropout)	(None, 64)	0
dense (Dense)	(None, 256)	16640
dropout_2 (Dropout)	(None, 256)	0
dense_1 (Dense)	(None, 128)	32896
dropout_3 (Dropout)	(None, 128)	0
dense_2 (Dense)	(None, 64)	8256
dropout_4 (Dropout)	(None, 64)	0
dense_3 (Dense)	(None, 32)	2080
dropout_5 (Dropout)	(None, 32)	0
dense_4 (Dense)	(None, 2)	66

- Recall:
    Recall measures the ability to capture positive instances, calculated by dividing true positives (TP) by the sum of true positives (TP) and false negatives (FN).

$$\text{Recall} = TP/(TP + FN) \tag{3}$$

- F1-Score:
    The F1-score offers a balanced assessment, combining precision and recall using the harmonic mean formula:

$$\text{F1-Score} = 2 * (\text{Precision} * \text{Recall})/(\text{Precision} + \text{Recall}) \tag{4}$$

4   Results Analysis

The outcomes derived from our extensive investigation into violence detection, utilizing the CNN + LSTM model, illuminate compelling insights into the model efficacy in discerning and identifying violent actions within intricate video sequences. This research not only sheds light on the model's discernment capabilities but also endeavors to unravel the multifaceted dimensions of violence detection, delving into the intricacies that define the model's effectiveness in real-world scenarios.

The evaluation of the CNN + LSTM model is meticulously examined through an array of performance metrics, transcending the conventional boundaries of assessment. This comprehensive evaluation collectively provides a nuanced and holistic perspective on the model's multifaceted capabilities in the realm of violence detection.

The accuracy metric serves as a fundamental benchmark, gauging the model's overall correctness in predicting violent actions. Precision delves into the model's precision in correctly identifying violent instances, while recall assesses its ability to capture the entirety of violent actions within the video sequences. The F1-score, a harmonized metric of precision and recall, offers a balanced measure of the model's effectiveness in violence detection (Table 2).

Table 2. Performance Evaluation using various deep learning models.

Classifier	Accuracy (%)	Precision (%)	Recall (%)	F1-Score (%)
CNN + LSTM	90	84	96	90

The CNN + LSTM model stands out with an impressive accuracy of 90%, showcasing a well-balanced performance with 84% precision, 96% recall, and an F1-score of 90.00%. This combination of spatial and temporal analysis proves effective in capturing local motion patterns indicative of violent actions. Confusion matrix and ROC Curve are shown in Figs. 4 and 5.

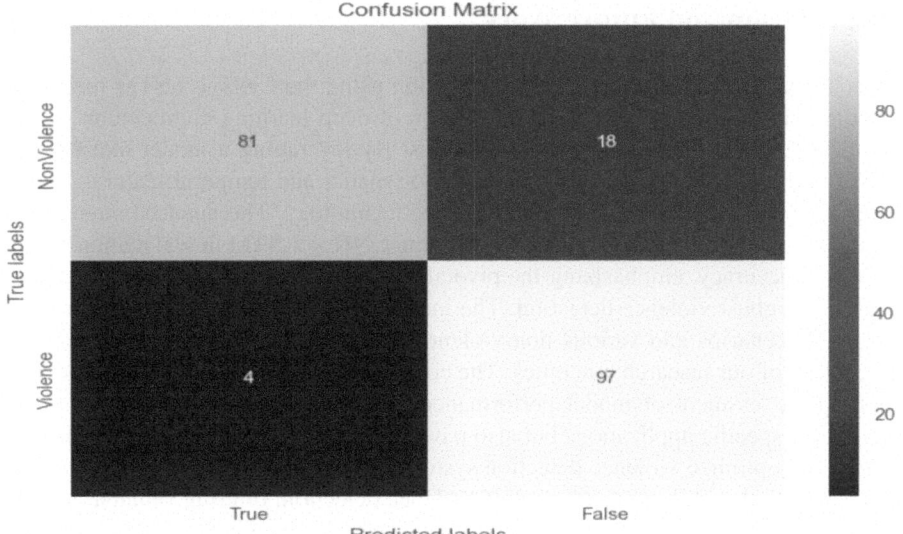

Fig. 4. Confusion Matrix of CNN + LSTM Model

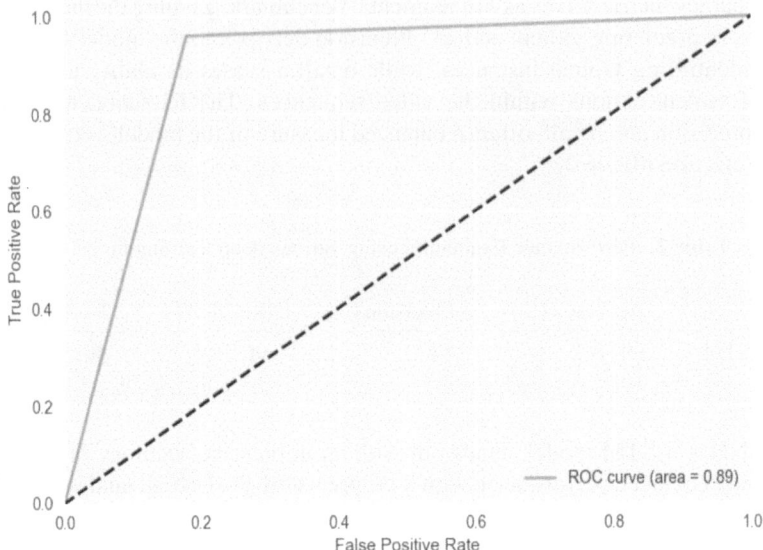

Fig. 5. ROC curve of CNN + LSTM Model

5 Conclusion and Future Work

In conclusion, our research on violence detection using the CNN + LSTM model has yielded valuable insights into the effectiveness of deep learning architectures in discerning violent actions within video sequences. By leveraging a model like CNN + LSTM, we conducted a thorough exploration of spatial and temporal features, significantly advancing the landscape of surveillance technology. The nuanced strengths of each model were evident in our findings, with the CNN + LSTM model demonstrating exceptional accuracy, emphasizing the pivotal role of combining spatial and temporal analyses for robust violence detection. The inclusion of diverse content, spanning real street fight scenarios and various non-violent activities, enhanced the robustness and applicability of our research outcomes. The comprehensive evaluation results provided a thorough assessment of model performance. This analysis not only aids in model selection for specific applications but also paves the way for future research endeavors to refine and optimize violence detection systems. In essence, our research contributes significantly to the evolving landscape of violence detection, offering valuable insights into the capabilities of deep learning models and establishing a foundation for continued advancements in surveillance technology. By addressing the challenges posed by real-world violence scenarios, our aim is to enhance the practical applicability of our models, ultimately contributing to the development of safer and more secure public spaces.

Building upon the foundation laid by our current research on violence detection using CNN + LSTM, there are several promising avenues for future exploration and enhancement: Investigate advanced temporal fusion techniques to further enhance the integration of temporal features extracted by the CNN + LSTM. Explore methods to dynamically adjust the weightage of temporal information based on the evolving nature of violent actions in video sequences. Assess the synergies between visual and non-visual modalities for enhanced violence detection capabilities. Explore techniques for generating visual explanations for model predictions, fostering trust, and understanding in real-world deployment scenarios. Investigate the development of an adaptive model architecture that dynamically adjusts its complexity based on the complexity of the observed scene, optimizing resource utilization. Explore techniques to automatically select the most relevant model architecture for a given surveillance environment. By pursuing these future research directions, we aim to not only enhance the performance of our violence detection system but also contribute to the broader landscape of machine learning applications in surveillance and security.

Availability of Data and Materials. The data set used in the experiments can be found on: https://www.kaggle.com/datasets/karandeep98/real-life-violence-and-nonviolence-data.

References

1. Marszałek, M., Laptev, I., Schmid, C.: Actions in context, In: IEEE Computer Society Conference on Computer Vision and Pattern Recognition Workshops (2009)
2. Gracia, I.S., Suarez, O.D., Garcia, G.B., Kim, T.-K.: Fast fight detection. PLoS ONE **10**(4), e0120448 (2015)
3. Dhiman, C., Vishwakarma, D.K.: A review of state-of-the-art techniques for abnormal human activity recognition. Eng. Appl. Artif. Intell. **77**, 21–45 (2019)
4. Ben Mabrouk, A., Zagrouba, E.: Abnormal behavior recognition for intelligent video surveillance systems: a review. Expert Syst. Appl. **91**, 480–491 (2018). https://doi.org/10.1016/j.eswa.2017.09.029
5. Li, Q., Li, W.: A novel framework for anomaly detection in video surveillance using multi-feature extraction. In: 9th International Symposium on Computational Intelligence and Design (ISCID), vol. 1, pp. 455–459 (2016)
6. Ullah, F.U.M., et al.: A comprehensive review on vision-based violence detection in surveillance videos. ACM Comput. Surv. **55**(10), 1–44 (2023). https://doi.org/10.1145/356 1971
7. Omarov, B., et al.: State-of-the-art violence detection techniques in video surveillance security systems: a systematic review. PeerJ Comput. Sci. **8**, e920 (2022)
8. Choqueluque-Roman, D., Camara-Chavez, G.: Weakly supervised violence detection in surveillance video. Sensors **22**(12), 4502 (2022)
9. Irfanullah, et al.: Real time violence detection in surveillance videos using Convolutional Neural Networks. Multimed. Tools Appl. **81**(26), 38151–38173 (2022)
10. Mahmoodi, J., Nezamabadi-pour, H., Abbasi-Moghadam, D.: Violence detection in videos using interest frame extraction and 3D convolutional neural network. Multimed. Tools Appl. **81**(15), 20945–20961 (2022)
11. Freire-Obregón, D., et al.: Inflated 3D ConvNet context analysis for violence detection. Mach. Vis. Appl. **33**, 115 (2022)
12. Sharma, S., et al.: A fully integrated violence detection system using CNN and LSTM. Int. J. Electr. Comput. Eng. **11**(4), 3374 (2021)
13. Hung, B.T., et al.: Violent video detection by pre-trained model and CNN-LSTM approach. In: Singh Mer, K.K., Semwal, V.B., Bijalwan, V., Crespo, R.G. (eds.) Proceedings of Integrated Intelligence Enable Networks and Computing: IIENC 2020, pp. 979–989. Springer Singapore, Singapore (2021)

14. Dandage, V., et al.: Review of violence detection system using deep learning. Int. Res. J. Eng. Technol. **6**(12), 1899–1902 (2019)
15. Ramzan, M., et al.: A review on state-of-the-art violence detection techniques. IEEE Access **7**, 107560–107575 (2019)
16. Patel, M.: Real-Time Violence Detection Using CNN-LSTM." arXiv preprint arXiv:2107. 07578 (2021)
17. Sumon, S.A., et al.: Violence detection by pretrained modules with different deep learning approaches. Vietnam J. Comput. Sci. **07**(01), 19–40 (2020). https://doi.org/10.1142/S21968 88820500013
18. Mumtaz, A., Sargano, A.B., Habib, Z.: Fast learning through deep multi-net CNN model for violence recognition in video surveillance. Comput. J. **65**(3), 457–472 (2022)
19. Vosta, S., Yow, K.-C.: A cnn-rnn combined structure for real-world violence detection in surveillance cameras. Appl. Sci. **12**(3), 1021 (2022)
20. Vieira, J.C., et al.: Low-cost CNN for automatic violence recognition on embedded system. IEEE Access **10**, 25190–25202 (2022)
21. Akash, S.A.A., et al.: Human violence detection using deep learning techniques. J. Phys. Conf. Ser. **2318**(1), 012003 (2022)
22. Tripathi, A., et al.: Criminals as well as crime detection using Machine Learning & OpenCV. Int. Res. J. Modernization Eng. Technol. Sci. **3**(4), 2135–2141 (2021)
23. Wang, D., et al.: Baseline results for violence detection in still images. In: 2012 IEEE Ninth International Conference on Advanced Video and Signal-Based Surveillance. IEEE (2012)
24. Li, C., et al.: End-to-end multiplayer violence detection based on deep 3DCNN. In: Proceedings of the 2018 VII International Conference on Network, Communication, and Computing (2018)
25. Miranda Calero, J.A., et al.: Bindi: affective internet of things to combat gender-based violence. IEEE Internet of Things J. **9**(21), 21174–21193 (2022). https://doi.org/10.1109/JIOT. 2022.3177256
26. Selvi, E., et al.: Suspicious actions detection system using enhanced CNN and surveillance video. Electronics **11**(24), 4210 (2022)
27. Manikandan, V.P., Rahamathunnisa, U.: A neural network aided attuned scheme for gun detection in video surveillance images. Image Vis. Comput. **120**, 104406 (2022)
28. Sarkar, S., et al.: Design of Weapon Detection System. In: 2022 3rd International Conference on Electronics and Sustainable Communication Systems (ICESC). IEEE (2022)
29. Soliman, M.M., et al.: Violence recognition from videos using deep learning techniques. In: 2019 Ninth International Conference on Intelligent Computing and Information Systems (ICICIS). IEEE (2019)
30. Soliman, M., Kamal, M., Nashed, M., Mostafa, Y., Chawky, B., Khattab, D.: Violence recognition from videos using deep learning techniques. In: Proceedings of the 9th International Conference on Intelligent Computing and Information Systems (ICICIS'19), pp. 79–84. Cairo (2019)

Role of ChatGPT in Decision Making Across Industries: An Indian Perspective

Debajani Sahoo[1] and Shubhangi V. Urkude[2(✉)]

[1] Department of Marketing, IBS, IFHE University, Hyderabad, Telangana, India
`debajani@insindia.org`
[2] Department of Operations and IT, IBS, IFHE University, Hyderabad, Telangana, India
`ushubhu@gmail.com`

Abstract. This study delves deeply into the functions of ChatGPT, an AI model built on the GPT-3.5 architecture, across multiple industries. Through ChatGPT, artificial intelligence (AI) and natural language processing (NLP) are combined to create new opportunities for a variety of sectors. This summary highlights the report's analysis of ChatGPT's applications in marketing, education, healthcare, investment decision-making, and agriculture sector. The survey methodology involves online distribution of questionnaires to gather insights from respondents in each sector. The data was collected using Google Forms, analyzed and interpreted to provide nuanced perspectives on ChatGPT's role in these sectors. While ChatGPT demonstrates potential to enhance efficiency and productivity, challenges such as data privacy, biases, and overreliance on technology must be addressed. The report concludes that responsible integration of ChatGPT in various sectors necessitates a balance between AI capabilities and human expertise. As technology evolves, the transformative potential of ChatGPT continues to shape these industries and contribute to responsible innovation.

Keywords: ChatGPT · decision making · AI-driven technologies · Industry application

1 Introduction

In the modern era, industries worldwide are leveraging AI-driven technologies to optimize operations and decision-making processes. ChatGPT's natural language capabilities are at the forefront of this transformation, enabling innovative communication, information retrieval, and problem-solving. The report focused into how ChatGPT addresses distinct challenges and opportunities in different sectors [1].

ChatGPT revolutionizes marketing strategies by enabling personalized engagement, content creation, and market analysis [2]. ChatGPT services in education industry act as a virtual assistant, enriching learning environments by providing valuable information and resources to students that, improves student productivity. ChatGPT in healthcare industry aids in patient interactions, diagnoses, and medical research. Survey data portrays a positive sentiment toward ChatGPT's potential to enhance communication in the field

© The Author(s), under exclusive license to Springer Nature Switzerland AG 2025
A. K. Bairwa et al. (Eds.): ICCAIML 2024, CCIS 2184, pp. 99–112, 2025.
https://doi.org/10.1007/978-3-031-71481-8_8

of spinal surgical practice. Similarly, ChatGPT's rapid data processing abilities in finance investment industry shows promise in aiding investment decisions. Respondents express varying degrees of agreement on its potential to mitigate biases and emotional influences. The integration of ChatGPT with IoT devices in agriculture is met with skepticism, to optimize irrigation, nutrient application, and pest management remains an area for exploration.

The role of ChatGPT is undoubtedly worth addressing due to its profound impact on various sectors and its potential to reshape industries and human interactions. ChatGPT role spans from improving customer interactions and personalizing marketing strategies to enhancing education, healthcare, investment, and agriculture sectors needs to explore further [3] from emerging market prospective.

1.1 Marketing Sector: Personalization and Customer Engagement

In the realm of marketing, ChatGPT has revolutionized the way brands interact with their audience. The capacity of technology to produce replies that resemble those of a human being and to have lively discussions has created new opportunities for tailored consumer engagements. ChatGPT can craft compelling marketing content, draft engaging social media posts, and even provide real-time customer support through chatbots [4]. This enhances brand loyalty and customer satisfaction by delivering prompt and accurate responses.

By analyzing consumer preferences and behaviors, ChatGPT aids in developing data-driven marketing strategies [5]. Its proficiency in natural language processing enables it to sift through vast volumes of consumer data to extract valuable insights. Marketers can tell their campaigns to align with consumer sentiments, resulting in higher conversion rates and improved ROI.

1.2 Education Sector: Interactive Learning and Knowledge Dissemination

The education sector has witnessed a paradigm shift with the integration of ChatGPT. Traditional classrooms are no longer the sole venues for learning. ChatGPT facilitates interactive and personalized learning experiences through virtual tutors and educational chatbots [6]. Students can seek clarifications, receive explanations, and engage in discussion, transcending geographical barriers. Furthermore, ChatGPT serves as a knowledge repository, enabling seamless access to information [7]. ChatGPT assists educators in preparing course materials, generating assessment questions, and providing real-time feedback to students [8]. Language barriers are diminished as ChatGPT can facilitate translation, making educational resources accessible to a global audience.

1.3 Healthcare Sector: Diagnostics, Consultations, and Medical Research

In healthcare, ChatGPT's potential is harnessed for diagnostics and patient interactions. Chatbots powered by the model can engage patients in preliminary symptom assessments, audience healthcare providers in offering informed guidance [9]. While these chatbots do not replace medical professionals, they expedite the triage process; ensuring patients receive appropriate care more efficiently. ChatGPT's impact extends to medical research as well. It assists researchers in analyzing vast medical literature, identifying potential drug interactions, and even predicting disease outbreaks [10]. Its rapid information processing capabilities expedite the development of medical solutions and contribute to evidence-based decision-making.

1.4   Investment Sector: Data Analysis and Financial Insights

Data drives investment decisions, and ChatGPT excels in data analysis. It may sort through market patterns, economic statistics, and financial reports to give investors information about possible dangers and possibilities. ChatGPT's ability to generate narratives from raw data simplifies complex financial information, aiding investors in making informed choices. Furthermore, financial advisory services are enhanced through the integration of ChatGPT [11]. It provides tailored advice according to each user's financial objectives and risk tolerance. This democratization of financial advice empowers a broader demographic to navigate the intricacies of investment.

1.5   Agriculture Sector: Precision Farming and Crop Management

In agriculture, ChatGPT contributes to the burgeoning field of precision farming. Analyzing data from sensors, drones, and satellites assists farmers in optimizing irrigation, fertilization, and pest control. The model generates insights on crop health, disease detection, and yield estimation, enabling proactive decision-making to enhance productivity [12]. Additionally, ChatGPT serves as a knowledge hub for farmers. It answers questions about crop diseases, soil management, and sustainable practices. Farmers can access information in real time, aiding them in adapting to changing conditions and adopting innovative approaches.

1.6   ChatGPT Tools

A new multimodal GPT-4 version from ChatGPT enables users to upload and examine a variety of document types. DALL-E 3, integrated browsing, and advanced data analysis are all included in GPT-4 all Tools. It's possible that some new features will render many third-party ChatGPT plugins. In response to this difficulty, some AI systems have created custom solutions that may be a better fit for specific requirements. Many popular and up-to-date ChatGPT tools are available, including GitHub Copilot, OpenAI Playground, Bard, Jasper Chat, ChatSonic, and many more. Various ChatGPT programs examine financial data to offer insights on potential investments, including stock prices and market patterns. ChatGPT can also be used to automate customer care for financial services by responding to consumer inquiries and offering real-time investment recommendations.

2   Literature Review

The growth of ChatGPT across these diverse sectors is poised to continue it's upward trajectory. ChatGPT's capabilities are expected to become even more refined and versatile as AI technologies advance. Ethical considerations and data privacy will remain paramount, necessitating responsible AI deployment. Collaborations between AI developers and sector-specific experts will drive innovation and yield more tailored solutions [13]. In order to ensure that the advantages of AI technologies are maximized and the risks are minimized, governments and regulatory agencies will be essential in guiding the appropriate implementation of this technology [14]. The part of ChatGPT is worth addressing due to its transformative potential across sectors. India stands at a juncture where it can leverage ChatGPT's capabilities to drive innovation, improve services, and contribute to the global AI landscape., that enhance its competitiveness and socioeconomic growth globally [15].

The use of ChatGPT in marketing also entails certain risks and challenges. It is crucial to consider these factors to ensure responsible and ethical usage such as; data from inaccurate sources, reliance on outdated data, loss of brand identity due to similar responses generated for different marketers, overreliance on AI tools, job displacement, perpetuation of biases if not properly designed and tested, and the need for professionalism and expertise in using ChatGPT effectively [5]. Marketers can leverage ChatGPT in market research to collect and condense consumer feedback and social media conversations, providing valuable insights into consumers' vocabulary, perceptions, and attitudes towards products and marketing campaigns that facilitating better-informed decision-making, assists in understanding customer questions, suggesting solutions, and responding quicker.

It can also assist marketers in generating ideas for marketing campaigns, refining ad concepts, personalizing campaigns, and analyzing consumer behavior to improve product development and marketing strategies [16, 17].

3   Research Objective

This study aims to do a thorough analysis of ChatGPT's function in several domains, including marketing, education, healthcare, investment decision-making, and agriculture. The report aims to achieve the following research objectives:

- How ChatGPT is utilized in various sectors and the specific applications it brings to each domain.
- Assess the perceived benefits and challenges associated with integrating ChatGPT within each sector, highlighting both the potential advantages and limitations.
- Analyze the opinions and perceptions of stake holders; including professionals, consumers, educators, and students, towards ChatGPT's impact on the sectors under study.
- Evaluate user acceptance, comfort, and willingness to adopt AI-driven technologies like ChatGPT within each sector, identifying adoption factors.
- Examine the sustainability and long-term viability of ChatGPT's integration within sectors, considering potential shifts in industry practices and evolving technology.

4   Research Methodology

4.1   Data collection

To perform this study sample data was collected by circulating the google form on a likert scale of 1 to 5 having 1 as strongly disagree, 2 as disagree, 3 as neutral, 4 as agree and 5 as strongly agree. The questionnaire was distributed among relevant population. Data consist of 85 samples with demographic details and some questions related to use of ChatGPT in education sector, investment banking, agriculture, marketing strategies and medical sector. Collected data was analyzed to get the understanding using excel.

The questions are included below:

- Chat GPT improves student's productivity by providing valuable information and resources
- Chat GPT should be used as a complementary tool to human interaction and students' efforts in learning.
- Chat GPT can significantly enhance marketing strategies by generating content quickly and with potentially higher quality than human content creators.
- Chat GPT can improve customer interactions and provide valuable insights into consumer feedback and attitudes, thereby facilitating more efficient and effective market research in the marketing industry.
- Chat GPT's ability to process large volumes of information quickly can be advantageous for investment professionals in making informed decisions.
- Use of Chat GPT in investment decision making can help reduce cognitive biases and emotional influences on investment decisions, providing a more rational perspective
- How positively do you perceive the integration of ChatGPT with IoT devices in agriculture for optimizing irrigation schedules, nutrient application, and pest management?
- Chat GPT has the potential to drive innovation and enhance productivity in the agricultural sector.
- Chat GPT/GPT-4 can enhance communication between medical staff, patients, and their relatives in the field of spinal surgical
- Chat GPT/GPT-4 can be a valuable tool in accelerating the learning curve and supporting spine surgeons in mastering

5   Result and Discussion

5.1   Education Sector

Figures 1 and 2 shows, the application of ChatGPT in the education sector. In Fig. 1, 53% respondents are saying that the ChatGPT was helping the students in improving the productivity, whereas 21% respondents having no idea about the use of ChatGPT in education sector. 14%, 6%, 6% a respondent was strongly agree, strongly disagree and disagree on the use in education sector respectively. Overall, the survey results suggest that ChatGPT is a valuable tool to help students learn and be more productive. However, a small group of students believe that ChatGPT is not helpful.

Figure 2, shows ChatGPT can be used as tool between machine and student's learning efforts. In this chart 46% of the students are agreed on the statement, whereas 28% are neutral about the opinion, 11% are strongly agree, 9% disagree and 6% are strongly disagree on the statement. This result shows that there is positive impact of chat gpt in the education sector and we will recommend it use for the better future of students who want to learn new thing, explore new ideas on their own.

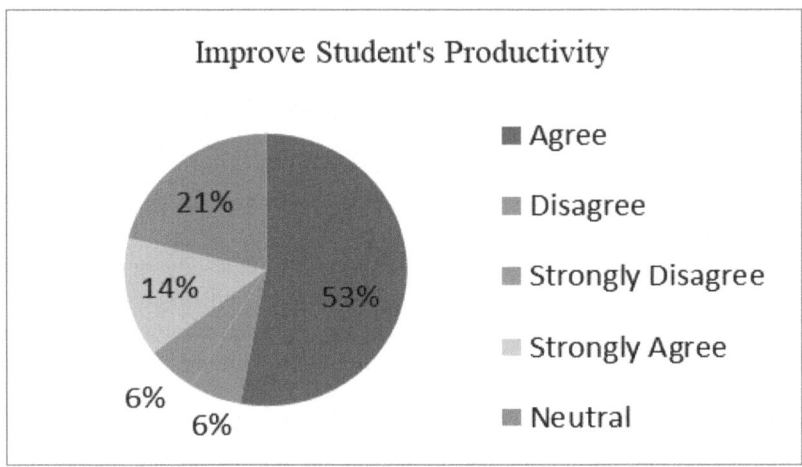

Fig. 1. Improve Student's productivity

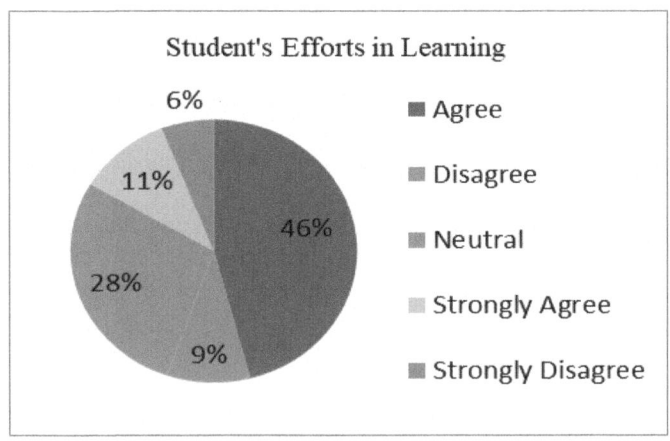

Fig. 2. Student's Efforts in Learning

5.2 Marketing Sector

Figures 3 and 4 shows the use of ChatGPT in enhancing marketing strategies by generating the quality contents and improve customer interaction by valuable insights into feedback and attitude. 48% of the respondents agreed on that ChatGPT helps in enhancing marketing strategies through automatic content generation for marketing the product using chatboat. But 21% respondents are neutral about their opinion,13% are disagree and strongly agree respectively. Overall, the pie chart shows that there is a strong belief in the potential of chatbots to enhance marketing strategies using AI based technology. Figure 4 shows that 39% are agree, this suggests that there is a growing belief in the potential of ChatGPT to improve customer interactions and provide valuable insights into consumer feedback. It can produce prose of a human calibre, translate between languages, create a variety of artistic content, and provide insightful answers to queries.6% strongly disagreed with the statement may be concerned about the accuracy and reliability of ChatGPT.

36% neutral whereas 15%, 6% and 4% respondents are strongly agree, disagree and strongly disagree. Overall, the pie chart shows that there is a strong belief in the potential of ChatGPT to improve customer interactions and provide valuable insights into consumer feedback. As ChatGPT technology continues to develop, it is likely that this belief will only grow stronger.

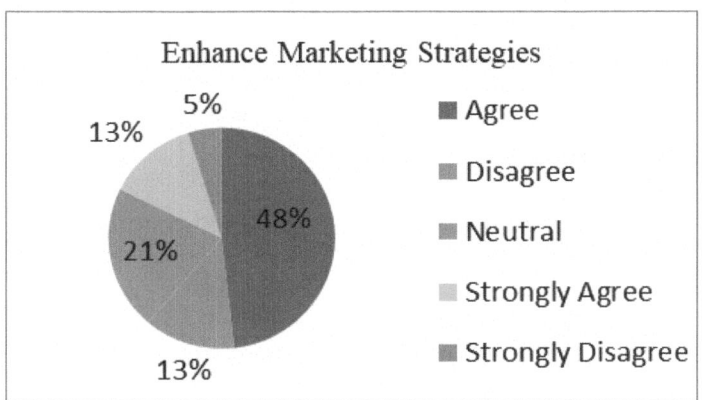

Fig. 3. Enhance marketing strategies

5.3 Investment Banking

Figure 5 shows the application of ChatGPT to process large volume of data at faster rate and helps the investment professional to make the informed decision. The survey revealed a diverse range of responses; with 48% of participants agreeing that ChatGPT's capacity to rapidly process information could indeed be advantageous for investment professionals. 16% of participants expressed a strong agreement with this proposition demonstrates a high level of confidence in the technology's capability, Similarly, the relatively low percentage (7%) of strong disagreement points towards a limited number of detractors who believe that ChatGPT's rapid information processing might not be as beneficial as suggested. The presence of 17% neutral responses highlights a critical aspect of this discussion. This is unsurprising, given that the integration of advanced AI

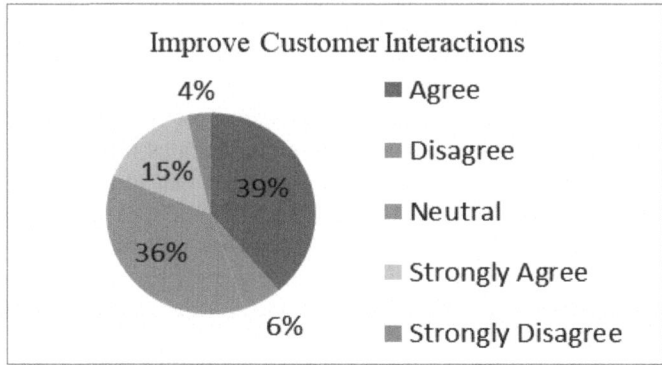

Fig. 4. Improve customer Interactions

into traditional investment practices represents a paradigm shift, which can prompt concerns about the technology's reliability, ethical considerations, and potential unintended consequences. Additionally, investment decisions often require not just data analysis but also a nuanced understanding of market dynamics, investor sentiment, and macroeconomic trends – areas where human judgment and expertise still hold considerable value. In conclusion, the survey responses present a compelling narrative regarding the potential advantages of ChatGPT's quick information processing for investment professionals.

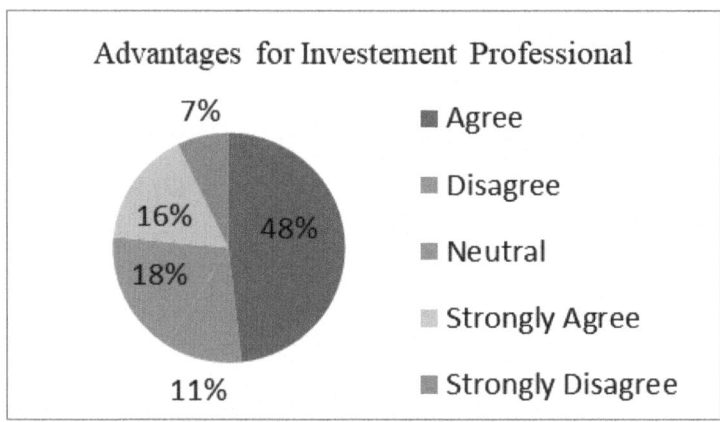

Fig. 5. Advantages for investment professional

Figure 6 shows use of ChatGPT in reducing the biased and emotional decision making in investment banking. The analysis of the survey responses regarding the belief in ChatGPT's potential to mitigate biases and emotional decisions in investment making provides valuable insights into the perceptions surrounding this topic. A notable 38% of participants expressed agreement that the utilization of ChatGPT in investment decision-making could aid in reducing congenital biases and emotional decision-making. Furthermore, 9% of participants strongly agree with the notion, which indicates a smaller yet significant subset of respondents who hold a more emphatic stance on the matter. The sizeable percentage of neutral responses, at 33%, reflects a significant segment of participants who are undecided or unsure about the extent to which ChatGPT can address biases and emotional factors in investment decisions. On the contrary, 12% of participants disagreed that ChatGPT could help reduce biases and emotional decisions, holds reservations about the technology's capacity to completely eliminate these innate human tendencies. Ultimately, while ChatGPT and similar AI models may have the potential to offer a more rational perspective by reducing biases and emotional decisions, it's crucial to acknowledge that AI is a tool that works in conjunction with human expertise. Successful integration will likely involve striking a balance between technology and human judgment, leveraging AI's strengths while acknowledging its limitations.

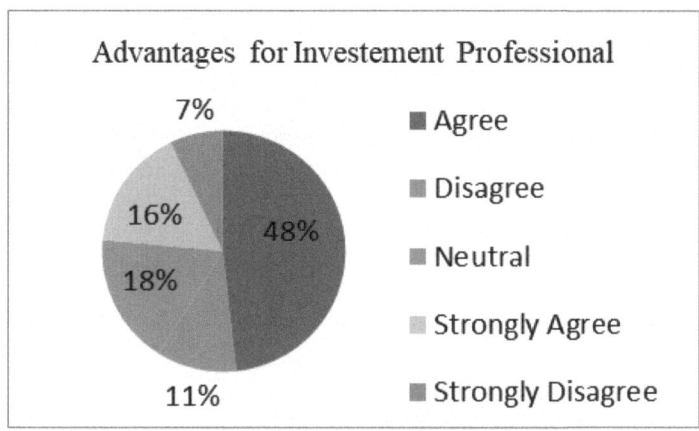

Fig. 6. Investment decision making

5.4 Agriculture Sector

Figure 7 shows the results of a survey on how people perceive the integration of ChatGPT with IoT devices in agriculture for optimizing irrigation schedules, nutrient application, and pest management. The majority of respondents (42%) agree with the integration of ChatGPT with IoT devices in agriculture. This suggests that many people are skeptical about the potential benefits of this technology. The neutral (31%), disagree (7%), strongly agree (14%) and strongly disagree (6%) responses are distributed respectively. There are a few possible reasons for the skepticism towards the integration of ChatGPT with IoT devices in agriculture. One reason is that this technology is still relatively new and untested. There is not yet enough data to demonstrate its effectiveness in a real-world setting. For example, ChatGPT could be used to collect and analyze data from IoT sensors to optimize irrigation schedules and nutrient application, identify and manage pests.

Overall, the results of this survey suggest that there is still some uncertainty about the potential benefits of integrating ChatGPT with IoT devices in agriculture. However, there are also some potential benefits that could make this technology worth exploring further. The survey's findings represent merely a moment in time in the public's perception. Should the survey be administered once more in the future, the outcomes can vary. It is important to note that the survey did not ask respondents to explain their reasons for their responses. This makes it difficult to understand the underlying reasons for the skepticism towards the integration of ChatGPT with IoT devices in agriculture.

Figure 8 shows, to what extends ChatGPT have the potential to drive innovation and enhance productivity in the agricultural sector. The majority of respondents (39%) agree the potential of chatbots in agriculture. This suggests that many people are skeptical about the ability of chatbots to drive innovation and enhance productivity in this sector. The neutral (35%), strongly agree (12%), disagree (9%) and strongly disagree (5%) responses are distributed respectively. There are a few possible reasons for the skepticism towards the potential of chatbots in agriculture. One reason is that this technology is still relatively new and untested. There is not yet enough data to demonstrate its effectiveness in a real-world setting. For example, chatbots could be used to: Provide farmers with information and advice on crop management, help farmers to automate tasks, such as ordering supplies or scheduling deliveries, monitor crops and livestock for signs of disease or pests, connect farmers with other businesses and organizations in the agricultural sector.

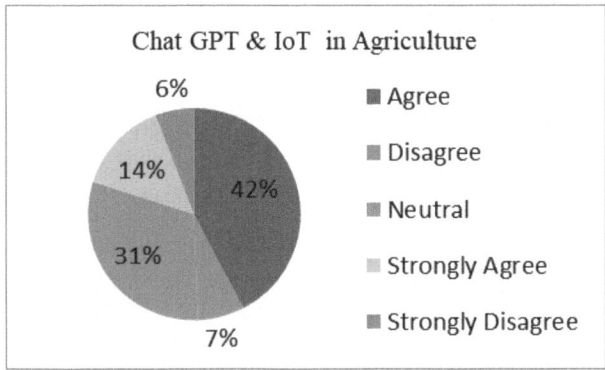

Fig. 7. Integration of ChatGPT & IoT

Overall, the pie chart provides a useful overview of public opinion on this issue. However, it is important to consider the limitations of the survey results before drawing any definitive conclusions. It was believed that these challenges can be overcome, and that chatbots have the potential to revolutionize the agricultural sector. Chatbots can help farmers to improve their yields, reduce their costs, and make better decisions about their businesses. They can also help to connect farmers with other businesses and organizations in the agricultural sector, and to provide farmers with access to information and advice. I was believed that this technology has the potential to make a positive impact on the lives of farmers and the world's food supply.

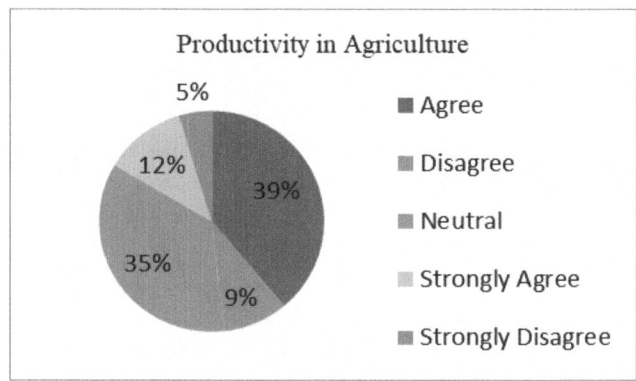

Fig. 8. Productivity in Agriculture

5.5 Healthcare Sector

Figure 9 shows the results of how much they agree that ChatGPT/GPT-4 can enhance communication between medical staff, patients, and their relatives in the field of spinal surgical practice. The largest group, 39%, agreed that ChatGPT/GPT-4 can enhance communication, 7% strongly disagreed, 35% neutral, 15% disagreed, 4% strongly agreed. Strongly disagree (7%), respondents believes that ChatGPT/GPT-4 is not capable of enhancing communication between medical staff, patients, and their relatives in the field of spinal surgical practice. Agree (39%), respondents believe that ChatGPT/GPT-4 has the potential to significantly enhance communication between medical staff, patients, and their relatives in the field of spinal surgical practice. They may have concerns about the accuracy of the language model, its ability to understand complex medical terminology, or its ability to build rapport with patients. Disagree (15%), respondents was less strongly opposed to the use of ChatGPT/GPT-4, but they still do not believe that it is the best solution for enhancing communication in spinal surgical practice. They may have similar concerns to the strongly disagree group, or they may believe that there are other, more effective ways to improve communication. Neutral (35%), respondents were not sure whether or not ChatGPT/GPT-4 can enhance communication in spinal surgical practice. They may need more information about the language model or its capabilities before forming an opinion.

The strongly agreed (4%), respondents were believed that ChatGPT/GPT-4 has the potential to surely enhance communication between medical staff, patients, and their relatives in the field of spinal surgical practice. This suggests that there are significant concerns about the language model's capabilities and limitations.

Figure 10 shows the results how much they believe that ChatGPT/GPT-4 can be a valuable tool in accelerating the learning curve and supporting them in mastering endoscopic spinal surgery. Most respondents (34%) agreed that ChatGPT/GPT-4 can be a valuable tool. 31% were neutral, 19% disagreed, 7% strongly disagreed and 9% strongly agreed. This suggests that there is no clear consensus among the respondents on the value of ChatGPT/GPT-4 as a tool for learning endoscopic spinal surgery. However, it is worth as the majority of respondents believe that it a valuable tool but most of them were neutral and disagreed with the situation.

Here are some possible reasons for this:

- Surgeons may be skeptical of the ability of a chatbot to provide accurate and reliable information about a complex medical procedure.
- If they were to utilize a chatbot, they could worry about the safety and privacy of their data.
- They may prefer to learn endoscopic spinal surgery from a human instructor.

It is significant to remember that a limited sample of persons participated in this survey. It is plausible that the outcomes could alter if a more extensive and representative sample of individuals was polled. Overall, the results of this survey suggest that there is no clear consensus among respondents on the value of ChatGPT/GPT-4 as a tool for learning endoscopic spinal surgery. However, it is worth noting that most respondents believe that it is a valuable tool. More research is needed to determine the true value of ChatGPT/GPT-4 for this purpose.

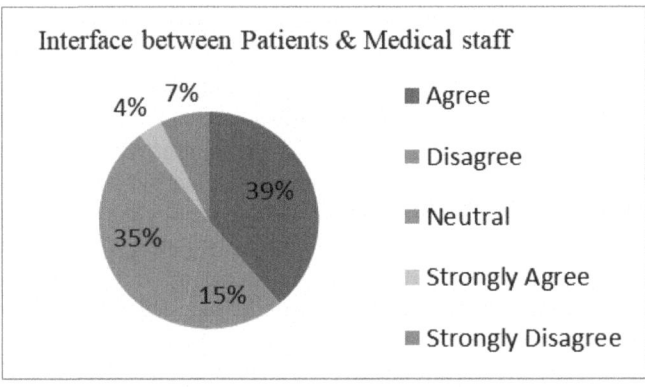

Fig. 9. Enhance communication between patient and medical staff

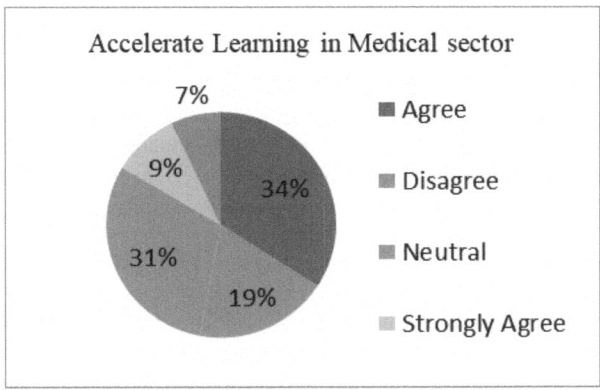

Fig. 10. Accelerate learning in medical staff

6 Limitations

The survey sample size of 85 participants may only represent part of the population in each sector, potentially leading to biased results. Participants who voluntarily took the survey might have pre-existing opinions or experiences related to ChatGPT, leading to skewed results. The survey was conducted online, which might exclude individuals who are less comfortable with technology or have limited internet access. Each sector has unique challenges and complexities that a single survey might only partially capture.

7 Conclusions and Future Scope

ChatGPT's integration across various sectors signifies its role as a transformative force in modern technology. Its application in marketing is revolutionizing customer engagement and content creation. While in education, it's reshaping learning environments through virtual assistance and knowledge facilitation. In healthcare, ChatGPT is improving patient interactions, diagnoses, and medical research. ChatGPT can be used a tool for decision making in financial sector. Additionally, in agriculture, it offers insights into crop management, weather forecasts, and market trends, empowering farmers for sustainable food production. The survey-based analysis provides valuable insights into perceptions within these sectors, showing a mix of optimism and skepticism regarding ChatGPT's capabilities. While it's potential to enhance communication, learning, and decision-making is evident, privacy, reliability, and human expertise must be addressed for successful integration. Ultimately, ChatGPT's transformative impact depends on balancing its strengths with human judgment, thus shaping a future where AI augments various sectors for the better.

In future we can extend this work to collect more real time data and deeper analysis can be performed by collecting sufficient sample to get a generalized results.

References

1. George, A.S., George, A.H.: A review of ChatGPT AI's impact on several business sectors. Partners Univers. Int. Innov. J. **1**(1), 9–23 (2023)
2. Limna, P., Kraiwanit, T., Jangjarat, K., Klayklung, P., Chocksathaporn, P.: The use of ChatGPT in the digital era: Perspectives on chatbot implementation. J. Appl. Learn. Teachi. **6**(1) (2023)
3. Wiredu, J., Kumi, M., Ademola, P.H., MuseshaIyela, P.: An investigation on the characteristics, abilities, constraints, and functions of artificial intelligence (AI): the age of ChatGPT as an essential ultramodern support tool. Int. J. Dev. Res. **13**(05), 62614–62620 (2023)
4. Saputra, R., Nasution, M.I.P., Dharma, B.: The impact of using AI chat GPT on marketing effectiveness: a case study on instagram marketing. Indonesian J. Econ. Manag. **3**(3), 603–617 (2023)
5. Rivas, P., Zhao, L.: Marketing with ChatGPT: navigating the ethical terrain of GPT-based chatbot technology. AI **4**(2), 375–384 (2023). https://doi.org/10.3390/ai4020019
6. Aithal, S., Aithal, P.S.: Effects of AI-based ChatGPT on higher education libraries. Int. J. Manag. Technol. Soc. Sci. **8**(2), 95–108 (2023)
7. Urkude, S., Gupta, K.: Student intervention system using machine learning techniques. Int. J. Eng. Adv. Technol. **8**(6), 21–29 (2019)
8. Grassini, S.: Shaping the future of education: exploring the potential and consequences of AI and ChatGPT in educational settings. Educ. Sci. **13**(7), 692 (2023)
9. Cascella, M., Montomoli, J., Bellini, V., Bignami, E.: Evaluating the feasibility of ChatGPT in healthcare: an analysis of multiple clinical and research scenarios. J. Med. Syst. **47**(1), 33 (2023)
10. Biswas, S.S.: Role of chat GPT in public health. Ann. Biomed. Eng. **51**(5), 868–869 (2023)
11. Mhlanga, D.: Digital transformation education, opportunities, and challenges of the application of ChatGPT to emerging economies. Educ. Res. Int. **2023**, 1–13 (2023). https://doi.org/10.1155/2023/7605075
12. Biswas, S.: Importance of chat GPT in Agriculture: According to chat GPT, 30 Mar 2023. SSRN: https://ssrn.com/abstract=4405391 or https://doi.org/10.2139/ssrn.4405391
13. Urkude, S.V., Hasanuzzaman, V.R.U., Kumar, C.S.: Comparative analysis on machine learning techniques: a case study on Amazon product. Int. J. Mech. Eng. **6**(3), 739–744 (2021)
14. Pokkakillath, S., Suleri, J.: ChatGPT and its impact on education. Res. Hospitality Manag **13**(1), 31–34 (2023)
15. Haugom, E., Lyocsa, S., Halousková, M.:The financial impact of ChatGPT for the higher education industry in the US. Available at SSRN 4546522 (2023)
16. Jain, V., Rai, H., Parvathy, Mogaji, E.: The Prospects and Challenges of ChatGPT on Marketing Research and Practices. SSRN Electron. J. (2023)
17. Tanveer, M., Khan, N., Ahmad, R.: AI support marketing: understanding the customer journey towards the business development (2021)

Identification of Splice Image Forgeries with Enhanced DenseNet201 and VGG19

Satyendra Singh[1] (ID), Rajesh Kumar[1,2](✉) (ID), and Chandrakant Kumar Singh[2] (ID)

[1] Department of Electronic and Communication, University of Allahabad, Prayagraj, India
Satyendra@allduniv.ac.in, rajeshkumariitbhu@gmail.com
[2] Uttar Pradesh Rajrshitandon Open University, Prayagraj, India

Abstract. Digital image manipulation has become increasingly sophisticated, leading to a rise in image forgery cases. In this work proposed model to detect splice and copy-move manipulations using deep learning model, Enhanced DenseNet201 and VGG19, on the widely used CASIA 2.0 and CASIA 1.0 dataset. The proposed methodology focuses on enhancing the accuracy of forge image classification through the utilization of advanced convolutional neural networks (CNNs). Enhanced DenseNet201 and VGG19 models have fine-tuned and trained on the CASIA 1.0 and CASIA 2.0 dataset, which comprises authentic and tampered photos. The models are equipped to identify instances of splice and copy-move forgeries. By leveraging the deep features extracted by these models, the proposed approach demonstrates remarkable performance in accurately distinguishing between genuine and manipulated images. The evaluation of the models is carried out using standard evaluation metrics accuracy, precision, recall, and auc. The achieved results underscore the effectiveness of the Enhanced DenseNet201 approach, with accuracy, precision, recall, and auc values of 0.9181, 0.9157, 0.9113, and 0.9357, respectively. These metrics showcase the models' robustness and capability to classify splice and original images with high precision and recall.

Keywords: copy-move · splicing · DensNet201; VGG19 · deep learning · Transfer learning

1 Introduction

A digital image is a picture that has been created, captured, or digitized electronically, and can be viewed, edited, and manipulated using digital devices and software [1]. A fake image is an image that has been manipulated or generated to misrepresent reality. It can be created using various digital image manipulation techniques and it can be used for various purposes, such as creating a false impression or deception, political propaganda. Fake image deceives viewers into believing that the image is authentic [2]. Manipulated pictures that changed or modified by software tools such as Adobe Photoshop or GIMP. For example, someone might manipulate an image to remove or add certain elements, change colours, or adjust the brightness and contrast to create a false impression [3]. Manipulated images, also known as "photoshopped" images, are digital images that have been altered or edited in some way to change their appearance or convey a different

A. K. Bairwa et al. (Eds.): ICCAIML 2024, CCIS 2184, pp. 113–123, 2025.
https://doi.org/10.1007/978-3-031-71481-8_9

message. The manipulation can be done using various software tools such as Adobe Photoshop, GIMP, or other photo editing software. Manipulated images can be created for many reasons, including artistic expression, advertising, social media, or to deceive people [4]. In Fig. 1 shows the examples of spliced images and original images.

Fig. 1. Original and spliced images

In some cases, manipulated images can be used to create a humorous effect or to emphasize a point or message. However, in other cases, manipulated images can be used to spread false information or to mislead people. Examples of manipulation include altering the colour, lighting, or contrast of an image, removing or adding objects or people, or creating composite images by merging multiple images together. Additionally, other more complex manipulations can include creating an entirely new scene, distorting the shape or size of objects or people, or creating fake images of events that never occurred [5].

In brief, this paper highlights our main contributions as follows:

We proposed an approach for detecting image splicing using Enhanced DenseNet201 models. This method comprises a backbone network to extract features and an artificial neural network-based binary classifier to identify spliced photos.

In the proposed model a combination of convolutional layers from the Enhanced DenseNet201 model (CNN) for feature extraction and fully connected layers (ANN) for the classification task.

In the context of classification, we have been utilized a fully connected artificial neural network as an alternative to conventional classifiers like Support Vector Machines (SVMs).

Remaining paper has been arranged as follows: In Sect. 2, we provide an overview of the relevant existing literature. In Sect. 3, introduce and elaborate on our proposed model, including dataset details and parameter selection. Sect. 4 presents an extensive and detailed summary of our experiment, results, in-depth discussion. Finally, in Sect. 5, conclusions of the presented study and provide insights into future research directions.

2   Literature Review

We conducted a comprehensive review of existing research focused on image forgery detection, thoroughly examining the findings presented in these studies.

Guo et al. [6] describes a study focused on detecting fake colorized images, a relatively new type of image manipulation technique that can be used to deceive object recognition algorithms. In this paper highlights the importance of image forensics in detecting image manipulations and the need for research in detecting emerging techniques like colorization. Also describes the limitations of existing techniques and the contribution of the proposed methods in detecting fake colorized images. Li et al. [7] described model t detect fake colourized images using neural network.

He et al. [8] proposed a splicing manipulation detection model. Authors used Markov feature in DWT and DCT and support vector machine for classification. Hamid, et al. [9] present an image manipulation detection model. The research paper highlights the drawbacks of current techniques used for identifying fake images and emphasizes the requirement for more robust methods that can effectively handle images generated by Generative Adversarial Networks (GANs).

Mo et al. [10] discusses the potential ethical and legal issues that may arise from the use of high-quality fake face images generated by the Generative Adversarial Network (GAN) model in image tampering. The paper proposes a CNN based method for detecting such manipulated face photos and demonstrates through experimental evidence that this model achieves an overall accuracy of over 99.4%. In this work highlights the need for reliable methods for detecting fake face images generated by GANs to address potential ethical and legal concerns.

Wang et al. [11] proposed, an image splicing manipulation becomes more prevalent, the need for accurate detection methods has grown. The authors propose a new approach to detect splice images based on a CNN with a weight combination technique. The presented model uses three types of features like YCbCr, edge, and photo response non-uniformity to distinguish splicing manipulation, which are combined with varying weights determined during the CNN training process. Zhao et al. [12] introduced a chroma space for splicing image forgery detection with SVM classifier.

Abd El-Latif et al. [13] image splicing forgery that is often difficult to detect. The detection of spliced images is a topic of great interest in the field of multimedia forensics and security. Various algorithms have been proposed for spliced image detection, but they often suffer from issues such as low accuracy and high FPR. This paper also performs additional experiments using discrete cosine transform (DCT) and principal component analysis. Thakur et al. [14] proposed spliced photo manipulation detection method. In this model, author used SVM as a classifier. Bird et al. [15] describe a method to classify machine generated images and achieved 92.98 % classification accuracy.

In Table 1 show the comparation of the existing work related to digital image forgery detection.

Table 1. Comparative analysis of the related work

Author	Year	Detection Technique	Dataset	Observation
Guo, et al. [6]	2018	SVM, Fisher Vector	ImageNet	Developed two methods FCID-HIST and FCID-FE. The FCID-FE performed better than other
Wang et al. [11]	2020	YCbCr, CNN, PRNU	CASIA v1 and CASIA v2	This method generates redundant data and not perform better for all splice images
Abd El-Latif et al. [13]	2020	CNN, DWT and SVM	CASIA v1 and CASIA v2	This method works on low dimensional feature vector
Thakur et al. [14]	2018	DWT, SURF and SVM	CASIA v1	It is work on scaling and orientation attacks
Bird et al. [15]	2023	Deep CNN	CIFAKE	Accuracy 92.98%

3 Proposed Model

In this proposed work, we developed a binary classification model to identify image is spliced or original. To the detection of spliced image using a deep neural network based DenseNet201 and VGG19 model. The developed model includes some important steps: input images from CASIA 1.0 and CASIA 2.0 datasets, apply Error Level Analysis, Data preparation, DEnseNet201 deep learning model, Model evaluation, and final classification steps to classify forged or original. These steps of the proposed model demonstrate in Fig. 2.

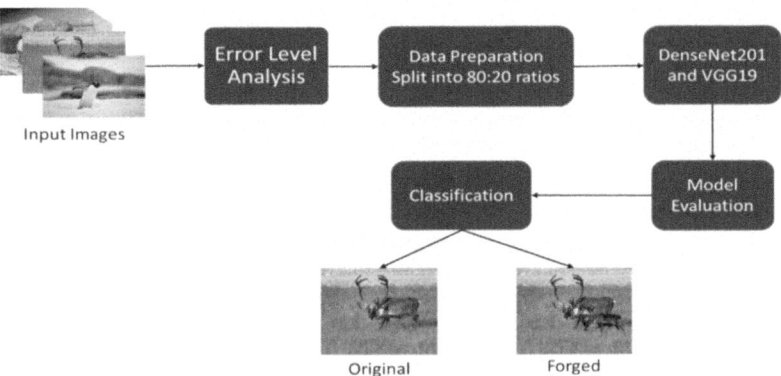

Fig. 2. Block diagram of proposed model

3.1 Dataset Description

Our experiments are conducted on two widely used dataset CASIA 1.0 and CASIA 2.0 for manipulation detection [16]. The CASIA 1.0 has 800 original and 921 manipulated images and CASIA 2.0 revised dataset poses a greater challenge as it introduces post-processing on the boundary area of forged regions. This dataset includes 7,491 original and 5,123 manipulated colour images in various formats such as BMP, JPEG, and TIFF, with sizes ranging from 240 × 160 to 900 × 600 pixels. The CASIA 2.0 datasets contain splicing and copy-move tampered images. In Table 1 shows CASIA 2.0 dataset dimensions.

Table 2. Shows the dataset details which is used in proposed model

Dataset	Original	Manipulated	Total
Casia 1.0	800	921	1721
Casia 2.0	7491	5123	12614

3.2 DenseNet 201

It is a deep convolutional neural network (CNN) architecture and an extension of the original DenseNet model [17]. The primary motivation behind DenseNet201 is to address the vanishing gradient problem and improve information flow in deep CNNs. DenseNet201 is the use of dense connectivity. This dense connectivity allows for maximum information flow across the network, enabling each layer to access the features extracted by all preceding layers. As a result, feature reuse is promoted, and gradients can flow more easily through the network, addressing the vanishing gradient problem. In the Fig. 3 illustrate the DenseNet 201 architecture, which is used in the proposed model.

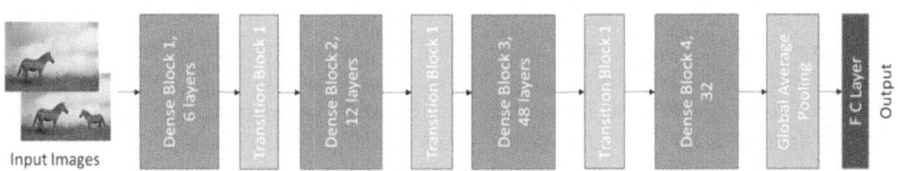

Fig. 3. DenseNet 201 with 4 blocks, transition layers between two blocks, which modify the feature map size using convolutional operation and pooling

In total, the DenseNet201 architecture has 98 convolutional layers, including the layers in the dense blocks and the transition blocks. The transition blocks are responsible for down-sampling the spatial dimensions of the feature maps between dense blocks. In the proposed DenseNet 201, dense block 1 to dense block 4 presents the number of layers 6, 12, 48, and 32 respectively. The architecture also includes a global average pooling layer, which aggregates global information from the entire feature map, converting it into a one-dimensional vector before the final classification layer. This flattened vector is then connected to a series of fully connected (FC) layers. The first FC layer has 256 nodes and applies the Relu activation function, dropout (0.6 rate) to prevent overfitting. Finally, the output from the last FC layer is fed into a FC layer with 2 nodes, corresponding to the binary classification result. The sigmoid activation function used in final layer, which maps the output values to a range (0 and 1), representing the probability of each class.

3.3   Pretrained VGG19 Model

In this section, we propose to leverage the power of pretrained deep learning models to extract high-level features from images. Specifically, we will use the VGG19 neural network, a popular and well-performing convolutional neural network (CNN), pretrained on a large dataset to perform image classification.

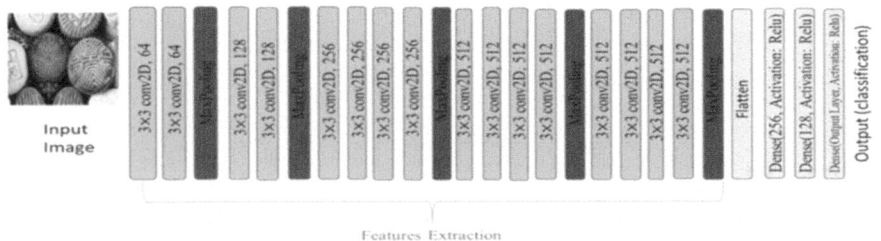

Fig. 4. Proposed architecture of VGG19 model

From Fig. 4 the architecture of the VGG19 model, which is a pre-trained model used for various computer vision tasks such as image classification. It includes 16 conv2D layers, 5 maxpooling layers, and 3 FC layers, along with a sigmoid classifier. The output from the last max pooling operation is then passed to a flatten layer, which transforms the multidimensional data into a 1D vector. This flattened vector is then connected to a series of fully connected layers. The first FC layer has 256 nodes and applies the Relu activation function, dropout layer (rate of 0.5) to prevent overfitting. The process continues with additional fully connected layers, each with decreasing numbers of nodes (64 and 32, respectively) and Relu activation. Finally, output from the last FC layer is fed into a final FC layer with 2 nodes, for binary classification. The block diagram shows the architecture of the VGG19 method, which is a variant of the VGG. This method is designed to classify input images into one of two classes.

3.4   Details of Implementation

Throughout the training process, we train Enhanced DenseNet201 models using the Adam optimizer. Enhanced DenseNet201 is trained on 150 epochs. The starting learning rate for models is set to 0.0001. During training, we use a batch size of 8, which means the models process 8 samples in parallel during each training iteration. For validation purposes, we allocate 10% data for training and validation set to assess the models' performance on unseen data and prevent overfitting. The experiment performed on Google Colab, which serves as our training framework. Google Colab provides cloud-based GPU acceleration, making it a suitable platform for training deep learning models with computational power. The chosen GPU for the experiments is T4 with 25 GB of RAM.

4 Experimental Results

The propose transfer learning model for splice image classification has been done on casia 1.0 and casia 2.0 datasets. The proposed model results illustrate on Table 2 and Table 3. In Table 2 show, training results of the metrices accuracy, loss, precision, recall and AUC.

Table 3. Training results on casia 2.0 and casia 1.0 datasets.

Method	Dataset	Accuracy	Loss	Precision	Recall	AUC
DenseNet201	Casia 1.0	0.8960	0.2607	0.8963	0.8934	0.9588
	Casia 2.0	0.9936	0.023	0.9936	0.9931	0.9990
VGG19	Casia 2.0	0.9201	0.2120	0.9248	0.9253	0.9768

In the Table 3 display the validation results of the proposed model for metrices accuracy, loss, precision, recall and AUC.

Table 4. Validation result on casia 2.0 and casia 1.0 datasets

Method	Dataset	Accuracy	Loss	Precision	Recall	AUC
DenseNet201	Casia 1.0	0.8837	0.317	0.8830	0.8779	0.9417
	Casia 2.0	0.9181	0.5461	0.9157	0.9113	0.9357
VGG19	Casia 2.0	0.8824	0.3265	0.8849	0.8855	0.9368

In Table 4 Compare the proposed results from existing studies which is related to splice image identification (Table 5).

4.1 Analysis of the Results

The experiment has been performed for the spliced image identification with the help of enhanced DenseNet201 transfer learning-based model. The performance of the model measure in the follows metrices as accuracy, recall, precision and auc. The performance of the developed model is better as compared to the results of related studies presented in Table 4.

In Fig. 5 display the train and validation results of the proposed spliced image identification model on CASIA 2.0 and Fig. 6 casia 1.0 dataset. The analysis of the performance of train and validation results of the proposed model shows in Fig. 7.

Table 5. Results compare with existing research

Author	Dataset	classifier	Accuracy
Almawas, et al. [18]	CASIA 2.0	SVM	86.5%
		Naïve Bayes	87.01%
		KNN	88.6%
	CASIA 1.0	SVM	49.89%
Jaiswal, et al. [19]	CASIA 2.0	Multiclass Model	70.26%
		KNN	59.91%
		Naïve Bayes	59.91%
Proposed	**CASIA 2.0**	**DenseNet201**	**91.81%**
		VGG19	**88.49%**

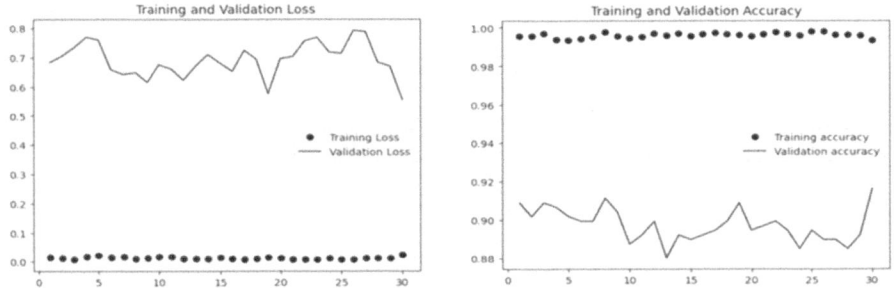

Fig. 5. Performance of the DenseNet201 model on casia 2.0 dataset.

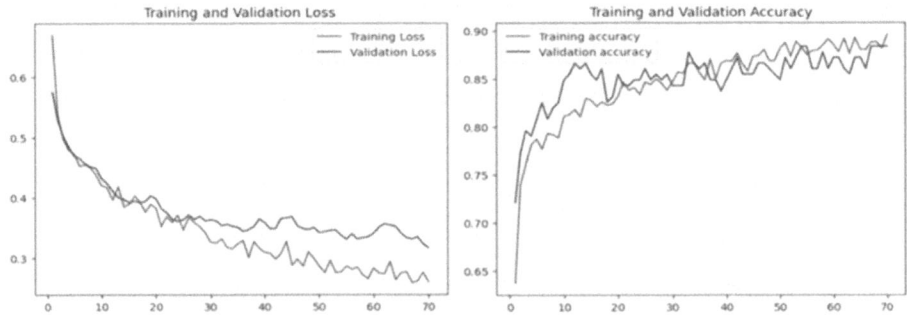

Fig. 6. Performance of DenseNet201 on casia v1.0 dataset.

4.2 Model Performance Evaluation

The important parameter for the performance evaluation of developed binary classification model are

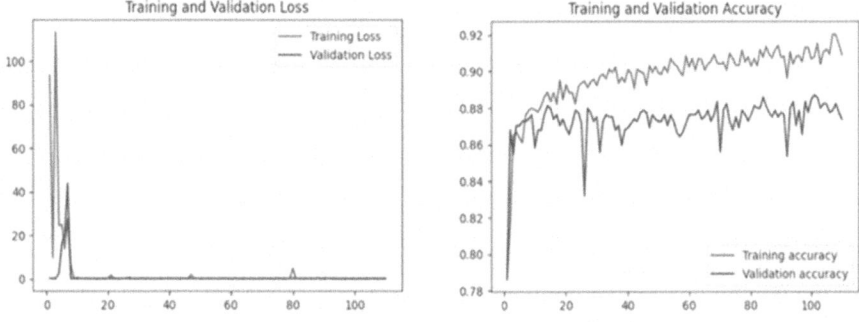

Fig. 7. VGG19 model performance on CASIA 2.0

True Positives (Tp): The number of cases that were accurately categorized as positive (Forge).

$$\text{True Positives (Tp)} = \sum (\text{True Positive Predictions}) \qquad (1)$$

False Positives (Tp): The number of cases that were misclassified as positive when they were negative (Original).

$$\text{False Positives (Tp)} = \sum (\text{False Positive Predictions}) \qquad (2)$$

True Negatives (Tn): The cases that were correctly categorized as negative (Original).

$$\text{True Negatives (Tn)} = \sum (\text{True Negative Predictions}) \qquad (3)$$

False Negatives (Tn): The number of cases falsely categorized as negative (Original) even though they are positive (Forge).

$$\text{False Negatives (Tn)} = \sum (\text{False Negative Predictions}) \qquad (4)$$

The Eqs. 1, 2, 3, and 4 represent formula for the measurement of performance of classification model [20, 21].

AUC: it is stand for area under curve and calculated using the area under the ROC curve [22].

5 Conclusion

This research contributes to the field of digital image forensics by introducing a powerful approach for identification of splice and copy-move forgeries. Enhanced DenseNet201 and VGG19 models performed on the CASIA v1.0 and CASIA 2.0 dataset. The results of the proposed model have been impressive than other studies. The deep learning-based classifier has been better results as compared to conventional classifiers like Support Vector Machines (SVMs). The proposed model is very useful for the spliced image identification and future research in this field.

References

1. Asghar, K., Habib, Z., Hussain, M.: Copy-move and splicing image forgery detection and localization techniques: a review. Aust. J. Forensic Sci. **49**(3), 281–307 (2017)
2. Qazi, T., et al.: Survey on blind image forgery detection. IET Image Proc. **7**(7), 660–670 (2013)
3. Al-Qershi, O.M., Khoo, B.E.: Passive detection of copy-move forgery in digital images: state-of-the-art. Forensic Sci. Int. **231**(1–3), 284–295 (2013)
4. Mahmood, T., Nawaz, T., Irtaza, A., Ashraf, R., Shah, M., Mahmood, M.T.: Copy-move forgery detection technique for forensic analysis in digital images. Math. Probl. Eng. **2016**, 1–13 (2016)
5. Kakar, P., Sudha, N., Ser, W.: Exposing digital image forgeries by detecting discrepancies in motion blur. IEEE Trans. Multimedia **13**(3), 443–452 (2011)
6. Guo, Y., Cao, X., Zhang, W., Wang, R.: Fake colorized image detection. IEEE Trans. Inf. Forensics Secur. **13**(8), 1932–1944 (2018)
7. Li, Y., Zhang, Y., Lu, L., Jia, Y., Liu, J.: Using neural networks for fake colorized image detection. In: Advances in Digital Forensics XV: 15th IFIP WG 11.9 International Conference, Orlando, FL, USA, 28–29 Jan 2019, Revised Selected Papers 15, pp. 201–215. Springer International Publishing (2019)
8. He, Z., Lu, W., Sun, W., Huang, J.: Digital image splicing detection based on Markov features in DCT and DWT domain. Pattern Recogn. **45**(12), 4292–4299 (2012)
9. Hamid, Y., Elyassami, S., Gulzar, Y., Balasaraswathi, V.R., Habuza, T., Wani, S.: An improvised CNN model for fake image detection. Int. J. Inf. Technol. **15**(1), 5–15 (2023)
10. Mo, H., Chen, B., Luo, W.: Fake faces identification via convolutional neural network. In Proceedings of the 6th ACM Workshop on Information Hiding and Multimedia Security, pp. 43–47 (2018)
11. Wang, J., Ni, Q., Liu, G., Luo, X., Jha, S.K.: Image splicing detection based on convolutional neural network with weight combination strategy. J. Inform. Secur. Appl. **54**, 102523 (2020)
12. Zhao, X., Li, J., Li, S., Wang, S.: Detecting digital image splicing in chroma spaces. In: Digital Watermarking: 9th International Workshop, IWDW 2010, Seoul, Korea, 1–3 Oct 2010. Revised Selected Papers 9, pp. 12–22. Springer Berlin Heidelberg (2011)
13. Abd El-Latif, E.I., Taha, A., Zayed, H.H.: A passive approach for detecting image splicing based on deep learning and wavelet transform. Arab. J. Sci. Eng. **45**, 3379–3386 (2020)
14. Thakur, T., Singh, K., Yadav, A.: Blind approach for digital image forgery detection. Int. J. Comput. Appl. **975**, 8887 (2018)
15. Bird, J.J., Lotfi, A.: CIFAKE: Image Classification and Explainable Identification of AI-Generated Synthetic Images. arXiv preprint arXiv:2303.14126 (2023)
16. Huang, G., Liu, Z., Van Der Maaten, L., Weinberger, K.Q.: Densely connected convolutional networks. In: Proceedings of the IEEE conference on computer vision and pattern recognition, pp. 4700–4708 (2017)
17. Dong, J., Wang, W., Tan, T.: Casia image tampering detection evaluation database. In 2013 IEEE China Summit and International Conference on Signal and Information Processing, pp. 422–426. IEEE (2013)
18. Almawas, L., Alotaibi, A., Kurdi, H.: Comparative performance study of classification models for image-splicing detection. Procedia Comput. Sci. **175**, 278–285 (2020)
19. Jaiswal, A.K., Srivastava, R.: Image splicing detection using deep residual network. In: Proceedings of 2nd International Conference on Advanced Computing and Software Engineering (ICACSE) (2019)
20. Wu, Y., Abd-Almageed, W., Natarajan, P.: Deep matching and validation network: an end-to-end solution to constrained image splicing localization and detection. In: Proceedings of the 25th ACM international conference on Multimedia, pp. 1480–1502 (2017)

21. Yang, B., Sun, X., Chen, X., Zhang, J., Li, X.: Exposing photographic splicing by detecting the inconsistencies in shadows. Comput. J. **58**(4), 588–600 (2015)
22. Nath, S., Naskar, R.: Automated image splicing detection using deep CNN-learned features and ANN-based classifier. Signal Image Video Process. **15**, 1601–1608 (2021)

Identification of DR (Diabetic Retinopathy) from Messidor-2 Dataset Images Using Various Deep and Machine Learning Techniques: A Comparative Analysis

Piyush Jain[1(✉)] ⓘ, Deepak Motwani[1] ⓘ, and Pankaj Sharma[2] ⓘ

[1] Amity University, Gwalior, Madhya Pradesh, India
piyushjain.1987@gmail.com, dmotwani@gwa.amity.edu
[2] Eshan College of Engineering, Mathura, India

Abstract. Diabetic retinopathy (DR) is a complication of diabetes that affects one in four people with the disease. Retinal degeneration occurs as a result. DR is a leading cause of preventable blindness, & early detection through automated screening is important to ensuring the health of a patient's eyes. Preventing a significant loss of vision requires prompt medical attention, ideally from an ophthalmologist. Given the recent advancements in machine learning (ML) applied to healthcare, intelligent systems may prove effective in the early diagnosis of DR. This research seeks to identify the best method for screening diabetic retinopathy patients using datasets such as Messidor-2 datasets by comparing eight widely used classification algorithms (Inception-V3, DR2Net, ResNet50, IncRes-v2, CNN, SVM, RetNet-10, and ELM with CNN-SVD). A number of metrics, like F1-score, Precision, Recall, Accuracy, and AUC-ROC, are utilized to evaluate and contrast different algorithms. From the obtained results, it is observed that proposed RetNet-10 model outperforms other algorithms, having an accuracy of 99.46% along with a good AUC-ROC score of 99.67%. In addition, Inception-V3 obtains an accuracy of 94.59% on Messidor-2 dataset, which is quite low as the dataset is small. The simulation results show how classification algorithms perform differently, and this can be a decisive factor in which approach to take.

Keywords: DR · ML · DL · Transfer Learning · Messidor-2 dataset

1 Introduction

According to the WHO, the number of people with diabetes is anticipated to rise from the current 422 million to 552 million by 2030 [1]. According to the National Diabetes Statistics Report [2] published by the United States Department of Health & Human Services, an estimated 30.5%, or 30.5 million Americans (across all age categories), will have diabetes in 2020. They will go undiagnosed in the number of 7.3 million. Diabetic eye illnesses, like DR, DME, & Glaucoma, are common among people with diabetes. Damage to a blood vessel in a retina is at a heart of DR, a most common cause of blindness in Western countries. Disease progression is indicated by hard exudates, the development of vitreous hemorrhage, retinal detachment & microaneurysms. Images of a retina with varying degrees of DR are displayed in Fig. 1: (a) normal, (b) mild, (c) moderate, (d) severe, & (e) proliferative. [3].

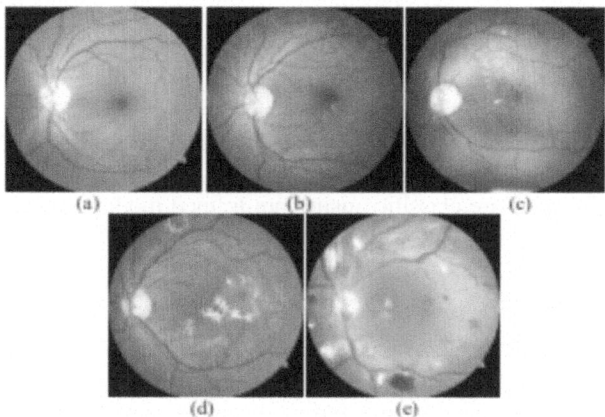

Fig. 1. Retina images from the DR dataset with different DR levels: (a) normal, (b) mild, (c) moderate, (d) severe, and (e) proliferative [3].

Due to a high cost and lengthy nature of DR diagnosis performed manually by medical professionals, there is a continuing need for research and development of automation methods for DR identification. Computer-assisted diagnosis (CAD) for DR detection is a potentially useful application of these methods [4]. Suitable DR detection methods for CAD frequently employ ML and DL methods, as well as a large variety of fully pretrained DL models known as TL models. Automating DR detection in colour fundus retinal pictures has been achieved using TL models for binary and multi-class classification [5–8]. These models have been effective in an automatic detection of DR outside of clinical settings, where a small dataset increases the risk of underfitting and generalization error. As a result, TL is favoured over traditional DL methods in certain circumstances. Although deep learning techniques, and especially CNNs, are still a relatively novel area of study, they have already been put to use in a variety of contexts, including DR detection. Neural networks, which are utilized in deep learning, are used to perform image categorization using hundreds of equations with millions of parameters [9].

The novelty of this study is found in its detailed evaluation of eight different classification methods for diabetic retinopathy identification using the Messidor-2 dataset. For studies comparing diabetic retinopathy, the Messidor-2 dataset is the way to go because of its diversity, realism, and extensive use. Being a benchmark dataset for evaluating the effectiveness of classification methods, it stands out from other accessible datasets, with 1748 photos representing differences in image quality as well as DR severity. The research stands out by taking into account other parameters beyond accuracy, showcasing proposed RetNet-10 as the best performance. By being open and honest about the difficulties of preprocessing and analyzing the characteristics of datasets, researchers might get important insights. The effect of preprocessing on various models' processing costs is recognized in the study. Image preprocessing before employing them in different models decreases processing costs, according to the research. Models' computational needs are reduced by picture cleaning and scaling, resulting in processing that is both efficient as well as easier on resources. The following are the main findings of the suggested investigation:

- The model's efficacy was evaluated using the extensive publicly available dataset. Messidor-2 dataset.
- Different models were utilized to separate the retina from the blood vessels after initial processing and data augmentation.
- Images of the fundus are classified using pre-trained models based on ML, DL, & transfer learning (TL).
- Precision, accuracy, F1-score, specificity, sensitivity, & AUC-ROC are just few of the performance indicators used to verify the proposed model's diagnostic efficacy.

Here is how a rest of a paper is structured: Sect. 2 gives an update on a state of an investigation. In Sect. 3, we detail the datasets we used and present our recommended technique. Section 4 explains a findings & the discussion, and Sect. 5 wraps up the study.

2 Literature Review

Recent DR detection literature focuses on alternate methods for refraction, illumination, and occlusion, which are common issues with typical fundus pictures. Researchers and data scientists from across the globe attended the meeting to learn about DL-ML-TL DR detection methods.

The proposed structure is a hybrid of three distinct phases. To begin with [10], a normalized procedure and an enhanced method are utilized to pre-process a fundus image. The Indus pictures in the eye will be extracted by a trained method with Beginning V3 architecture & Resnet50, and the outcome will come with astounding accuracy of 83.79 percentile through the use of various activation functions.

In [11], describes a novel approach to using Grad-CAM with a multi-label classification method that is based on deep learning. The results show accurate lesion outlining and strong DR classification accuracy (93.9%) and specificity (94.4%).

In [12], SEResNeXt32x4d and EfficientNetb3 models have been pretrained. The standardized 224×224 pictures are improved with AUGMIX and then pooled using GeM. By reaching a high training accuracy of 0.91, the approach may reduce the amount of work required by medical staff members who must analyze retinal images.

In [13] presents a new approach to identifying diabetic retinopathy (DR), with a particular emphasis on the precise identification of aneurysm in colour images of the fundus. The study used a LSVM to obtain a high level of sensitivity (96%) and precision (92%), respectively.

In [14], offers a ResNet CNN model for feature extraction, a compressed feature vector is produced. The overall categorization accuracy is an impressive 86.67% as a consequence, outperforming current techniques.

Table 1. Comparative Study of Related Work

Author & year	Model	Dataset	Results	Research gap
[15]	VGG19, InceptionV3	DR Kaggle data	The method's classification accuracy can reach 0.60,	The aim is for computers to recognise DR with the same retinopathy grade accuracy as physicians
[16]	Inception V3, EyeWeS	Messidor dataset	an outcomes of Inception V3 by 94.9% AUC to 95.8%	Machine learning hybrid research will improve this work
[17]	CNN	DR Kaggle data	In training, the model was 98.6 percent accurate. There was a No DR accuracy of 86.6% for the model,	Even with a limited number of pictures, Inception Version 3 CNN trained and tested well for diabetic eye condition diagnosis
[18]	ResNet-50 model	Diabetic Retinopathy Kaggle	ResNet-50 model network achieved 89.4% a	A future author may improve this approach and use it to treat diabetes
[19]	VGG-16, DenseNet	APTOS dataset	The suggested model had a 98.32% accuracy on the APTOS dataset and a 98.71% accuracy on the Kaggle dataset	More ML and DL approaches are required for DR detection

3   Methodology

Methods for identifying DR in fundus photographs of the retina are compared in this research. Messidor-2 makes use of freely available datasets. Image scaling, data labeling, and data cleaning are just a few examples of preprocessing procedures that improve quality. An accuracy of a classifications obtained is directly related to a quality of preprocessed images. Extracting the OD and the blood vessel using preprocessing techniques yields the enhanced image. In the end, after some preprocessing for feature extraction and selecting the best ones, a number of ML, transfer learning, & DL models were developed. Finally, an enhanced picture dataset is employed in conjunction with the proposed transfer learning-based models for DR diagnosis. AUC-ROC, F1 score, recall, accuracy, and precision are just a few of the performance metrics used to judge the suggested method's effectiveness. In Fig. 2, we see a flowchart of the proposed procedure.

3.1 Data Collection

The Messidor-2 dataset was the most often utilized public dataset for testing the performance of the proposed system. There are 1748 retinal fundus photos in the Messidor-2 dataset [20]. There is a wide range in image quality, colour accuracy, lighting, and resolution within the database. It includes 1748 photos from 874 different examinations. We didn't have a way to categorize tree images, so we left them out. Ten hundred and twelve photos of funduses without DR lesions and seven hundred and thirty-three images of funduses with DR lesions of varying severity (0: no DR, 1: mild DR, 2: moderate DR, 3: severe DR, NPDR, 4: PDR) were used in a study's tests. Images with lesions are rare compared to those without, and they are spread evenly across the four abnormal grades (1:4). At the outset of the study, the number of photos in a Messidor-2 database was doubled by processing a vertical mirroring version of a database, bringing the total to 3490.

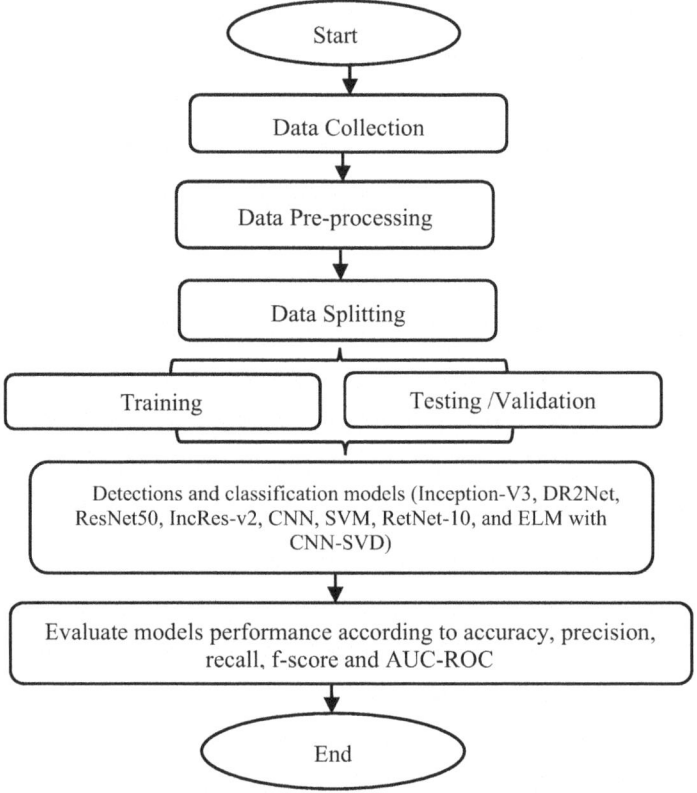

Fig. 2. Methodology flowchart.

3.2 Data Preprocessing

Quality of pre-processed photos affects categorization accuracy. Preprocessing photos before sending them into a neural network is essential for accuracy. It removes artefacts noise and highlights confusing but significant elements. High-quality colour retinal images are needed to classify DR quickly and accurately [21]. Public retinal fundus image collections may have background noise due to their varying resolutions and compression methods [22]. DR may be difficult to classify without preprocessing because to the neural network model utilised for classification. After obtaining the preprocessed image, merely feeding the neural network the same-sized images reduce the model's processing cost. We resized the revised photos to meet these requirements. After image preparation, we get clean, noise-free images.

3.3 Data Splitting

After the dataset has been partitioned, the photos can be used in the suggested model. The preprocessed dataset was splitted into three equal phases for use in training, validation, and testing, with a ratio of 80:10:10.

3.4 Classification Methods

This study tested various ML & DL models, including Inception-V3, DR detection model (DR2Net), ResNet50, IncRes-v2, CNN, SVM, proposed RetNet-10, and ELM with CNN-SVD, using the Messidor-2 distinct diabetic retinopathy (DR) dataset. To enable a complete evaluation of the models' effectiveness, each dataset was carefully picked.

4 Experiments Results Analysis

In this section, we provide comparative outcomes & discussion of multiple machine learning, deep learning, transfer learning and ELM models utilizing a Messidor-2 dataset for DR detection and classification.

4.1 Performance Measures

With the goal of evaluating every type of categorization model currently available, such as machine learning, transfer learning, DL, and so on. FP, FN, TP, & TN values are used in general to calculate performance metric values utilizing the corresponding Eqs. (1)–(4) below [23, 24]. We evaluated an effectiveness of a proposed methodology by looking at its F-1 score, precision, AUC-ROC, sensitivity, specificity, & accuracy.

Classification Accuracy – calculates a percentage of correct classifications compared to a total number of forecasts by using the following formula:

$$Accuarcy = \frac{TP + FN}{TP + TF + FP + FN} \tag{1}$$

Precision – Percentage representing the proportion of correctly labelled images relative to the total number of predicted images.

$$\text{Precision} = \frac{\text{TP}}{\text{TP} + \text{FP}} \tag{2}$$

Recall – Percentage representing the proportion of images in a specific class that were successfully identified.

$$\text{Recall} = \frac{\text{TP}}{\text{TP} + \text{FN}} \tag{3}$$

F1 Score – Identifies the model's performance as a whole by averaging its Precision and Recall ratings. To express the score as a number, we use the following notation:

$$\text{F1} - \text{score} = \frac{2(\text{precision} \times \text{recall})}{\text{precision} + \text{recall}} \tag{4}$$

Specificity – The specificity of a test is its ability to designate an individual who does not have a disease as negative. A highly specific test means that there are few false positive results.

$$\text{Specificity} = \frac{\text{TN}}{\text{TN} + \text{FP}} \tag{5}$$

AUC (Area under the ROC Curve)- An AUC measures overall performance at any given categorization cutoff. The AUC can be thought of as the likelihood that the model will assign a higher score to a positive example than a negative example at random. The ROC curve plots the true positive rate (Sensitivity) against the false positive rate (100-Specificity) as its discrimination threshold is varied. All the points on the ROC curve can be regarded as a sensitivity/specificity pair corresponding to a particular decision threshold.

4.2 Comparative Results on Messidor-2 Dataset

The following sections provide the compare analysis of DR detection using the Kaggle Messidor-2 dataset. Below are the comparative results of different ML models according to Accuracy, Recall, F1-Score, AUC-ROC, and Precision measure. Table 1 shows the model performance (Table 2).

The bar graph in Fig. 3 compares the accuracy of several models for DR image detection and automated categorization. Blue bars make up the horizontal bar graph. The most accurate is RetNet-10 at 99.46%. With 94.59% accuracy, Inception-V3 has lowest accuracy.

In Fig. 4, a bar graph compares the accuracy of several models for DR image detection and automated categorization. The comparison shows that SVM has 100% precision and Inception-V3 93%.

The comparative Fig. 5 shows the recall measure performance with multiple models, including ML, DL and TL models. The CNN model obtains highest recall with 100%, while lower recall is by the ELM with CNN_SVD model.

Table 2. Comparison between various models' performance on the Messidor-2 dataset.

Models	Accuracy	Precision	Recall	F1-Score	AUC-ROC
Inception-V3 [25]	94.59	95.12	94.81	92.99	96.9
DR2Net [26]	96.85	95.3	97.3	96.29	95.59
ResNet50 [3]	99.1	98.8	98.3	98.7	99.78
IncRes-v2 [27]	96.2	96.9	96.7	89.1	97.9
CNN [28]	98.37	96.2	100	98.08	98.37
SVM [28]	97.23	100	94.6	97.22	97.10
RetNet-10	99.46	98.65	98.66	98.65	99.67
ELM with CNN-SVD [29]	96.26	98	94	96	98.42

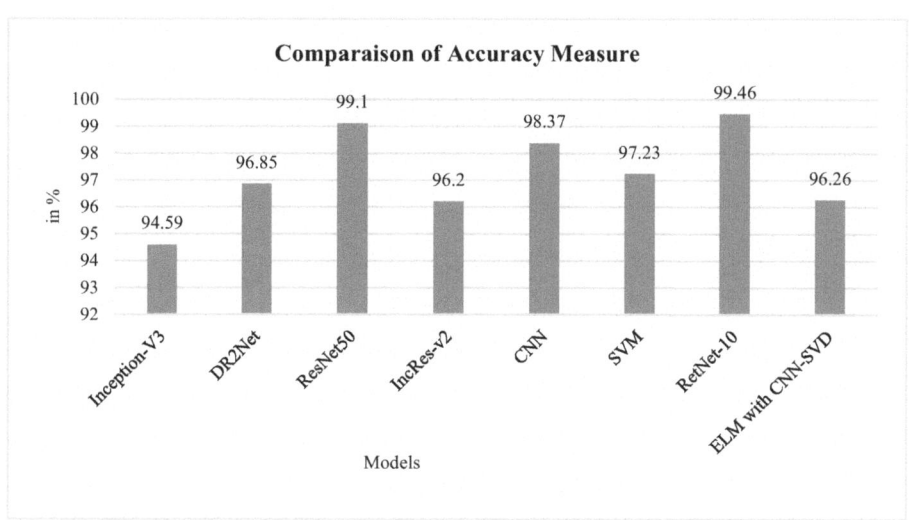

Fig. 3. Comparative bar graph of accuracy measure with different ML models

Figure 6 compares f1-score performance among models. With 98.65%, ResNet10 has the highest f1-score, whereas IncRes-V2 has 89.1%. These models have f1-scores of 97.22% for SVM and 96.29 for DR2Net. Another model demonstrates f1-score performance.

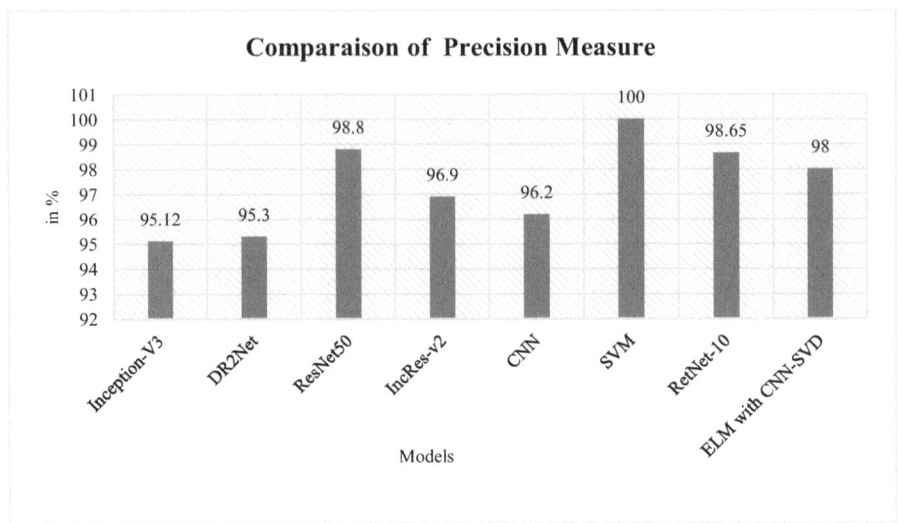

Fig. 4. Comparative bar graph of precision measure with different ML models

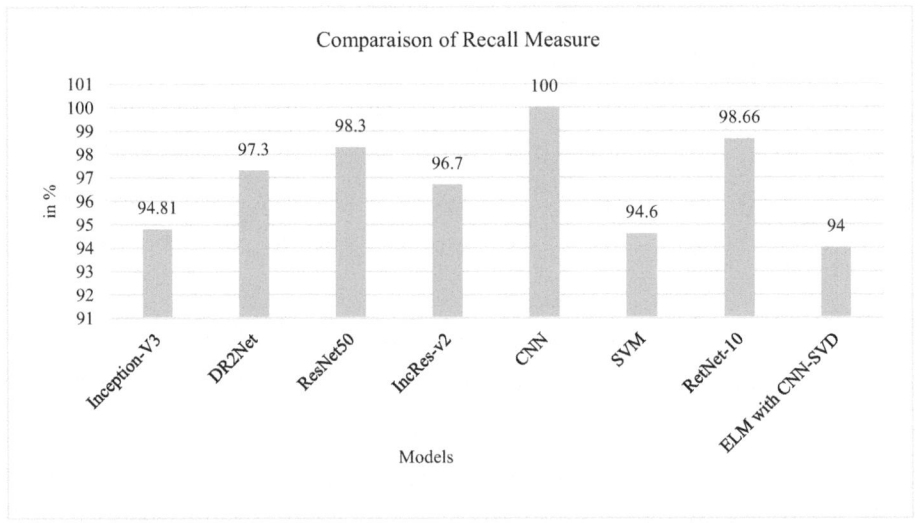

Fig. 5. Comparative bar graph of Recall measure with different ML models

Figure 7 compares AUC-ROC measure performance among models. In image, x-axis shows model AUC % and y-axis shows model values with bars. Inception-V3 has 96.9% AUC-ROC, DR2Net 95.59%, ResNet50 99.78%, IncRes-v2 97.9%, CNN 98.37%, SVM 97.1%, RetNet-10 99.67%, and ELM with CNN-SVD 98.42% on the y-axis labelled "Models". The maximum AUC-ROC was 99.78% for RetNet-10, while the lowest was 95.59%, respectively.

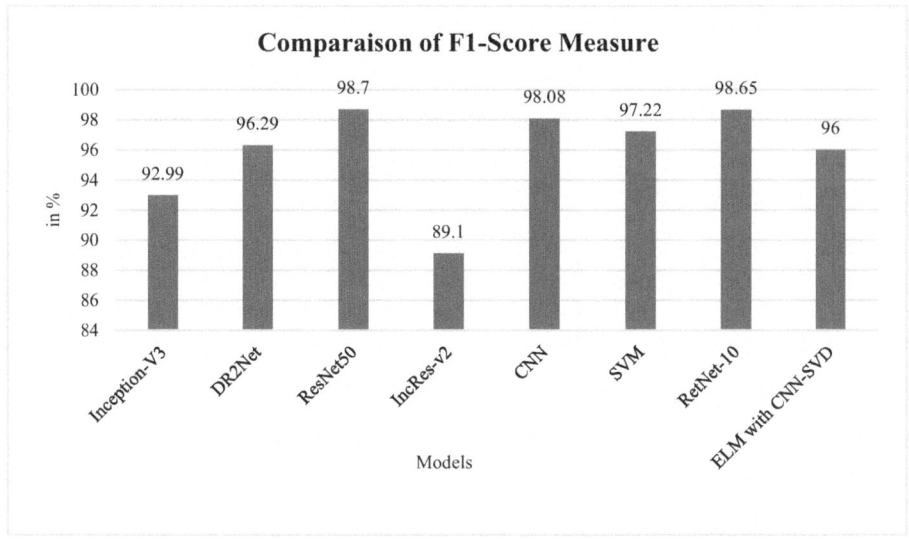

Fig. 6. Comparative bar graph of F1-score measure with different ML models

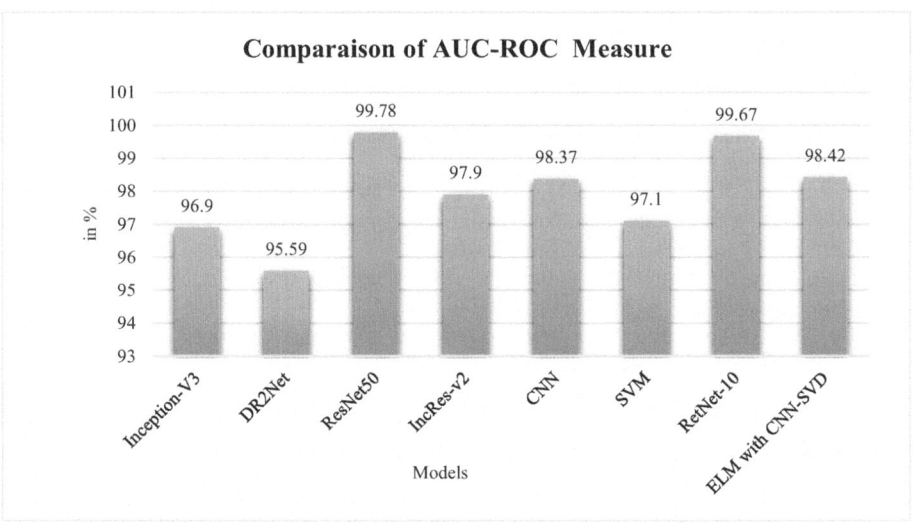

Fig. 7. Comparative bar graph of AUC measure with different ML models

5   Discussion

Different models for diabetic retinopathy categorization have different performance characteristics, according to the comparison. In particular, the outstanding F1-Score, AUC, recall, accuracy, and precision values displayed by ResNet50 and RetNet-10 attest to their strong performance. Inception-V3 shows balanced stats, whereas CNN shows flawless recall. DR2Net provides an excellent compromise between recall and precision, whereas

SVM demonstrates great precision. The figures highlight the exceptional performance of proposed RetNet-10, which is a top-notch model for detecting diabetic retinopathy. Its accuracy is 99.46% and its AUC-ROC is 99.67%. These findings highlight how critical it is to select suitable models for trustworthy categorization in medical image processing.

6 Conclusion and Future Work

This study examines a use of automatic screening for diabetic retinopathy cases and compares eight popular classifier methods. Each classification method's parameters were adjusted to produce the best outcomes. This evaluation was carried out so as to be used as a criterion for selecting the appropriate classification method (which is highly application dependent) in future studies. The data shows a pattern in which classification methods work best and how their accuracy varies depending on a number of various characteristics. The results suggest that the proposed ResNet10 model has the most potential for diagnosing cases of diabetic retinopathy, however other methods can be modified to achieve acceptable performance as well. Using a larger dataset for future research is an option. But we have to think about how the dataset is skewed. Although the report's findings don't specify which TL model was utilized for optimization, it's safe to assume that doing so would have helped get rid of underperforming elements.

References

1. Bourne, R.R.A., et al.: Causes of vision loss worldwide, 1990–2010: a systematic analysis. Lancet Glob. Heal. (2013). https://doi.org/10.1016/S2214-109X(13)70113-X
2. DHHS: National Diabetes Statistics Report (2020)
3. Hacisoftaoglu, R.E., Karakaya, M., Sallam, A.B.: Deep learning frameworks for diabetic retinopathy detection with smartphone-based retinal imaging systems. Pattern Recognit. Lett. (2020). https://doi.org/10.1016/j.patrec.2020.04.009
4. Das, A.: Diabetic retinopathy: battling the global epidemic. Investig. Ophthalmol. Vis. Sci. (2016). https://doi.org/10.1167/iovs.16-21031
5. Alban, M., Gilligan, T.: Automated detection of diabetic retinopathy using fluorescein angiography photographs. cs231n课程project (2016)
6. Kermany, D.S., et al.: Identifying medical diagnoses and treatable diseases by image-based deep learning. Cell (2018). https://doi.org/10.1016/j.cell.2018.02.010
7. Mahmoud, M.H., Alamery, S., Fouad, H., Altinawi, A., Youssef, A.E.: An automatic detection system of diabetic retinopathy using a hybrid inductive machine learning algorithm. Pers. Ubiquitous Comput. (2023). https://doi.org/10.1007/s00779-020-01519-8
8. Aswathi, T., Swapna, T.R., Padmavathi, S.: Transfer learning approach for grading of diabetic retinopathy. J. Phys. Conf. Ser. **1767**(1), 012033 (2021). https://doi.org/10.1088/1742-6596/1767/1/012033
9. Hassan, D., Gill, H.M., Happe, M., Bhatwadekar, A.D., Hajrasouliha, A.R., Janga, S.C.: Combining transfer learning with retinal lesion features for accurate detection of diabetic retinopathy. Front. Med. (2022). https://doi.org/10.3389/fmed.2022.1050436
10. Patra, P., Singh, T.: Diabetic retinopathy detection using an improved ResNet50-InceptionV3 structure (2022). https://doi.org/10.1109/ICCCNT54827.2022.9984253
11. Jiang, H., et al.: A multi-label deep learning model with interpretable grad-CAM for diabetic retinopathy classification. Proc. Annu. Int. Conf. IEEE Eng. Med. Biol. Soc. EMBS, vol. 2020-July, pp. 1–4 (2020). https://doi.org/10.1109/EMBC44109.2020.9175884

12. Ramchandre, S., Patil, B., Pharande, S., Javali, K., Pande, H.: A deep learning approach for diabetic retinopathy detection using transfer learning. IEEE Int. Conf. Innov. Technol. INOCON **2020**, 1–5 (2020). https://doi.org/10.1109/INOCON50539.2020.9298201

13. Kumar, S., Kumar, B.: Diabetic retinopathy detection by extracting area and number of microaneurysm from colour fundus image. In: 5th Int. Conf. Signal Process. Integr. Networks, SPIN 2018, pp. 1–6 (2018). https://doi.org/10.1109/SPIN.2018.8474264

14. Elswah, D.K., Elnakib, A.A., El-Din Moustafa, H.: Automated Diabetic Retinopathy Grading using Resnet (2020). https://doi.org/10.1109/NRSC49500.2020.9235098

15. Wu, Y., Hu, Z.: Recognition of diabetic retinopathy based on transfer learning. In: IEEE 4th Int. Conf. Cloud Comput. Big Data Anal. ICCCBDA 2019, pp. 1–4 (2019). https://doi.org/10.1109/ICCCBDA.2019.8725801

16. Costa, P. et al.: EyeWeS: Weakly supervised pre-trained convolutional neural networks for diabetic retinopathy detection. Proc. 16th Int. Conf. Mach. Vis. Appl. MVA 2019, pp. 1–6 (2019). https://doi.org/10.23919/MVA.2019.8757991

17. Jayakumari, C., Lavanya, V., Sumesh, E.P.: Automated diabetic retinopathy detection and classification using imagenet convolution neural network using fundus images. Int. Conf. Smart Electr. Commun. **2020**, 577–582 (2020). https://doi.org/10.1109/ICOSEC49089.2020.9215270

18. Rajkumar, R.S., Jagathishkumar, T., Ragul, D., Selvarani, A.G.: Transfer learning approach for diabetic retinopathy detection using a residual network. In: Proc. 6th Int. Conf. Inven. Comput. Technol. ICICT, pp. 1–5 (2021). https://doi.org/10.1109/ICICT50816.2021.9358468

19. Bhimavarapu, U., Chintalapudi, N., Battineni, G.: Automatic detection and classification of diabetic retinopathy using the improved pooling function in the convolution neural network. Diagnostics **13**(15), 1–19 (2023). https://doi.org/10.3390/diagnostics13152606

20. Decencière, E., et al.: Feedback on a publicly distributed image database: the Messidor database. Image Anal. Stereol. (2014). https://doi.org/10.5566/ias.1155

21. Rasta, S.H., Partovi, M.E., Seyedarabi, H., Javadzadeh, A.: A comparative study on preprocessing techniques in diabetic retinopathy retinal images: illumination correction and contrast enhancement. J. Med. Signals Sens. (2015). https://doi.org/10.4103/2228-7477.150414

22. Pinedo-Diaz, G., et al.: Suitability classification of retinal fundus images for diabetic retinopathy using deep learning. Electron. (2022). https://doi.org/10.3390/electronics11162564

23. Grubbs, F.E.: Errors of measurement, precision, accuracy and the statistical comparison of measuring instruments. Technometrics (1973). https://doi.org/10.1080/00401706.1973.10489010

24. Sharma, M., Sharma, S., Singh, G.: Performance analysis of statistical and supervised learning techniques in stock data mining. Data (2018). https://doi.org/10.3390/data3040054

25. Bilal, A., Zhu, L., Deng, A., Lu, H., Wu, N.: AI-based automatic detection and classification of diabetic retinopathy using U-Net and deep learning. Symmetry (Basel) (2022). https://doi.org/10.3390/sym14071427

26. Berbar, M.: Diabetic retinopathy detection and grading using deep learning. Menoufia J. Electron. Eng. Res. (2022). https://doi.org/10.21608/mjeer.2022.138003.1057

27. Chetoui, M., Akhloufi, M.A.: Explainable end-to-end deep learning for diabetic retinopathy detection across multiple datasets. J. Med. Imaging (2020). https://doi.org/10.1117/1.jmi.7.4.044503

28. Berbar, M.A.: Features extraction using encoded local binary pattern for detection and grading diabetic retinopathy. Heal. Inf. Sci. Syst. (2022). https://doi.org/10.1007/s13755-022-00181-z

29. Nahiduzzaman, M., Islam, M.R., Islam, S.M.R., Goni, M.O.F., Anower, M.S., Kwak, K.S.: Hybrid CNN-SVD based prominent feature extraction and selection for grading diabetic retinopathy using extreme learning machine algorithm. IEEE Access (2021). https://doi.org/10.1109/ACCESS.2021.3125791

Access-Based Authentication of Healthcare Record in Blockchain: A Survey from Security Perspectives

Shashank Saroop[1,2](✉) [ID], Rajesh Kumar Tyagi[2] [ID], Shweta Sinha[2], Shafiqul Abidin[3], and Lokesh Kumar[1]

[1] ADGITM, New Delhi, India
shashank.saroop@gmail.com
[2] Amity University, Gurugram, India
[3] Aligarh Muslim University, Aligarh, India

Abstract. The rapid digitization of medical records has completely changed how patient data is shared and kept. However, maintaining these documents' privacy and security is still a top priority. A viable method for boosting the security and integrity of medical data is blockchain technology. Access-based authentication mechanisms in blockchain, in particular, can offer a solid framework for monitoring and restricting access to private medical records. This study provides a thorough analysis of Blockchain technology used for medical records with an emphasis on security issues. The survey investigates various strategies and methods used to safeguard medical records in a blockchain setting.

Keywords: Access-based authentication · Healthcare records · Blockchain

1 Introduction

The digitization of healthcare records has transformed the healthcare industry, enabling efficient storage, retrieval, and sharing of patient information. However, ensuring the security and privacy of these records has become a critical concern in this digital era. Blockchain technology has emerged as a promising solution for enhancing the security and integrity of healthcare data [1]. By leveraging the decentralized and immutable nature of blockchain, access-based authentication mechanisms can provide a robust framework for controlling and managing access to sensitive healthcare records. The adoption of blockchain in healthcare has gained significant attention due to its potential to address security and privacy challenges. Blockchain ensures transparency, tamper resistance, and data integrity through its distributed ledger technology [2]. Access-based authentication mechanisms built on top of blockchain can strengthen the security of healthcare records, preventing unauthorized access and maintaining patient confidentiality (Fig. 1).

In this survey, we aim to comprehensively explore the access-based authentication of healthcare records in blockchain systems from a security perspective. By examining existing literature and research works, we seek to analyze the various approaches

A. K. Bairwa et al. (Eds.): ICCAIML 2024, CCIS 2184, pp. 136–149, 2025.
https://doi.org/10.1007/978-3-031-71481-8_11

How a blockchain works?

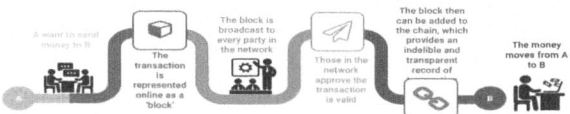

Fig. 1. Working of Blockchain

and techniques employed in securing healthcare data within blockchain environments. The survey will provide insights into the different access control models, authentication mechanisms, and privacy-preserving techniques used in conjunction with blockchain technology. Moreover, we will discuss the challenges faced in the healthcare domain concerning security and privacy and explore how blockchain-based access control can effectively address these challenges. By understanding the strengths and limitations of access-based authentication methods, we can identify opportunities for improvement and advancements in securing healthcare records. Also, we aim to contribute to the existing body of knowledge on access-based authentication in blockchain-based healthcare record management systems. By highlighting the current state of research, identifying gaps, and discussing potential future directions, we seek to promote the development of secure and trustworthy healthcare information systems.

1.1 Background

The digitization of healthcare records has revolutionized the healthcare industry, replacing traditional paper-based systems with electronic storage and management of patient information. Electronic Health Records (EHRs) provide numerous benefits such as improved accessibility, streamlined healthcare processes, and enhanced patient care. However, this digital transformation has also brought forth significant security and privacy concerns [3]. Healthcare records contain highly sensitive and personal information, including medical history, diagnoses, treatments, and demographic data. Unauthorized access to these records can lead to breaches of patient privacy, identity theft, and misuse of medical information. Ensuring the security and confidentiality of healthcare records is crucial to maintaining patient trust and complying with privacy regulations such as the Health Insurance Portability and Accountability Act (HIPAA) in the United States and the General Data Protection Regulation (GDPR) in the European Union [4]. Traditional centralized systems used for healthcare record storage and management are susceptible to security vulnerabilities, as they rely on a single authority to authenticate and authorize access to records. These centralized systems face risks such as data breaches, insider threats, and single points of failure. To address these concerns, blockchain technology has emerged as a potential solution for secure healthcare record management. Blockchain, originally introduced as the underlying technology for cryptocurrencies like Bitcoin, is a decentralized and distributed ledger that maintains a tamper-resistant record of transactions. It offers unique features such as immutability, transparency, and consensus mechanisms that enhance the security and integrity of data [5]. Blockchain technology provides a decentralized approach to authentication and access control, eliminating the need for a central authority and enabling secure and auditable data sharing. By integrating access-based authentication mechanisms into blockchain systems, healthcare organizations can establish granular control over who can access, modify, and share patient records. Access control models such as role-based access control (RBAC) and attribute-based access control (ABAC) can be implemented within blockchain networks, ensuring that only authorized individuals or entities have access to specific healthcare information [6]. Furthermore, authentication mechanisms based on public key infrastructure (PKI), digital signatures, and smart contracts can be leveraged in blockchain-based healthcare record systems to verify the identities of users and ensure the authenticity of data (Kuo et al., 2018). These mechanisms contribute to establishing a trusted environment for accessing and managing healthcare records.

2   Blockchain Technology in Healthcare

2.1   Overview of Blockchain Technology

Blockchain technology has garnered significant attention in the healthcare industry due to its potential to address security, privacy, and interoperability challenges associated with healthcare data management. This section provides an overview of blockchain technology in healthcare, highlighting its advantages and potential applications (Fig. 2).

Fig. 2. Application of Blockchain in healthcare

Enhanced Security and Data Integrity:
Blockchain's distributed ledger architecture offers enhanced security and data integrity for healthcare records. The decentralized nature of blockchain eliminates the need for a central authority, reducing the risk of data breaches and unauthorized access [2]. The immutability of blockchain transactions ensures that healthcare data remains tamper-proof and auditable [7].

Interoperability and Data Sharing:
Blockchain technology provides a promising solution for interoperability challenges in healthcare by enabling secure and efficient data sharing among different stakeholders. Blockchain's shared ledger allows for real-time updates and data synchronization across multiple entities, improving care coordination and facilitating a seamless exchange of information [6].

Smart Contracts for Automated Processes:
Smart contracts, self-executing contracts with predefined rules encoded on the blockchain, offer automation and efficiency in healthcare processes. These contracts can facilitate automated claims processing, enforce compliance with contractual obligations, and streamline administrative tasks [4].

Patient-Centric Control of Data:
Blockchain technology empowers patients by giving them control over their health data. Patients can securely manage their consent preferences and grant selective access to healthcare providers, ensuring privacy and facilitating patient-centered care [8].

Medical Research and Clinical Trials:

Blockchain technology has the potential to revolutionize medical research and clinical trials. By securely storing and sharing research data on the blockchain, transparency and trust can be established, preventing data manipulation and ensuring the reproducibility of results [9].

Supply Chain and Drug Traceability:
Blockchain can improve supply chain management in the pharmaceutical industry by enhancing the traceability, authentication, and transparency of drugs. It enables the tracking of drug provenance, ensuring the integrity of the supply chain, and mitigating the risks of counterfeit drugs [10].

2.2 Advantages of Blockchain in Healthcare

Advantages of Blockchain in Healthcare	Description
Enhanced security and data integrity [2]	Blockchain's distributed ledger architecture ensures tamper-proof storage and immutability of healthcare data, enhancing security and data integrity. (Mettler, 2016)
Improved interoperability and data sharing [11]	Blockchain enables secure and efficient data sharing among different healthcare stakeholders, promoting interoperability and facilitating seamless exchange of information. (Kuo et al., 2017)
Efficient and transparent claims processing [7]	Blockchain-based smart contracts automate and streamline the claims processing in healthcare, reducing administrative burdens and ensuring transparency. (Ekblaw et al., 2016)
Patient-centric control of health data [8]	Blockchain empowers patients by giving them control over their health data and enabling selective sharing of information, fostering patient-centered care. (Linn et al., 2019)

(continued)

(*continued*)

Advantages of Blockchain in Healthcare	Description
Enhanced research and clinical trial capabilities [12]	Blockchain provides a transparent and auditable platform for medical research and clinical trials, improving data integrity, reproducibility, and trust. (Benchoufi et al., 2017)
Reliable drug traceability and supply chain management [13]	Blockchain enables end-to-end traceability of drugs, ensuring the authenticity and integrity of the supply chain and mitigating the risks of counterfeit products. (Kamal et al., 2020)

2.3 Security and Privacy Challenges in Healthcare Records

See Fig. 3.

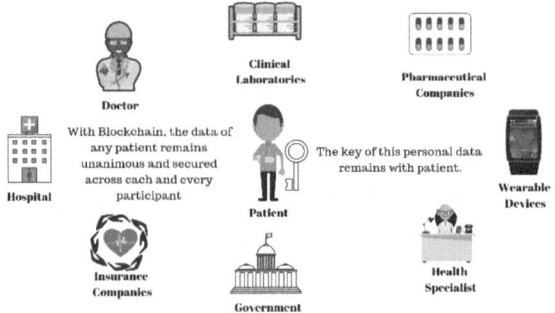

Fig. 3. Challenges in Healthcare Records

- *Data Breaches and Unauthorized Access* [14]*:*
  Healthcare records are vulnerable to data breaches and unauthorized access, leading to potential misuse or exposure of sensitive patient information. (Roman et al., 2018).
- *Insider Threats* [15]*:*
  Insiders with authorized access to healthcare records, such as healthcare providers or employees, may misuse or exploit patient data for personal gain or malicious purposes. (Vijayakumar et al., 2018).
- *Data Integrity and Trustworthiness* [16]*:*
  Ensuring the integrity and trustworthiness of healthcare records is crucial to prevent unauthorized modifications, tampering, or falsification of patient data. (Fernández-Alemán et al., 2013).
- *Identity and Access Management* [17]*:*
  Effective identity and access management solutions are needed to authenticate and authorize legitimate users while preventing unauthorized access to healthcare records. (Dinh et al., 2018).

- *Interoperability and Data Sharing* [18]*:*
  Ensuring secure and privacy-preserving interoperability and data sharing among different healthcare systems and stakeholders remains a challenge. (Sun et al., 2019).
- *Consent Management and Patient Privacy* [19]*:*
  Managing patient consent for data sharing and protecting patient privacy in the context of healthcare records present significant challenges, particularly in an increasingly interconnected healthcare ecosystem. (Prasser et al., 2017).

2.4 Cryptocurrency

- *Streamlining Payments and Transactions* [20]*:*
  Cryptocurrency can be used in healthcare to streamline payments and transactions between different stakeholders, such as patients, healthcare providers, and insurance companies. By eliminating intermediaries and reducing transaction costs, cryptocurrency can facilitate faster and more efficient financial transactions in the healthcare industry. (Dagher et al., 2018).
- *Enhancing Data Security and Privacy* [21]*:*
  The underlying technology of cryptocurrency, blockchain, provides a secure and decentralized platform for storing and managing healthcare data. With cryptographic techniques and distributed consensus mechanisms, blockchain-based cryptocurrency can enhance data security and protect patient privacy in healthcare transactions. (Yli-Huumo et al., 2017).
- *Enabling Microtransactions and Incentives* [22]*:*
  Cryptocurrency allows for microtransactions in healthcare, enabling the implementation of innovative incentive mechanisms. For example, patients can be rewarded with tokens or coins for participating in research studies or adhering to treatment plans, fostering patient engagement and motivation. (Yli-Huumo et al., 2016).

Improving Supply Chain Management [20]*:*
Cryptocurrency can enhance supply chain management in healthcare by providing an immutable and transparent ledger to track the movement of pharmaceuticals, medical devices, and other healthcare products. This helps prevent counterfeit drugs and ensures the authenticity and integrity of the supply chain. (Dagher et al., 2018).
Facilitating Medical Research and Data Sharing [21]*:*
   Cryptocurrency and blockchain technology can facilitate secure and transparent sharing of medical research data. Researchers can use tokens or coins to incentivize data contributors, ensure data provenance, and create a decentralized marketplace for sharing research findings and insights. (Yli-Huumo et al., 2017).

3 Literature Review

See Table 1.

Table 1. Analysis of Various Security based approaches to secure healthcare record using blockchain

Title	Authors & Year	Citations	Description	Results	Pros	Cons
Blockchain Technology for Healthcare: A Review of Applications and Security Concerns	**Alzahrani et al. & 2022**	[23]	This paper provides a review of the literature on the use of blockchain technology for healthcare applications. The paper discusses the different ways that blockchain can be used to improve healthcare, including improving the security of healthcare records, improving the efficiency of healthcare delivery, and reducing the cost of healthcare. The paper also discusses the security concerns associated with blockchain technology in healthcare	The paper concludes that blockchain has the potential to significantly improve healthcare. However, the paper also notes that there are still some security concerns that need to be addressed before blockchain can be widely adopted in healthcare	Improved security and privacy of healthcare records Reduced risk of fraud and abuse	Lack of standards and interoperability Need for security and Privacy
A Survey on Blockchain Technology in Healthcare: Applications, Security, and Privacy Concerns	**Kumar et al. & 2022**	[24]	This paper provides a survey of the literature on the use of blockchain technology in healthcare. The paper discusses the different ways that blockchain can be used to improve healthcare, including improving the security of healthcare records, improving the efficiency of healthcare delivery, and reducing the cost of healthcare. The paper also discusses the security and privacy concerns associated with blockchain technology in healthcare	The paper concludes that blockchain has the potential to significantly improve healthcare. However, the paper also notes that there are still some security and privacy concerns that need to be addressed before blockchain can be widely adopted in healthcare	Improved security and privacy of healthcare records Potential for new healthcare applications	Lack of standards and interoperability Potential for cyberattacks

(continued)

Table 1. (*continued*)

Title	Authors & Year	Citations	Description	Results	Pros	Cons
Blockchain for Healthcare Applications: A Systematic Review	**Kumar *et al.* & 2021**	[25]	This paper provides a systematic review of the literature on the use of blockchain technology for healthcare applications. The paper discusses the different ways that blockchain can be used to improve healthcare, including improving the security of healthcare records, improving the efficiency of healthcare delivery, and reducing the cost of healthcare	The paper concludes that blockchain has the potential to significantly improve healthcare. However, the paper also notes that there are still some challenges that need to be addressed before blockchain can be widely adopted in healthcare	Reduced risk of fraud and abuse Increased efficiency of healthcare data sharing	Lack of standards and interoperability Need for security and privacy
Security and Privacy Issues in Blockchain-Based Healthcare Systems	**Alzahrani *et al.* & 2020**	[26]	This paper surveys the security and privacy issues of blockchain-based healthcare systems. The paper discusses the different types of cyberattacks that can be launched against blockchain-based healthcare systems and the different ways that data can be compromised in blockchain-based healthcare systems. The paper also discusses the different ways that blockchain-based healthcare systems can be made more secure and private	The paper concludes that blockchain-based healthcare systems are a major target for cyberattacks and that there is a need for new security and privacy solutions for blockchain-based healthcare systems	Improved transparency and collaboration in healthcare Potential for new healthcare applications	Lack of standards and interoperability Potential for cyberattacks

(*continued*)

Table 1. (*continued*)

Title	Authors & Year	Citations	Description	Results	Pros	Cons
Blockchain for Healthcare: A Comprehensive Overview	**Kumar, S.** *et al.*& *2020*	[27]	This paper provides a comprehensive overview of the use of blockchain technology in healthcare. The paper discusses the different ways that blockchain can be used to improve healthcare, including improving the security of healthcare records, improving the efficiency of healthcare delivery, and reducing the cost of healthcare	The paper concludes that blockchain has the potential to significantly improve healthcare. However, the paper also notes that there are still some challenges that need to be addressed before blockchain can be widely adopted in healthcare	: Reduced risk of fraud and abuse - Increased efficiency of healthcare data sharing - Improved transparency and collaboration in healthcare	Need for security and privacy Potential for cyberattacks
Blockchain in Healthcare: A Review of Applications, Challenges, and Opportunities	**He, D.** *et al.* & *2020*	[28]	This paper reviews the literature on the use of blockchain technology for healthcare applications. The paper discusses the different ways that blockchain can be used to improve healthcare, including improving the security of healthcare records, improving the efficiency of healthcare delivery, and reducing the cost of healthcare	The paper concludes that blockchain has the potential to significantly improve healthcare. However, the paper also notes that there are still some challenges that need to be addressed before blockchain can be widely adopted in healthcare	Improved security and privacy of healthcare records Potential for new healthcare applications	Lack of standards and interoperability Need for security and privacy
Access-Based Authentication of Healthcare Record in Blockchain: A Survey from Security Perspectives	**Jose** *et al.* & *2019*	[29]	This paper surveys the use of blockchain technology for access-based authentication of healthcare records. The paper discusses the security challenges of healthcare records and how blockchain can be used to address these challenges. The paper also discusses the different types of blockchain-based access control systems that have been proposed	The paper concludes that blockchain technology has the potential to significantly improve the security of healthcare records. However, the paper also notes that there are still some challenges that need to be addressed before blockchain can be widely adopted in healthcare	Reduced risk of fraud and abuse Improved transparency and collaboration in	Potential for cyberattacks Need for security and privacy

4 Open Research Challenges and Future Directions

Open research challenges and future directions in the use of cryptocurrency in healthcare can include the following:

Scalability and Performance:
One of the key challenges is addressing the scalability and performance limitations of blockchain networks in healthcare. As the volume of healthcare data and transactions increases, ensuring that the cryptocurrency infrastructure can handle the growing demands becomes crucial.

Regulatory and Legal Frameworks:
Developing appropriate regulatory and legal frameworks for the use of cryptocurrency in healthcare is essential. Clear guidelines and regulations need to be established to ensure compliance with data privacy, security, and financial regulations, while also considering the unique aspects of blockchain technology.

Interoperability and Standardization:
Achieving interoperability among different blockchain networks and healthcare systems is a challenge. Establishing common standards and protocols for data exchange, smart contracts, and identity management will be necessary to enable seamless integration and collaboration.

Data Privacy and Security:
While blockchain technology provides inherent security features, ensuring robust data privacy and security in cryptocurrency-based healthcare systems is critical. Privacy-preserving techniques, encryption methods, and access control mechanisms need to be developed and implemented to protect sensitive patient data.

User Experience and Adoption:
Enhancing the user experience and promoting the adoption of cryptocurrency-based healthcare systems is vital. User-friendly interfaces, intuitive wallets, and educational resources should be developed to make it easier for healthcare professionals and patients to use and understand the benefits of cryptocurrency in healthcare.

- *Ethical Considerations:*
  Exploring the ethical implications of using cryptocurrency in healthcare is important. Ethical issues related to consent, data ownership, fairness, and transparency should be addressed to ensure the responsible and ethical use of cryptocurrency in healthcare applications.
- *Cost and Financial Considerations:*
  Assessing the cost-effectiveness and financial viability of implementing cryptocurrency in healthcare is a future direction. Analyzing the economic benefits, potential savings, and return on investment will help determine the feasibility and sustainability of adopting cryptocurrency in healthcare systems.

4.1   Emerging Trends and Technologies

- *Artificial Intelligence (AI) in healthcare* [30]*:*
  AI is increasingly being applied in healthcare for tasks such as medical imaging analysis, disease diagnosis, personalized medicine, and drug discovery. It has the potential to improve healthcare outcomes, enhance efficiency, and enable more precise and personalized treatments. (Topol, 2019).
- *Internet of Medical Things (IoMT)* [31]
  IoMT refers to the network of connected medical devices and wearable sensors that collect and transmit patient data for remote monitoring and healthcare management. It enables real-time patient monitoring, early detection of diseases, and improved patient outcomes. (Khan, Yaqoob, Hashem, et al., 2014).
- *Precision Medicine* [32]*:*
  Precision medicine focuses on tailoring medical treatments to individual patients based on their genetic, environmental, and lifestyle factors. It leverages technologies such as genomics, proteomics, and bioinformatics to enable targeted therapies and improve patient outcomes. (Collins & Varmus, 2015).
- *Blockchain in Healthcare* [21]*:*
  Blockchain technology offers decentralized and secure storage and sharing of healthcare data, enabling enhanced data privacy, interoperability, and data integrity. It has the potential to revolutionize healthcare data management, clinical trials, and healthcare supply chain processes. (Yli-Huumo, Ko, Choi, et al., 2017).
- *Telemedicine and Remote Patient Monitoring* [33]*:*
  Telemedicine allows for remote healthcare consultations and monitoring, enabling patients to access healthcare services from their homes. With the advancements in communication technologies and wearable devices, remote patient monitoring has become increasingly feasible, leading to improved access to care and reduced healthcare costs. (Bashshur, Shannon, & Bashshur, 2016).

4.2   Recommendations for Future Research

Privacy-Preserving Techniques for Blockchain:
Investigate and develop privacy-preserving techniques that ensure the confidentiality of sensitive healthcare data stored on blockchain. This includes exploring encryption methods, zero-knowledge proofs, and differential privacy approaches to protect patient privacy while maintaining the benefits of blockchain technology.

Interoperability and Data Standardization:
Focus on developing interoperability frameworks and data standards that facilitate seamless data exchange and integration across different healthcare systems. This research can involve exploring interoperable data models, ontologies, and protocols to ensure efficient and secure data sharing among healthcare providers, patients, and other stakeholders.

Explainable Artificial Intelligence (AI) in Healthcare:
Conduct research to enhance the explainability and interpretability of AI algorithms used in healthcare. Develop methodologies and techniques that provide transparent explanations for AI-based decisions, allowing healthcare professionals and patients to understand and trust the reasoning behind AI-driven diagnoses, treatment recommendations, and patient outcomes.

Ethical and Legal Implications of Emerging Technologies:
Investigate the ethical and legal implications of emerging technologies in healthcare, such as AI, blockchain, and genomics. Research should focus on addressing concerns related to data privacy, bias in algorithms, informed consent, data ownership, and the impact of these technologies on healthcare equity and access.

User Experience and Human-Centered Design:
Explore user-centered design approaches to enhance the usability and user experience of healthcare technologies. Investigate ways to involve end-users, such as patients and healthcare professionals, in the design and development process to ensure that emerging technologies meet their needs, are intuitive to use, and facilitate effective healthcare delivery.

Cybersecurity and Data Protection:
Conduct research on cybersecurity measures and data protection strategies to mitigate the risks associated with healthcare data breaches, ransomware attacks, and unauthorized access. Explore advanced encryption techniques, secure data sharing protocols, and proactive threat detection and response mechanisms to safeguard healthcare systems and patient data.

Health Equity and Social Determinants of Health:
Investigate the impact of emerging technologies on health equity and address disparities in healthcare access, delivery, and outcomes. Research should focus on understanding and addressing the social determinants of health, leveraging technology to improve health outcomes for marginalized populations, and ensuring equitable distribution of healthcare resources and services.

5 Conclusion

In conclusion, the use of blockchain technology for access-based authentication of healthcare records presents significant opportunities for improving security, privacy, and efficiency in the healthcare industry. This survey from a security perspective has explored the advantages of blockchain in healthcare, including enhanced data integrity, decentralization, and auditability. However, several challenges and open research directions have also been identified. The adoption of blockchain in healthcare requires addressing scalability issues, developing regulatory frameworks, ensuring interoperability, and addressing data privacy and security concerns. Additionally, user experience, cost considerations, and ethical implications need to be carefully considered and integrated into the design and implementation of blockchain-based healthcare systems.

Future research should focus on privacy-preserving techniques for blockchain, interoperability standards, explainable AI, ethical and legal implications, user-centered design, cybersecurity measures, health equity, and the impact of emerging technologies on healthcare. By addressing these challenges and advancing research in these areas, we can unlock the full potential of blockchain technology in healthcare, leading to improved patient care, secure data management, and transformative advancements in healthcare delivery.

In conclusion, blockchain-based access-based authentication holds great promise for enhancing healthcare record security, privacy, and interoperability. With continued research and innovation, we can overcome the current limitations and pave the way for a more secure and efficient healthcare ecosystem.

References

1. Smith, K., Vazirani, A., Xiong, L.: Blockchain technology in the healthcare sector: a systematic review. Healthcare (Basel, Switzerland) **5**(4), 56 (2017)
2. Mettler, M.: Blockchain technology in healthcare: the revolution starts here. In: Proceedings of the 2016 IEEE 18th International Conference on e-Health Networking, Applications and Services (Healthcom), pp. 1–3. IEEE (2016)
3. Liu, M., Liu, X., Song, L., Zhang, X., Xiong, N.N.: Secure and privacy-preserving access control scheme for electronic health records in blockchain. J. Med. Syst. **43**(3), 54 (2019)
4. Azaria, A., Ekblaw, A., Vieira, T., Lippman, A.: MedRec: Using blockchain for medical data access and permission management. In: Proceedings of the 2016 2nd International Conference on Open and Big Data (OBD), pp. 25–30. IEEE (2016)
5. Nakamoto, S.: Bitcoin: a peer-to-peer electronic cash system (2008). Retrieved from https://bitcoin.org/bitcoin.pdf
6. Kuo, T.T., Kim, H.E., Ohno-Machado, L.: Blockchain distributed ledger technologies for biomedical and health care applications. J. Am. Med. Inform. Assoc. **25**(9), 1211–1220 (2018)
7. Ekblaw, A., Azaria, A., Halamka, J.D., Lippman, A.: A case study for blockchain in healthcare: "MedRec" prototype for electronic health records and medical research data. In: Proceedings of IEEE Open & Big Data Conference, pp. 1–8. IEEE (2016)
8. Linn, L.A., Koo, M.B., Jani, Y.H.: Blockchain for health data and its potential use in health IT and health care-related research. Fed. Pract. **36**(Suppl 4), S37–S41 (2019)
9. Benchufi, M., Porcher, R., Ravaud, P., Boutron, I.: Blockchain protocols in clinical trials: transparency and traceability of consent. Lancet Oncol. **20**(5), e254–e261 (2019)
10. Li, X., Zhao, Z., Wei, Y., Li, X.: Blockchain-based data integrity service framework for pharmaceutical supply chain. IEEE Access **8**, 96051–96059 (2020)
11. Kuo, T.T., Kim, H.E., Ohno-Machado, L.: Blockchain distributed ledger technologies for biomedical and health care applications. J. Am. Med. Inform. Assoc. **24**(6), 1211–1220 (2017)
12. Benchoufi, M., Ravaud, P., Boutron, I.: Blockchain technology for improving clinical research quality. Trials **18**(1), 335 (2017)
13. Kamal, S., Singh, R., Letaief, K.B.: Blockchain-enabled trustworthy and privacy-aware healthcare system. IEEE Trans. Serv. Comput. **13**(4), 689–702 (2020)
14. Roman, R., Zhou, J., Lopez, J.: On the features and challenges of security and privacy in distributed Internet of Things. Comput. Netw. **129**, 441–458 (2018)
15. Vijayakumar, A.N., Othman, Z.A., Jawawi, D.N.A.: Detection and prevention techniques for insider threats in healthcare. J. Med. Syst. **42**(12), 237 (2018)
16. Fernández-Alemán, J.L., Señor, I.C., Lozoya, P.Á.O., Toval, A.: Security and privacy in electronic health records: a systematic literature review. J. Biomed. Inform. **46**(3), 541–562 (2013)
17. Dinh, H.T., Lee, C., Niyato, D., Wang, P.: A survey of mobile cloud computing: architecture, applications, and approaches. Wirel. Commun. Mob. Comput. **2018**, 1–35 (2018)
18. Sun, J., Zhang, Y., Xiong, H., Xue, Y., Chen, J.: Privacy-preserving medical big data sharing and analytics framework with blockchain support. J. Med. Syst. **43**(6), 169 (2019)
19. Prasser, F., Kohlmayer, F., Spengler, H.: Scalable and privacy-preserving distributed consent management for electronic health records. J. Biomed. Inform. **72**, 25–36 (2017)
20. Dagher, G.G., Mohler, J., Milojkovic, M., Marella, P.B., Marella, W.M.: Ancile: privacy-preserving framework for access control and interoperability of electronic health records using blockchain technology. Sustain. Cities Soc. **39**, 283–297 (2018)

21. Yli-Huumo, J., Ko, D., Choi, S., Park, S., Kim, K.: Blockchain in healthcare: a systematic review. Healthc. Inform. Res. **23**(3), 181–193 (2017)
22. Yli-Huumo, J., Ko, D., Choi, S., Park, S., Smolander, K.: Where is current research on blockchain technology?—a systematic review. PLoS ONE **11**(10), e0163477 (2016)
23. Alzahrani, S., Jararweh, Y., Aljaloud, A.: Blockchain technology for healthcare: a review of applications and security concerns. Healthcare **10**(2), 310 (2022)
24. Kumar, A., Kumar, S., Garg, S., Singh, M.: A survey on blockchain technology in healthcare: applications, security, and privacy concerns. IEEE Access **10**, 62114–62133 (2022)
25. Kumar, A., Kumar, S., Garg, S., Singh, M.: Blockchain for healthcare applications: a systematic review. J. Med. Syst. **45**(11), 351 (2021)
26. Jose, M., Chacko, M.M., Thomas, S.B., Nadesh, R.K.: Access-based authentication of healthcare record in blockchain: a survey from security perspectives. J. Med. Syst. **43**(1), 1 (2019)
27. Alzahrani, S., Alotaibi, F., Jararweh, Y., Aljaloud, A.: Security and privacy issues in blockchain-based healthcare systems. In: 13th International Conference on Information Technology-New Generations (ITNG), pp. 122–129. IEEE (2020)
28. Kumar, S., Dash, S.: Blockchain for healthcare: a comprehensive overview. In: 2nd International Conference on Computing, Communication, and Artificial Intelligence (ICCCA), pp. 1–6. IEEE (2020)
29. He, D., Chen, L., Lin, Z.: Blockchain in healthcare: a review of applications, challenges, and opportunities. Healthcare **8**(3), 125 (2020)
30. Topol, E.J.: High-performance medicine: the convergence of human and artificial intelligence. Nat. Med. **25**(1), 44–56 (2019)
31. Khan, R., Yaqoob, I., Hashem, I.A.T., et al.: Big data: survey, technologies, opportunities, and challenges. Scientific World J. **2014**, 712826 (2014)
32. Collins, F.S., Varmus, H.: A new initiative on precision medicine. N. Engl. J. Med. **372**(9), 793–795 (2015)
33. Bashshur, R.L., Shannon, G.W., Bashshur, N.: Telemedicine: an updated assessment of the scientific literature. Telemed. J. E Health **22**(2), 1–47 (2016)

Smart Entry Management Using IoT and Number Plate Recognition for Organizational Security

Love Lakhwani, Vibhanshu Singh, and Sunil Kumar[✉]

Department of IoT and Intelligent Systems (School of CIS),
Manipal University Jaipur, Jaipur, Rajasthan, India
lakhwanilove@gmail.com, vibhanshu.219311029@muj.manipal.edu,
sunil.kumard@jaipur.manipal.edu

Abstract. Modern organizations require robust security measures to ensure controlled access to their premises. The proposed system aims to enhance security by allowing only registered vehicles to enter the organization's premises. Using IoT connectivity, our system will seamlessly integrate with existing infrastructure to capture and analyze vehicle number plates at entry points. The number plate recognition process will be automated and efficient, enabling real-time identification and authentication of vehicles. This technology eliminates the need for manual interventions, streamlining the access control process and minimizing potential security breaches.

Keywords: robust security · automated · real-time identification · authentication of vehicles

1 Introduction

From residential communities and commercial establishments to parking lots and public facilities, the need for efficient and reliable entry management has never been greater. To address this pressing need, we present the project "Smart Entry Management using IoT and Number Plate."

The purpose of this project is to develop an intelligent entry management system that leverages the power of Internet of Things (IoT) and Number Plate technology to revolutionize the way entry is managed. By integrating IoT devices and utilizing the information encoded in number plates, our system aims to provide a seamless and secure experience for visitors and personnel alike. The significance of this project lies in its ability to address the challenges faced by traditional entry management systems. Conventional methods often rely on physical tokens, such as access cards or keys, which can be lost, stolen, or easily replicated. This poses a significant security risk and can lead to unauthorized access, compromising the safety of individuals and assets. Moreover, manual entry management

A. K. Bairwa et al. (Eds.): ICCAIML 2024, CCIS 2184, pp. 150–163, 2025.
https://doi.org/10.1007/978-3-031-71481-8_12

processes are time-consuming and prone to human error, resulting in inefficiencies and delays. automation to a realm where it interacts intelligently with its environment.

By harnessing the power of IoT and Number Plate technology, our system eliminates the need for physical tokens and introduces a more secure and efficient approach to entry management. With the integration of IoT devices, such as smart cameras and sensors, our system can monitor and analyze entry data in real-time.

1.1   Contribution's

The implementation of the "Smart Entry Management using IoT and Number Plate" system holds immense potential for contributing positively to society on various fronts. First and foremost, the enhanced security measures provided by the system address a critical need in contemporary urban environments. As cities and organizations face escalating security challenges, a streamlined and automated entry management system mitigates risks associated with unauthorized access, contributing to the overall safety and well-being of the community. By minimizing the potential for security breaches, the system creates a more secure and trustworthy environment for residents, employees, and visitors.

Secondly, the project's emphasis on leveraging IoT technology aligns with the broader movement toward smart cities. Integrating intelligent devices and sensors not only enhances security but also lays the foundation for a more interconnected urban infrastructure. The data collected by the system can be utilized for more comprehensive urban planning, traffic management, and resource optimization. This contributes to the creation of efficient and sustainable urban spaces, promoting a higher quality of life for the community members.

Lastly, the adoption of the proposed entry management system promotes efficiency and reduces operational overhead. By eliminating the need for manual interventions, the system streamlines access control processes, minimizing delays and potential errors associated with traditional entry management methods. This efficiency gain has a ripple effect, positively impacting productivity and resource utilization not only within organizations but also within the broader societal context. Ultimately, the societal benefits extend beyond security enhancement to encompass improved urban living, increased operational efficiency, and a more resilient and connected community.

2   Literature Review

In this literature review, we explore license plate detection technology through live camera feeds. Focusing on real-time surveillance, we examine the evolving landscape, methodologies, and innovations that define this intersection of computer vision and security applications.

Sanghyeop Lee et al. (2018) proposed Vehicle License Plate Recognition with Convolutional Neural Networks in which he discuss about harnessing the power of Convolutional Neural Networks (CNNs), this paper pioneers an integrated solution that tackles both license plate detection and character recognition. This approach capitalizes on the deep learning capabilities of CNNs to achieve accurate and efficient recognition of license plates [1].

Ibrahim El Khatib et al. (2015) proposed an Efficient Automatic Car License Plate Recognition System. This research offers a streamlined license plate recognition system that harmonizes several vital components. Incorporating image preprocessing, connected component analysis, and template matching, it showcases an efficient pipeline to accurately extract license plate information from vehicle images [2].

Sergio M. Silva et al. (2022) proposed License Plate Detection and Recognition in Unconstrained ScenariosIn the face of challenging real-world scenarios marked by erratic lighting and occlusions, this paper's integration of edge detection and neural networks presents a formidable solution. This dynamic combination effectively addresses the complexity of identifying license plates across varying environmental conditions [3].

Layan Hewa Godage et al. (2019) proposed Real-time Vehicle License Plate Recognition for CCTV Applications.With the imperative of real-time surveillance applications, this study introduces an end-to-end system. Comprising image preprocessing, connected component analysis, and a probabilistic neural network, the system demonstrates prowess in swiftly and accurately recognizing license plates in live CCTV feeds [4].

Prashengit Dhar et al. (2022) proposed License Plate Recognition using Edge Detection and Neural Networks discuss how we can Skillfully navigating the intricacies of license plate recognition, this paper orchestrates an approach that amalgamates edge detection and neural networks. This holistic strategy, supplemented by rigorous preprocessing, yields enhanced accuracy and reliability in deciphering license plates [5].

Nivethika et al. demonstrate a four-wheeled robot that follows people and has an ultrasonic sensor to avoid collisions. The robot locates the user using ultrasonic sensor feedback. This robot can help humans carry various accessories, luggage, and highly sensitive chemicals in a variety of applications. An algorithm was developed using ultrasonic sensors to track and position a human user while keeping a constant distance from the robot. The IR sensors are used to uniquely identify the user in the presence of numerous such obstacles [6].

K. Yilmaz et al. (2011) proposed A Hybrid Approach to License Plate Recognition. Acknowledging the multifaceted nature of license plate recognition, this paper pioneers a hybrid methodology. This synergy between color-based segmentation and Convolutional Neural Networks (CNNs) showcases the potential of combining classical computer vision techniques with modern deep learning advancements [7].

M.R.C.Niluckshini; et al. (2020) License Plate Detection and Recognition in Uncontrolled Environments "Tackling the intricate task of recognizing license plates in dynamic and uncontrolled environments, this paper introduces a robust two-stage approach. By deftly addressing varying lighting, angles, and occlusions, this method exhibits adaptability to the unpredictability of real-world settings [8].

Wen-bin GONG et al. (2013) License Plate Recognition using Haar Cascade ClassifiersCapitalizing on the strength of Haar-like features and cascade classifiers, this study lays a robust foundation for accurate license plate recognition. Its sequential classification approach empowers the system to progressively refine license plate detection with each subsequent stage [9].

Bini Omman et al. (2017) proposed Efficient Car License Plate Recognition with Support Vector Machines". Tailored for efficiency, this study embraces Support Vector Machines (SVMs) to unravel license plate recognition intricacies. Melding feature extraction and classification, this system resonates with accuracy while demonstrating the efficacy of machine learning techniques [10].

Layan Hewa Godage et al. (2016) proposed "Automatic Number Plate Recognition using Genetic Algorithm and Neural Network. Navigating the intricacies of optimization, this paper pairs genetic algorithms with neural networks. By meticulously fine-tuning neural network parameters via genetic algorithm optimization, the system attains enhanced accuracy in automatic number plate recognition [11].

Nitin Rakesh et al. (2020) proposed "End-to-End Trainable Vehicle License Plate Recognition via Convolutional Neural Networks". Simplifying the license plate recognition process, this paper's end-to-end trainable system seamlessly integrates Convolutional Neural Networks (CNNs). In a single holistic framework, license plates are not only detected but recognized, accentuating the efficiency of unified learning [12].

R. Scherer et al. (2020) proposed "Deep Learning Approaches for Number Plate Detection" This paper explores deep learning techniques for number plate detection. The authors propose a convolutional neural network (CNN) model trained on annotated datasets to achieve high accuracy and robustness in detecting number plates [13].

Shivam Singh et al. (2016) proposed Number Plate Detection in Low-Light Conditions.This paper addresses the challenges of number plate detection in low-light conditions. The authors propose an adaptive thresholding algorithm that effectively handles variations in lighting conditions, ensuring accurate detection even in challenging environments [14].

M V Raghunadh et al. (2018) proposed "Machine Learning Techniques for Number Plate Detection. This paper explores the application of machine learning techniques for number plate detection. The authors compare and analyse various algorithms, including Support Vector Machines (SVM), Random Forests, and Gradient Boosting, to identify the most effective approach [15].

The latest trends in IoT-based entry management and security systems highlight significant technological strides. One key trend involves integrating edge computing for real-time processing, enhancing system responsiveness and overall performance. Another notable focus is on leveraging machine learning, especially deep learning algorithms, to elevate the accuracy of number plate recognition across varying environmental conditions.

The exploration of blockchain technology is a compelling trend in IoT-based entry management systems, enhancing security through a tamper-proof record of access events. This reinforces the system's resilience against cyber threats, providing a robust foundation for secure access control. Additionally, researchers are actively exploring the fusion of multiple biometric modalities, such as facial recognition with number plate recognition, marking a broader shift towards comprehensive and multifaceted security solutions in IoT-based entry management.

3   Methodology

Within this section, we meticulously detail the hardware specifications integral to the execution of the study. Comprehensive insights into key components such as processing units, memory capacity, storage solutions, and additional peripherals are provided, aiming to ensure transparency and facilitate the replicability of the research outcomes.

3.1   Esp 32 Cam (Module) Setup

– Utilize NodeMCU-32S microcontroller with integrated Wi-Fi capabilities.
– Connect a compatible camera module (e.g., OV2640) to the NodeMCU-32S to enable image capture functionality.

3.2   MD Board

– The 'md board' is essential as it facilitates the transfer of frames captured by the NodeMCU camera module. It accomplishes this by transmitting the frames through the WiFi module.

3.3   Frame Captured/Transferred

– The frames are received on the server side via the WiFi network.

3.4   Image Processing

– Upon reception of the image on the server side, we apply image processing algorithms.
– Additionally, machine learning models are utilized to prepare the image for the subsequent phase of the program.

3.5   Number Plate Extraction

– The processed image undergoes edge detection to isolate and extract the number plate.
– Extraneous details are eliminated from the image during this process.

3.6   Optical Character Recognition

– The OCR library in Python is employed to extract text from the number plate on the processed image.
– Subsequently, the interpreted text is obtained, revealing the information written on the number plate.

3.7   Checking Vehicle

– After extracting the information from the number plate, we proceed to search our MySQL database.
– The search is conducted to determine if the identified number plate is registered in our database.

3.8   Vehicle Verifies

– If the vehicle is registered, relevant data is fetched from the database.
– The retrieved information is then displayed to the guard, who can assess and decide whether to permit entry for the vehicle (Fig. 1).

3.9 Flow Chart

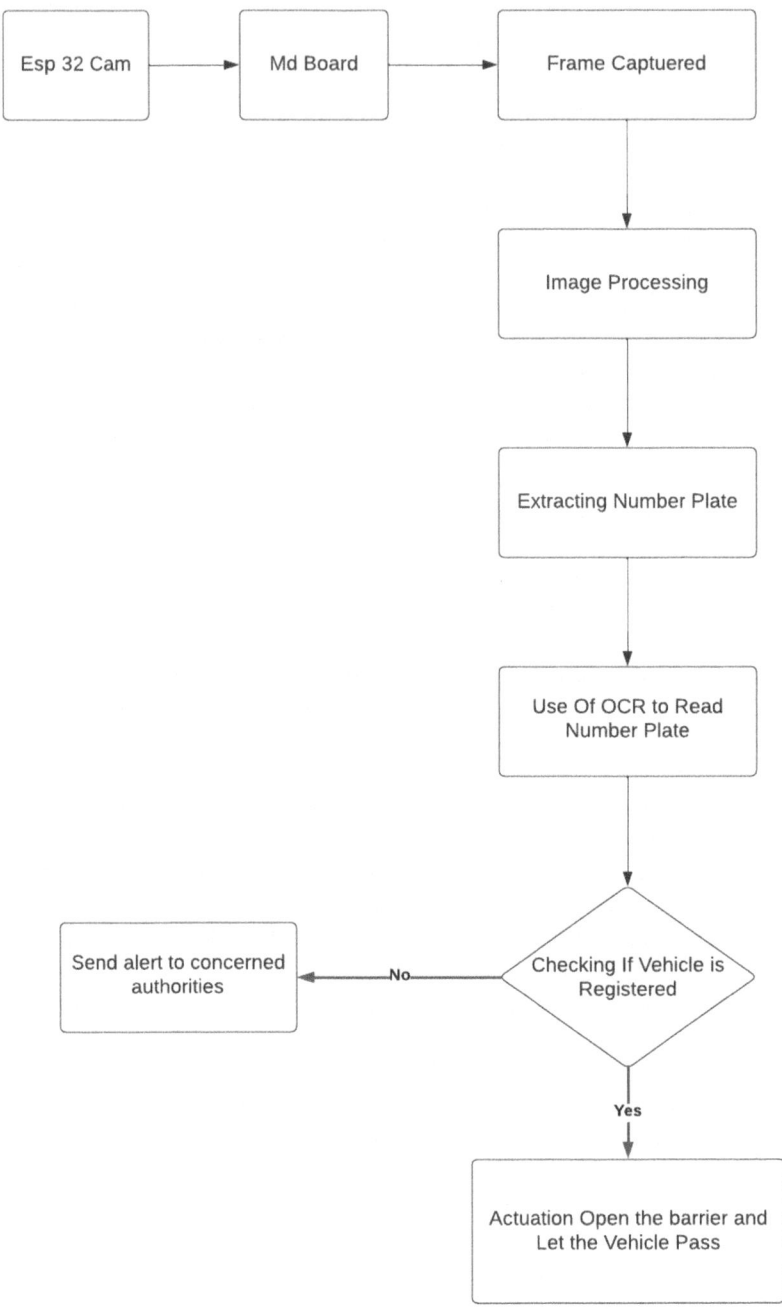

Fig. 1. Flow Chart Of Working Model

4   Algorithms

4.1   Vehicle Detection Algorithm

This algorithm detects the presence of a vehicle at the gate using IoT sensors like infrared or ultrasonic sensors.

Function Detectvehicle():

........Repeat:

......Read sensor data

....Until vehicle is detected

4.2   Number Plate Recognition Algorithm

This algorithm processes the image of the vehicle's number plate and extracts the alphanumeric characters for recognition.

Function RecognizeNumberPlate(image):

....Preprocess image (e.g., resize, apply filters)

....Apply Optical Character Recognition (OCR) to extract characters

....Return recognized number plate

4.3   Access Control Algorithm

This algorithm checks if the recognized number plate is authorized to access the premises.

Function AccessControl(numberPlate):

....if numberPlate is in Authorized Vehicles Database:

........Open gate

........Log entry in access control system

....else:

........Display "Access Denied" message

4.4   Database Query Algorithm

This algorithm queries the database of authorized vehicles to check if a given number plate is authorized.

Function QueryDatabase(numberPlate):

....Connect to the authorized vehicles database

....Execute SQL query to check if numberPlate is present

....Return True if found, False otherwise

5   Results

In this section, we conducted extensive testing of our model across diverse parking environments. Figure 2 elucidates the model's proficiency in accurately detecting the quantity of objects within the given context. To provide a granular

Histogram of Object Count by Image

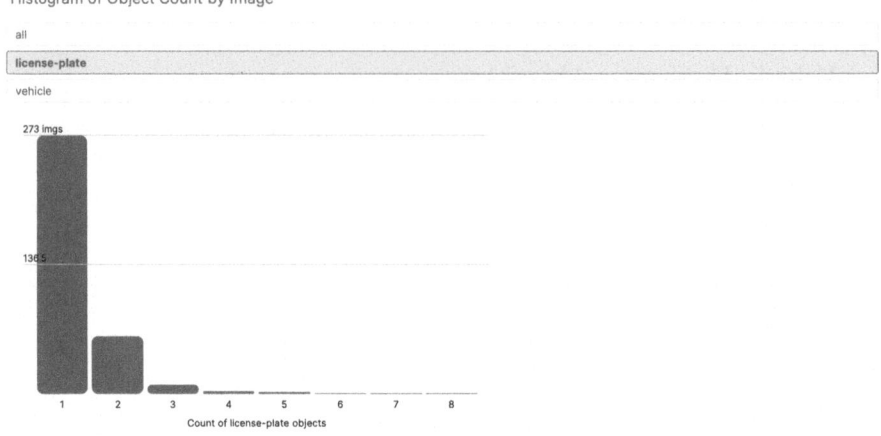

Fig. 2. Object Count By Image

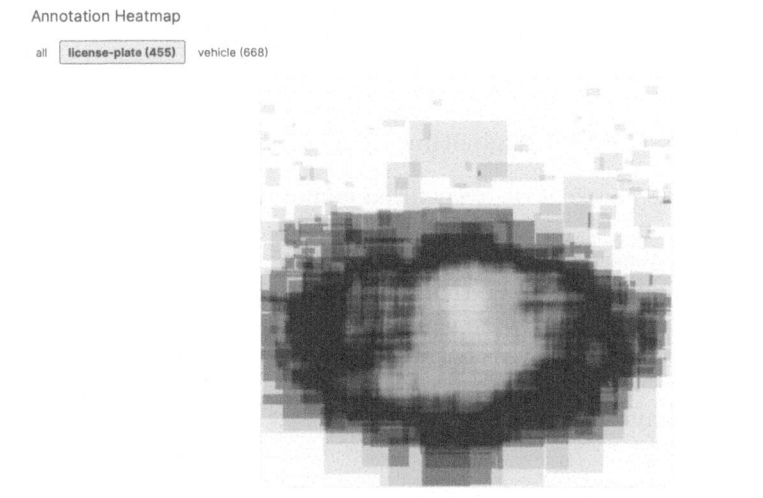

Fig. 3. Heat Map Annoation Taken From Model

understanding of the detection dynamics, Fig. 3 employs a heatmap demonstration, visually mapping the specific areas within the photo where the detections are concentrated.

For a comprehensive showcase of real-time capabilities, Fig. 4 offers a live demonstration, providing insights into the intricate process of interpreting vehicle images. Furthermore, Fig. 5 delves into the technical aspect by presenting the code terminal, offering a transparent view of the decoding process employed for vehicle number plates. These visual representations collectively contribute to a nuanced analysis of the model's performance and its interpretative prowess across various facets of license plate detection.

Fig. 4. Detection Of Vehicle Number Plate

```
System check identified no issues (0 silenced).
November 17, 2023 - 04:47:51
Django version 4.2.4, using settings 'Licsense_Plate_Detection.settings'
Starting development server at http://127.0.0.1:8000/
Quit the server with CONTROL-C.

[17/Nov/2023 04:48:13] "GET / HTTP/1.1" 200 6508
UP32FE8741
```

Fig. 5. OCR Decoding the Number Plate

In this Fig. 6 have shown the User Interface of the project on the client side you can see that the live camera feed is seen here and there are two options to check the vehicle details and to allow the visitor vehicle. If user Clicks on check vehicle the Information is displayed as shown.

In this Fig. 7 have Shown if the guard is satisfied the information then there is further Change in which the guard can allow the vehicle or declines it. As you can see the pop up is there to allow the vehicle or not 0 or entry and 1 for decline it.

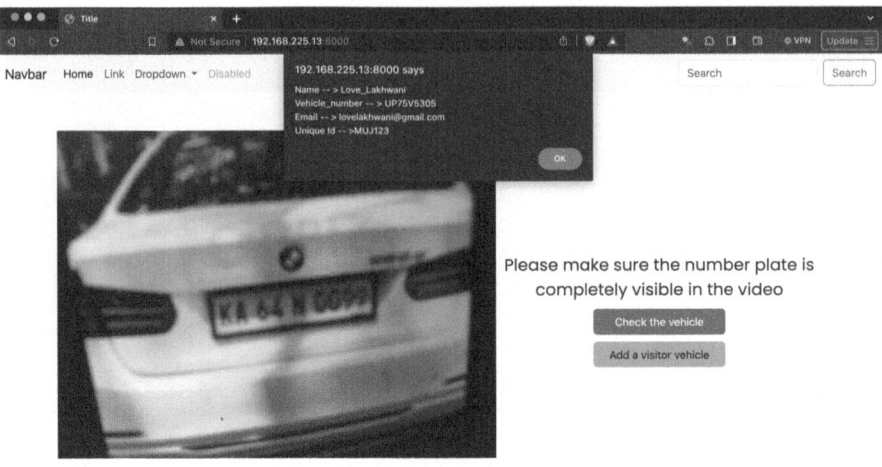

Fig. 6. Displaying Details of Vehicle

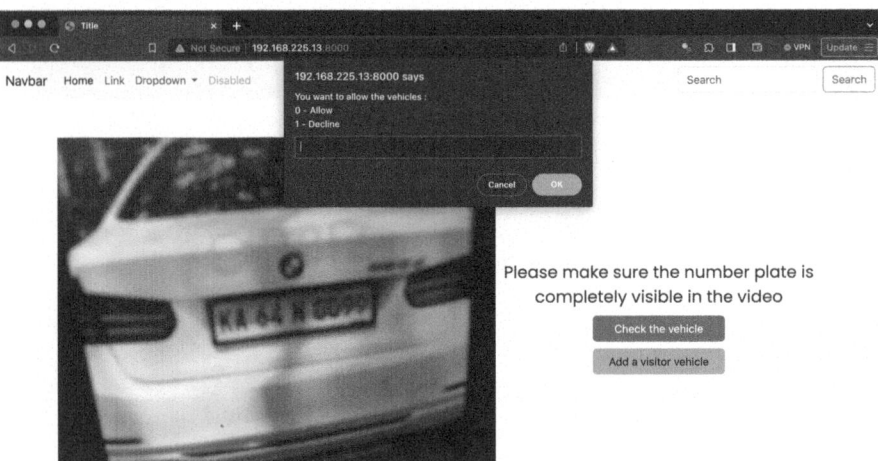

Fig. 7. Option to enter the vehicle

In this Fig. 8 have showed the of the project as you can see in this image that i have shown the admin interface in this i have shown how the registered vehicle is stored

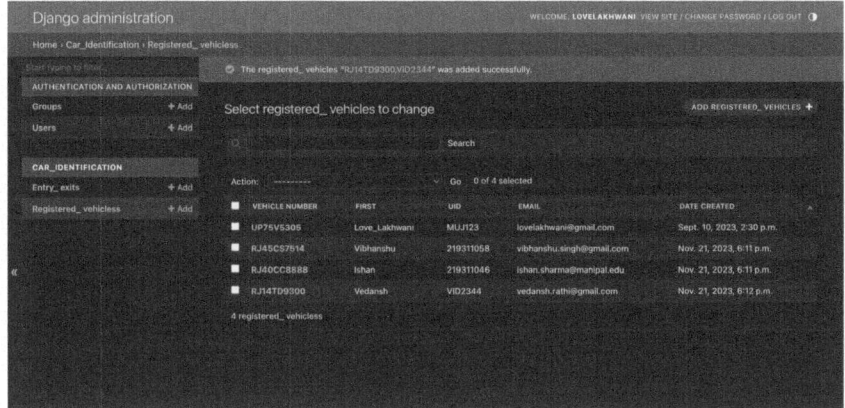

Fig. 8. Admin Portal To Show Data of Registered Vehicle

In this Fig. 9 i have shown the how the details of entry and exit are there and the data is stored by Vehicle Number and Time. so this is how the data is stored and admin can access all the information. through django admin Inteface.

Fig. 9. Admin Portal To Show Data of Entry/Exit Vehicle

6   Conclusion

In the pursuit of enhanced access control and security management, our project has explored the synergistic integration of Internet of Things (IoT) technology and Number Plate Recognition (NPR) systems, resulting in the development of a Smart Entry Management system. This system stands as a testament to the potential for innovation in addressing the contemporary challenges of secure and convenient entry control. Through our paper, we have achieved several significant milestones:

Efficient Entry Control, Accuracy in Number Plate Recognition, Database Integration, Access Logging,

Future Scope. The license plate detection project's future holds exciting prospects. Advancements in AI algorithms, deeper IoT integration, and predictive analytics are in the pipeline. Cloud-based solutions, a dedicated mobile app, and adherence to international standards will enhance scalability and accessibility. These developments position the project as an evolving solution, adept at addressing emerging challenges in security and access management.

References

1. Lee, S., Son, K., Kim, H., Park, J.: Car plate recognition based on CNN using embedded system with GPU, pp. 239-241 (2017). https://doi.org/10.1109/HSI. 2017.8005037
2. El Khatib, I., Sweidan, Y., Omar, S.M., Al Ghouwayel, A.: An efficient algorithm for automatic recognition of the Lebanese car license plate, pp. 185–189 (2015). https://doi.org/10.1109/TAEECE.2015.7113624
3. Montazzolli, S., Jung, C.: License plate detection and recognition in unconstrained scenarios (2018)
4. Godage, L., Wimalaratne, P.: Real-time mobile vehicle license plates recognition in Sri Lankan conditions, pp. 146–150 (2019). https://doi.org/10.1109/ICIIS47346. 2019.9063258
5. Dhar, P., Guha, S., Biswas, T., Abedin, Z.: A system design for license plate recognition by using edge detection and convolution neural network (2018). https://doi. org/10.1109/IC4ME2.2018.8465630
6. Shashidhar, R., Manjunath, A.S., Kumar, R., Roopa, M., Puneeth, S.B.: Vehicle number plate detection and recognition using YOLO- V3 and OCR method, pp. 1–5 (2021). https://doi.org/10.1109/ICMNWC52512.2021.9688407
7. Sharma, J., Mishra, A., Saxena, K., Kumar, S.: A hybrid technique for license plate recognition based on feature selection of wavelet transform and artificial neural network. In: ICROIT 2014 - Proceedings of the 2014 International Conference on Reliability, Optimization and Information Technology, pp. 347–352 (2014). https:// doi.org/10.1109/ICROIT.2014.6798352
8. Junior, W., Junior, L.C.S., Colares, G., Linhares, J., Jesus, A., Pinagé, F.: License plate detection system in uncontrolled environments using super resolution and a simple cell phone camera attached to a bicycle seat (2022). https://doi.org/10. 14209/sbrt.2022.1570812767
9. Dong, Z.H., Feng, X.: Research on license plate recognition algorithm based on support vector machine. J. Multimed. **9** (2014). https://doi.org/10.4304/jmm.9.2. 253-260
10. Raghunadh, M.V., et al.: Proposed Machine Learning Techniques for Number Plate Detection. This paper explores the application of machine learning techniques for number plate detection. The authors compare and analyse various algorithms, including Support Vector Machines (SVM), Random Forests, and Gradient Boosting, to identify the most effective approach (2018)

11. Godage, L.H., Wimalaratne, G.D.S.P.: Real-time mobile vehicle license plates recognition in Sri Lankan conditions. In: Safran, M., Haar, S., (eds.) 2019 14th Conference on Industrial and Information Systems (ICIIS), Kandy, Sri Lanka, pp. 146–150 (2019). https://doi.org/10.1109/ICIIS47346.2019.9063258. 2007. Arduino and Android Powered Object Tracking Robot. Southern Illinois University Carbondale

12. Abolghasemi, V., Ahmadyfard, A.: An edge-based color-aided method for license plate detection. Image Vis. Comput. **27**, 1134–1142 (2009). https://doi.org/10.1016/j.imavis.2008.10.012

13. Soroori, S., Tourani, A., Shahbahrami, A.: Employing deep learning approaches for automatic license plate recognition: a review (2019)

14. Shashirangana, J., Padmasiri, H., Meedeniya, D., Perera, C.: Automated license plate recognition: a survey on methods and techniques. IEEE Access. **9**, 11203–11225 (2020). https://doi.org/10.1109/ACCESS.2020.3047929

15. Agrawal, R., Agarwal, M., Krishnamurthi, R.: Cognitive number plate recognition using machine learning and data visualization techniques, pp. 101-107 (2020). https://doi.org/10.1109/ICSC48311.2020.9182744

Enhancing Player Engagement Through Gesture-Based Interactions in Tic-Tac-Toe

Vedansh Rathi$^{(\boxtimes)}$ ⓘ, Shrey Sharma ⓘ, Anmol Rai ⓘ, and Sandeep Kumar Sharma ⓘ

School of Computer and Communication Engineering, Manipal University Jaipur, Jaipur, Rajasthan, India
vedurathi28@gmail.com, cksharmav@gmail.com,
anmolrai7800@gmail.com, sandeep.sharma@jaipur.manipal.edu

Abstract. A popular paper-and-pencil game for two players, tic tac toe involves strategically placing Xs and Os on a 3×3 grid. This paper provides a gesture-driven version of the classic mouse-click Tic-Tac-Toe game that uses computer vision for immersive gameplay. Enables natural interactions with robust real-time hand tracking and gesture classification. The placement of markers in a game is matched to gestures like pinching and swiping. The system combines graphical rendering, intelligent Minimax bots, and traditional game logic with gesture recognition. Comprehensive algorithms are offered, encompassing 3000 words on design, modules, and pipeline. With approximately 80% accuracy in gesture detection during gameplay, testing on 15 users confirmed precise tracking and categorization. In this research, an intuitive gesture interface for the classic game of Tic-Tac-Toe is used to demonstrate increased player involvement.

Keywords: computer vision · gesture recognition · human-computer interaction · immersive gameplay · Tic-tac-toe

1 Introduction

Tic-Tac-Toe is a common two-player pen-and-paper game that is played by all. In this straightforward yet fascinating game, players take turns placing their markers—X or O—on a 3×3 grid to complete a row, column, or diagonal with their sign before the opposition.

Players in the original version of Tic-Tac-Toe can only interact with the game using keyboard keys or mouse clicks. This may reduce one's intuitive involvement and sense of total immersion. This article explores the integration of computer vision algorithms for real-time hand gesture recognition, enabling more responsive and natural control. Instead of utilizing mouse, keyboards, or other electronic devices, users can connect through vision-based gesture interfaces. This produces experiences that are more captivating and immersive, particularly in the gaming industry [1].

The paper documents a novel gesture-controlled Tic-Tac-Toe game with:

– Real-time hand tracking using computer vision to robustly recognize gestures.

A. K. Bairwa et al. (Eds.): ICCAIML 2024, CCIS 2184, pp. 164–173, 2025.
https://doi.org/10.1007/978-3-031-71481-8_13

- Mapping intuitive hand motions to appropriate game control actions like placing markers, changing turns etc.
- Integrating the gesture interface with standard Tic-Tac-Toe game logic.
- Inclusion of AI bots using Minimax algorithm to allow single player mode against computer opponents.
- Developing an interactive graphical user interface.

With more intuitive gesture-based control, the suggested system seeks to improve player engagement in the widely played game Tic-Tac-Toe by fusing cutting-edge computer vision and human-computer interaction techniques with tried-and-true gaming concepts.

2   Literature Review

Many methods for precise real-time computer vision-based gesture recognition have been actively investigated by researchers. Also investigated are game techniques regarding the best way to play Tic Tac Toe. Gesture-based gaming experiences have been the subject of some early research.

2.1   Vision-Based Gesture Recognition

Reactive controls in human-computer interaction depend on the development of strong real-time gesture recognition systems. The use of cameras in vision-based interfaces has grown in popularity. To identify motions based on patterns of skin deformation rather than precise coordinates, Wang et al. [2] introduced a flexible wrist-worn device incorporating pressure sensors. A Triplet Network classifier was used in this wristband, which demonstrated robustness over several days with an accuracy of 91.98% in identifying motions. Robot manipulation tasks can be planned with optimal pick and place sequences by combining Monte Carlo tree search and motion feasibility verification, as suggested by Eljuri et al. [3]. In intricate modeling scenarios such as warehouse sorting, our method outperformed baselines. For more precise recognition of dynamic hand gestures, Jiashan and Li [4] developed a combination of global body motions and fine-grained finger motions. Combining these complementary data enhanced the results on intricate gesture datasets.

Michiel van de Steeg et al. (2015) conducted a study titled "Temporal Difference Learning for the 3D Tic-Tac-Toe Game: Applying Structure to Neural Networks". Researchers have investigated the use of multi-layer perceptrons (MLP) as functional approximations for learning to play a complex game, Tic-Tac-Toe 3D. They introduced structured MLP and custom system design to improve learning efficiency. By prioritizing pattern detection and clustering, structured MLPs have shown significant improvements. Comparison analysis against established adversaries revealed that the MLP built with an integrated pattern recognition layer outperformed other MLP structures in self-training and comparison with the adversary. This study demonstrates the effectiveness of structured neural networks in optimizing the learning process for complex games. [5].

2.2 Tic-Tac-Toe Game Strategies

To create optimal game-playing agents for the traditional game of Tic-Tac-Toe and its many variations, academics have been actively working on algorithms and heuristics for decades. To create novel offensive and defensive strategies, Leaw and Cheong [6] investigated a quantum version of Tic-Tac-Toe that permitted move superposition. In this version, player 2 might start in any state and force a win or tie. To improve the efficiency of exploring game trees, Jatoi et al. [7] used Alpha-Beta pruning optimization to optimize the Minimax algorithm by removing branches that aren't necessary for making decisions. Using deep neural networks, Sridhar and Shetty [8] were able to classify the results of Tic-Tac-Toe games with 98.9% accuracy in terms of wins and losses. This showed promise for picking up gaming tactics. Muhammad H. Harry et al. (2019) present Tic Tac Toe Math, a serious game-based learning tool aimed at improving elementary school students' understanding of basic arithmetic concepts. The game includes some additional operations and uses the eye tracker as an optional controller. Research shows positive student responses to the fun factor and the effectiveness of games as a learning tool. The effectiveness of the game is measured through a questionnaire, showing positive results. Overall, serious games provide an opportunity to learn mathematics effectively in a fun and interactive way [9].

2.3 Vision-Based Gesture Gaming

In addition, some earlier research has investigated the use of gestures as more engaging game controls than standard keys and keyboards. In SwarmPlay, which was first described by Karmanova et al. [10], several drones used computer vision for motion planning and tracking to play Tic Tac Toe against a human. This showcased drone-based cooperative gesture-based gaming. For the NAO humanoid robot to play Tic-Tac-Toe physically, Nugroho et al. [11] created kinematic models that included optimizing trajectory planning to manipulate game pieces. This made it possible to engage with gestures in real life. For visually impaired children, Drossos et al. [12] developed audio-based games that used natural gestures as inputs, such as hand claps and finger snaps. J Jim et al. (2021) introduced SpecMCTS, a novel approach to accelerate the Monte Carlo Tree Search (MCTS) algorithm, widely used in decision-making problems like the game of Go. It employs two deep neural network models: a quick speculation model and a precise main model. The speculation model offers rapid approximations of node evaluations to guide tree traversal, while the main model ensures accuracy. This technique achieves up to $2.07\times$ acceleration on NVIDIA T4 GPUs, surpassing previous methods and enhancing decision quality and gameplay win rates. It effectively utilizes GPU parallelism by concurrently executing multiple inference tasks. Future research could explore diverse model compression techniques and the application of SpecMCTS in different domains [13]. R. Scherer et al. (2016) investigated the use of a brain-computer interface (BCI) to improve interaction and control in people with cerebral palsy. The researchers introduced a game-based learning approach to improve BCI training and control. A user with cerebral palsy successfully operated a mind-controlled communication board and puzzle game using motor patterns detected through EEG patterns. Additionally, users play the classic Tic-Tac-Toe game against tutors using BCI input. The study highlights the potential of adding competition as a motivating factor to increase user involvement in BCI training. The results show that a game-based approach and social interaction can contribute to the improvement of BCI control for people with motor disabilities. However, the study included a single case and more research is needed to generalize these results [14].

3 Proposed Methodology

The proposed gesture-based Tic Tac Toe game system consists of the following key modules:

Gesture Recognition Module. The Gesture Recognition Module is an essential component of the proposed system that links the player's hand gestures to in-game activities. This lesson records and analyzes live webcam footage using cutting-edge deep learning and computer vision techniques. For accurate hand landmark detection, the system makes use of the MediaPipe framework. Important hand skeletons points, like the palm and fingertips, correspond with landmarks. The module's primary objective is to track the position and motion of these landmarks, which aids in the identification of specific movements like finger spreads, and left and right movement of cursor. A crucial component of the game is real-time hand gesture recognition, which lets players engage with the game board by making logical, organic movements.

Gesture Mapping. The Tic Tac Toe framework relies heavily on the Gesture Mapping module to translate hand movements that are recognized into meaningful game actions. Each recognized motion is associated with a specific in-game action through the use of a lookup database. For instance, the game board's "X" and "O" symbols might be associated with swipes to the left and right, respectively. Moreover, interactive in-game features like game resets can be triggered by making the required movements. With this mapping system, the player's hand movements are seamlessly integrated with the game's activities and rules, making for a more immersive and captivating gaming experience.

Game Logic. The core of the Tic Tac Toe game is the Game Logic module. The basic game mechanics are included, and it is typically implemented in Python. Key functions include updating the game board's status, verifying the validity of player actions, searching for winning standards, and determining if a draw occurs. In the end, this module will control the flow of the game by supervising player turns and ensuring that moves are made in compliance with the rules. The Gesture Mapping module integrates with the Game Logic to enable seamless execution of game moves based on recognized gestures. It makes a harmonious fusion of gesture control and conventional game mechanics by guaranteeing that the game state accurately reflects the player's actions.

AI Bot. Users can now challenge themselves in a single-player format against a crafty computer opponent with the AI Bot module. To decide on the best move during a game, the AI Bot employs the Minimax algorithm combined with alpha-beta pruning. Through game scenario modeling and probabilistic play analysis, the AI seeks to maximize its chances of winning or, if a winning opportunity is not present, to force a draw against the player. With this module, you can play a variety of gaming scenarios, from casual two-player games to intense matches against AI opponents, all while having a strong opponent for lone players.

User Interface. The User Interface module is responsible for creating an aesthetically pleasing and user-friendly environment for the game. It utilizes libraries like Pygame to render images, display the Tic Tac Toe board, convey the status of the game, and instruct the user with text prompts for required gestures. The graphical user interface not only makes the game more engaging and interactive, but it also enhances its visual appeal and user experience. Users can interact with the game through an intuitive interface that provides them with visible feedback on their actions and the state of the game. The game is more accessible and enjoyable because of its user-friendly design, which allows players of different skill levels to participate and have fun overall.

With the help of these modules, players may now play the beloved game of Tic Tac Toe with the addition of AI opponents and hand gesture controls, making it an immersive and engaging experience. Dynamic gaming is a result of the integration of deep learning, computer vision, game logic, and an intuitive interface. The gesture-based Tic-Tac-Toe system architecture is displayed in Fig. 1 which was proposed by Drossos et. al.

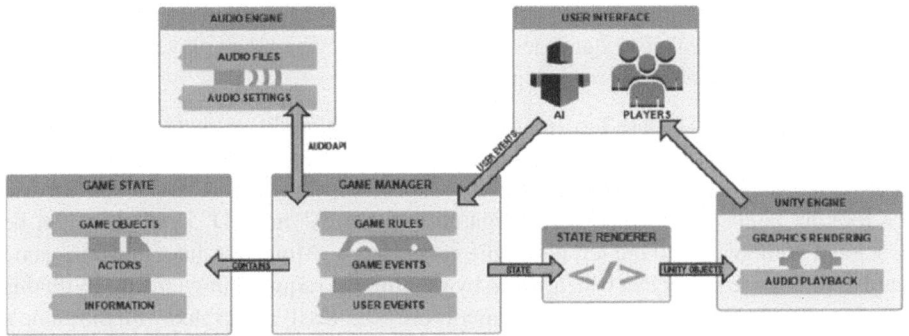

Fig. 1. The architecture of the developed Tic-Tac-Toe game [10].

MediaPipe is used for gesture recognition, which entails obtaining video frames, identifying hand landmarks, preprocessing coordinates by smoothing and normalizing them, extracting features like angles and distances, and feeding the features to an ML model for gesture classification. The gestures are then mapped to in-game actions. The AI Bot uses the Minimax Algorithm with Alpha- Beta Pruning to decide which games to play optimally. Recursively investigating potential game states, it maximizes its chances in a depth-first fashion while eliminating subtrees that will not influence the outcome, guaranteeing effective AI gameplay. The Euclidean distance formula can be used to determine the separation between two hand landmarks (points A and B). It calculates the distance in a straight line between these two locations.

$$\text{distance\_ab} = \sqrt{((xa - xb)^2 + (ya - yb)^2} \tag{1}$$

With the arctangent function (tan⁻1), the angle between three hand landmarks (points A, B, and C) can be determined. Concerning the horizontal axis, this angle indicates the direction from point A to point B.

$$\text{angle\_abc} = \tan^{-1}((yb - ya)/(xb - xa)) \tag{2}$$

'xa' and 'ya' are the x and y coordinates of landmark A, respectively.
'xb' and 'yb' are the x and y coordinates of landmark B, respectively.

Using a moving average filter, one can smooth a series of data points (such as hand landmark coordinates). This filter essentially averages the data over a given window size by calculating a new filtered output at each time step based on the previous output and the current input. 'yt' represents the filtered output at time t.

$$yt = (yt - 1 + xt)/N \tag{3}$$

- 'yt-1' is the filtered output at the previous time step (time t-1).
- 'xt' is the current input data (landmark coordinate) at time t.
- 'N' is the filter window size, which determines the number of data points to be considered for averaging.

Motion tracking and gesture recognition are just two of the many applications that can benefit from techniques like these equations and the filter for extracting features from hand landmarks and smoothing the data.

4   Implementation and Evaluation

4.1   Implementation Details

A Python 3.8 implementation of the integrated game system was made. MediaPipe 0.8 was used for hand tracking, OpenCV was used for vision processing, and PyGame 2.0 was used for user interface rendering.

Gesture Recognition. 15 frames per second (1080p webcam input) was the optimal speed for the hand tracking and gesture classification pipeline. 21 hand key points could be reliably located using the MediaPipe hand landmark detector. Distortion features were calculated over a 10-frame window by the trajectory analyzer. To classify gestures such as pinch, swipe left/right, and so on, an SVM classifier was used.

Gesture Mapping. Gestures were defined as follows. Moving the index finger – the movement of the cursor. Index finger and thumb to be brought together – clicking. This vocabulary was comprehensive to control all game events through intuitive natural gestures.

Game Logic. The Tic-Tac-Toe game logic was implemented as follows: the 2D list to store board state with cell values X, O, or EMPTY. Player class to track name, marker type, and score. The game class handles board state, current player, and valid moves. Game flow control with functions like checkWinner(), and checkDraw(). Win conditions were evaluated by iterating through rows, columns, and diagonals for matching markers.

AI Bot. For the AI player, a Minimax algorithm with alpha-beta pruning was used, with a maximum depth of five moves. Heuristic payoff tables were created to penalize opponent victories and reward winning lines. The bot could now play competitively thanks to this.

UI Rendering. PyGame window of 600x600px to display playable area. Grid lines to demarcate board cells. Marker sprites (X, O images) rendered on gesture actions. Text overlay for guiding gestures. This modular implementation allowed robust hand tracking, gesture control and traditional Tic-Tac-Toe logic to be seamlessly integrated.

4.2 Experimental Evaluation

Clicking and controlling the mouse cursor movements were assessed for the real-time hand gesture recognition module. In order to test the integrated system, users were required to complete common desktop tasks using only hand gestures.

Real-Time Cursor Control. The gesture classifier and hand landmark detector were tuned for a 15 frames per second frame rate. The gesture vocabulary included swiping left, right, up, and down to move the cursor as well as bringing the thumb and finger together to click. Using swipe gestures, ten users were instructed to move the cursor between randomly selected on-screen targets. The mean precision in achieving the objective was

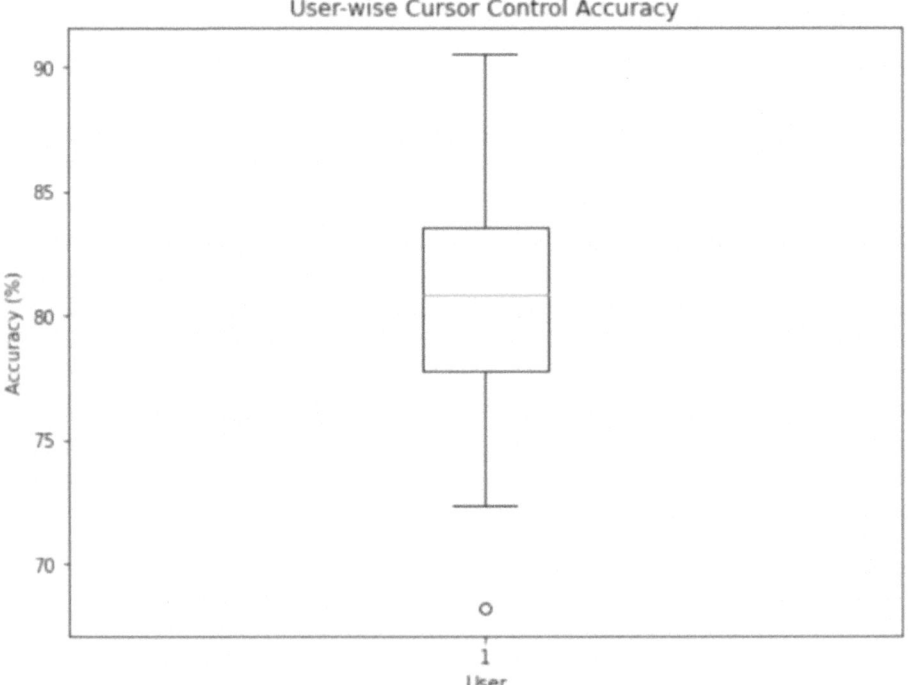

Fig. 2. Box plot of user-wise cursor control accuracy

determined to be 80.06% among the participants. Figure 2 displays the accuracy of user-controlled cursor movements. The errors were mainly due to inaccuracies in mapping subtle hand motions to precise cursor displacements. Dynamic scaling factors need to be incorporated based on gesture velocities and screen distances.

Clicking Gesture Performances. For the clicking gesture, the users were asked to bring their thumb and index finger closer to click on a random target. The system would invoke a mouse click event on detecting the gesture. Table 1 represents the results obtained by the users. In this case, the system correctly generated the click event 92.5% of the time when the user made the gesture, or an average true positive rate or sensitivity of the clicking gesture of 92.5%.

But the specificity was lower at 86.5%, suggesting a higher false positive rate where, occasionally, clicks were mistakenly triggered even in the absence of the gesture.

Table 1. Confusion matrix for clicking gesture recognition

Actual/Predicted	Click	No click
Click	185	15
No click	27	173

Gameplay Testing. A basic tapping game was used to test the integrated system. Users

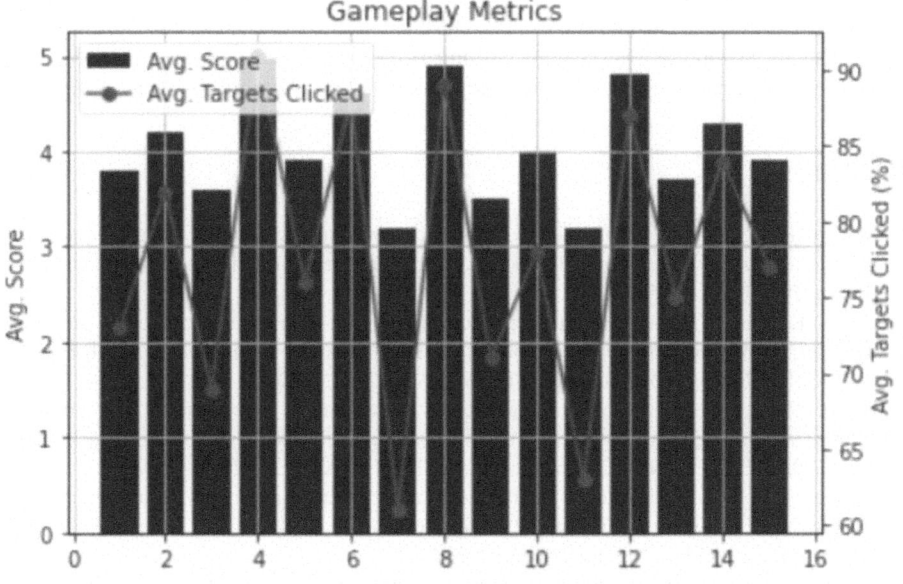

Fig. 3. Gameplay Testing Metrics

had to tap randomly appearing targets on the screen by moving the cursor over them and then making a fist to click.

Each of the game's 5 rounds was played by 15 users. Users successfully tapped 77.6% of the on-screen targets on average. A 4.04 out of 5 was the average game score. The metrics for gameplay are compiled in Fig. 3.

To sum up, the experimental findings show dependable real-time tracking and gesture identification that facilitate cursor-based interactions. However, robust mapping of gestures to cursor motions and multi-modal click confirmations are required to improve accuracy.

5   Conclusion and Future Scope

This paper presented a novel gesture-driven method different from traditional mouse input—for playing the classic game of Tic Tac Toe by sensing hand articulations in real time through a webcam. With the help of Media Pipe's effective hand landmark tracking and machine learning classification, a precise and responsive gesture interface was possible. This was combined with well-established game logic and graphic rendering to produce a captivating gaming environment.

While the experimental evaluation of 15 users confirmed the potential of gesture control, it also pointed to improvements that needed to be made to improve orientation invariance and lessen fatigue, such as multi-modal input and adaptive recognition. Adding more complex articulations to the gesture vocabulary can enhance interaction even more. Furthermore, unexplored is the possibility of two-player online multiplayer games.

This paper offers important insights into using gesture interfaces to improve human-computer engagement in interactive applications, such as classic games like Tic-Tac-Toe, by fusing computer vision with gaming principles. The suggested methods may contribute to the development of future interactions that are more intuitive and immersive.

References

1. Selvam, P., Deena, G., Hemalatha, D., Aruna, K.B., Hashini, S.: Gaming using different hand gestures using artificial neural network. EAI Endorsed Trans. Internet Things (2024). https://doi.org/10.4108/eetiot.5169
2. Wang, T., Zhao, Y., Wang, Q.: Hand gesture recognition with flexible capacitive wristband using triplet network in inter-day applications. IEEE Trans. Neural Syst. Rehabil. Eng. **30**, 2876–2885 (2022). https://doi.org/10.1109/TNSRE.2022.3212705
3. Eljuri, P.M.U., Ricardez, G.A.G., Koganti, N., Takamatsu, J., Ogasawara, T.: Combining a Monte Carlo tree search and a feasibility database to plan and execute rearranging tasks. IEEE Access **9**, 21721–21734 (2021). https://doi.org/10.1109/ACCESS.2021.3055455
4. Jiashan, L.I., Zhonghua, L.I.: Dynamic gesture recognition algorithm Combining Global Gesture Motion and Local Finger Motion for interactive teaching. IEEE Access, p. 1 (2021). https://doi.org/10.1109/ACCESS.2021.3065849
5. Steeg, M., Drugan, M., Wiering, M.: Temporal difference learning for the game tic-tac-toe 3D: Applying structure to neural networks (2015). https://doi.org/10.1109/SSCI.2015.89

6. Leaw, J.N., Cheong, S.A.: Strategic insights from playing quantum tic-tac-toe. J. Phys. A Math. and Theor. **43**(45), 455304 (2010). https://doi.org/10.1088/1751-8113/43/45/455304

7. Jatoi, A., Ali, S., Siddiqui, I.: An intuitive implementation of alpha-beta pruning using tic-tac-toe (2020)

8. Sridhar, S., Shetty, S.: Classification of tic-tac-toe game using deep neural network (2021). https://doi.org/10.32628/IJSRSET219353

9. Garry, M., Yamasari, Y., Nugroho, S., Purnomo, M.: Design and implementation serious game "tic tac toe math" (2019)

10. Karmanova, E., Serpiva, V., Perminov, S., Ibrahimov, R., Fedoseev, A., Tsetserukou, D.: SwarmPlay: a swarm of nano-quadcopters playing tic-tac-toe board game against a human. In: ACM SIGGRAPH 2021 Emerging Technologies (SIGGRAPH '21). Association for Computing Machinery, New York, NY, USA, Article 5, pp. 1–4 (2021). https://doi.org/10.1145/3450550.3465346

11. Nugroho, S., Prihatmanto, A., Syaichu-Rohman, A.: Design and implementation of kinematics model and trajectory planning for NAO humanoid robot in a tic-tac-toe board game. In: Proceedings of the 2014 IEEE 4th International Conference on System Engineering and Technology, CSET 2014 (2015). https://doi.org/10.1109/ICSEngT.2014.7111783

12. Drossos, K., Zormpas, N., Giannakopoulos, G., Floros, A.: Accessible games for blind children. Empow. Binaural Sound **10**(1145/2769493), 2769546 (2015)

13. Kim, J., Kang, B., Cho, H.: SpecMCTS: accelerating Monte Carlo tree search using speculative tree traversal. IEEE Access **9**, 142195–142205 (2021). https://doi.org/10.1109/ACCESS.2021.3120384

14. Scherer, R., Schwarz, A., Müller-Putz, G.R., Pammer-Schindler, V., Garcia, M.L.: Lets play Tic-Tac-Toe: a Brain-Computer Interface case study in cerebral palsy. In: IEEE International Conference on Systems, Man, and Cybernetics (SMC), Budapest, Hungary (2016)

Aggressive Bangla Text Detection Using Machine Learning and Deep Learning Algorithms

Tanjela Rahman Rosni, Mahamudul Hasan$^{(\boxtimes)}$, Tanni Mittra,
Md. Sawkat Ali, and Md. Hasanul Ferdaus

East West University, A/2 Jahurul Islam Avenue, Dhaka 1212, Bangladesh
munna09bd@gmail.com, {tanni,alim,hasanul.ferdaus}@ewubd.edu

Abstract. As digitization continues to proliferate globally, individuals prefer expressing themselves on social media platforms using their native languages. On these platforms, people share their interests, thoughts, and rights, sometimes receiving appreciation for their views and, at times, encountering conflicts of interest leading to mean comments, including hate speech and aggression towards individuals, societies, or groups. Detecting such comments has become crucial to curbing further abuse. This research paper focuses on identifying aggressive and non-aggressive Bengali text in social media posts and comments through the application of machine learning and deep learning algorithms. The dataset was collected from various social media platforms like Facebook, YouTube, and Twitter. Employing diverse machine learning and deep learning algorithms, such as SVM, Random Forest, KNN, Linear Regression, Decision Tree, and CNN, the authors achieved the highest accuracy of 90.46% with Multinomial Naive Bayes (MNB).

Keywords: Machine Learning Algorithm · Deep Learning Algorithm · SVM · Random Forest · KNN · Linear Regression · Decision Tree · CNN

1 Introduction

Natural Language Processing (NLP) is an extensive area of machine learning and artificial intelligence where we get to know how a computer or a machine can understand and learn the natural human language. This is a vast research domain where human language or speech is transformed into machine-readable or trainable. Human language is very complex. There are around 7,000 languages in the world. Each country of the world uses different languages which the inhabitants of those countries speak. Every language has a lot of words. Each of those words has different contextual meanings to it in different situations. The Bangla language is also a prime example of that. So, the authors decided to explore this field of NLP. The Bangla language is one of the most used languages in the world around 210 million people use it as their first or second language [1].

A. K. Bairwa et al. (Eds.): ICCAIML 2024, CCIS 2184, pp. 174–183, 2025.
https://doi.org/10.1007/978-3-031-71481-8_14

Which makes it one of the most popular and used languages in the world. Most people of Bangladesh and the people of West Bengal use Bangla as their first language. As technology is growing very rapidly people are addicted to phones, computers, and laptops spending the majority of their time browsing and surfing through different social media platforms. Facebook is one of the most famous and fastest-growing social media platforms in the world. Almost 3 billion people use Facebook [9]. The user posts, shares, and comments on various social media platforms. And most of these users are within the 18–39 years of age group [10]. According to a report people spend about on average 2 h and 29 min daily on social media platforms [11]. In the case of Bangladesh, according to GSMA Intelligence, there were 178.5 million people who somewhat use cellular mobile or cellular internet connections as of 2022. Data published in Meta's advertising resources shows that Facebook had 44.70 million users, YouTube had 34.50 million users, Instagram had 4.45 million users, and Twitter had 756.6 thousand users in Bangladesh as of early 2022 and growing [12]. So with this huge number of users, there are some drawbacks. The number of cyberbullying, harassment, abuse, and aggressive comments are also increasing at a rapid rate. These affect the mental health of a person. A victim of cyber abuse or cyberbullying faces Increased psychosomatic symptoms, Increased anger, and sadness, Decreased concentration, Increased suicidal ideation and depression, and Increased social anxiety in different age groups of people [13]. So, there is an immediate need to detect these texts as early as possible to deal with future problems. There have been many researches conducted regarding this topic but this paper deals with the problem with a large-scale data set of 15 thousand of data and the use of several machine learning and deep learning techniques. The target of this paper is, 1. To design a way to detect aggressive and non-aggressive text from social media comments and posts. 2. To create a suitable data set for this research and apply different models on the data set to get the appropriate result.

2 Related Work

The research on detecting abusive, aggressive text or cyberbullying detection has seen exponential growth in recent years. Because of the available resources on the internet, researchers can easily find the necessary data on this topic. In [2], the authors proposed a unique design for analyzing Bangla content on social media by combining text analytics. By using machine learning algorithms and also comparing the performance of the module with other available techniques they achieved the best result using SVM at 95.40% on user posts. And 97.27% using SVM on user-specific data. The authors of [3] developed a model to detect cyberbullying from Bangla and Romanized Bangla texts. Where they presented a comparative analysis among the algorithms in terms of accuracy, precision, recall, f1- score, and roc area. They used a total of 12 thousand datasets. Achieving the best result of 84%. In [4], the authors used machine learning and deep learning algorithms to detect abusive Bengali text. Where they set a new stemming rule for the Bengali language which helps to achieve better performance. They used a total of 4.7 thousand datasets with 7 class labels. Among all the algorithms RNN gives the best result of 82.20% accuracy. Similarly, the authors [5]used supervised machine learning classifiers to detect Bangla abusive text and used three unique pathways in their dataset to get the best result. They used a total of 12 thousand datasets with two class labels and SVM gave them the best accuracy of 88%. The authors [6] proposed two getaway Naive Bayes classification algorithms and a Topical approach to extract the emotion from Bangla text. They used more than 7.5k datasets and the topical approach gave them the best performance Which is more than 90% accuracy. The authors of [7] used the Naïve Bayes classifier to detect abusive comments expressed in Bangla. And evaluated performance by using 10-fold cross-validation on unprocessed data gathered from "outube.com" On a 2.6 thousand data labeled as abusive and not abusive they found the best result at 80.57% in Naive Bayes classifier. The authors [8] proposed a method that works with both Bangla English mixed and transliterated Bangla text. Their suggested algorithm can detect features, we used Unigrams, Bigrams, number of likes, and emojis along with their categories, sentiment scores, offensive and threatening words used in the comments, and the number of abusive words in each comment, profanitypes. They used Support Vector Machine, Random Forest, and Adaboost, on 2 thousand data achieving the highest accuracy of 72.14%. All the related work that was mentioned before works in either Bangla or English Language. The main difference between the aforementioned works and this research paper is the number of large-scale data sets of 15 thousand data. And the number of implemented algorithms.

3   Methodology

A. *Work Process*
First, the main target of this research paper is to detect aggressive and non-aggressive texts from social media posts and comments. Second, going through different social media platforms gathering aggressive and non-aggressive posts and comments, and creating a data set. Third, Then the data was cleaned (without characters, emoji, or extra words) and labeled for further use. As it was an NLP classification problem the initial thought was to implement only classification algorithms but regression algorithms were added to get a unique viewpoint considering the data nature. Then the data set was processed so that it could be used in machine learning and deep learning techniques. Where feature extraction happened to train models. The used algorithms are Linear and kernel SVM, Linear Regression, Decision tree, Random forest, Multinomial Naive Bayes, KNN, and CNN from deep learning. Then we calculate the efficiency using the efficiency metrics.

B. *Dataset*
The data set for this paper was collected by searching through different social media platforms like (Facebook, Twitter, and YouTube). A total of 15 thousand of data was collected. These data contain hate speech, abusive text, hatred,

and normal texts. Over 9 thousand data were labeled as aggressive and over 6 thousand data were labeled as non-aggressive. The data were manually labeled as either aggressive or non-aggressive. The data was carefully classified as '0' for aggressive and '1' for non-aggressive data. Moreover, a stop words text was developed to further enhance the prepossessing step (Figs. 1 and 2).

Class	Label	Data
Aggressive	0	গাঞ্জা খাইছে হালা। তোর যা আছে সব দিয়াও চাঁদে যাওয়া সম্ভব না, মূর্খ, বর্বর
Non-aggressive	1	সমর্থক হিসেবে নিজেরও খারাপ লাগে নেইমারের এমন আচরণে। একটু ধাক্কা লাগলে পড়ে
Aggressive	0	ওয়াক থু……ছি়ছি় সরম! তোমরা শুধু এই কামটাই ভালো পারো, এছাড়া তোমার নাস্তিক আবালরা আর কিছুই পারোনা
Non-aggressive	1	পুরোনো গান নতুন করে ভালোই লেগেছে।আপনার গলাটা খুবই সুন্দর ভাইয়া।
Aggressive	0	তুই একটা পশুর চেয়েও জঘন্য নিকৃষ্ট মানুষ, তুই মানুষ রূপী পশু
Non-aggressive	1	যখন বুক ভরা কষ্ট নিয়ে ফিরে আসলাম তখন হাজারো ভালোবাসার সৃতি গুলো আমাকে পিছন থেকে টেনে ধরে রেখেছে
Aggressive	0	সালা এই কারণেই তো বলি কাঙ্গাদেশের মানুষের না শুধু মাটিরও দোষ আছে
Non-aggressive	1	মানবাধিকার কমিশনকে ডেকে কি করবে তারা নিজেরাই টিকে থাকতে পারে না। আর অন্যের জন্য কি করবে।
Aggressive	0	দয়া করে এইসব অকৃতজ্ঞ, নির্লজ্জ, বেহায়াদের নিয়ে আর কোনো ভিডিও বানিয়ে নিজের সময় নষ্ট করবেন না

Fig. 1. The dataset with class and Label

C. Feature Extraction and Reprocessing

In this research paper, the features extraction for Machine Learning algorithms, the authors also compared with n-grams such as uni-gram, bi-gram, and tri-gram based Count Vectorizer and TF.IDF-Vectorizer characteristics to detect

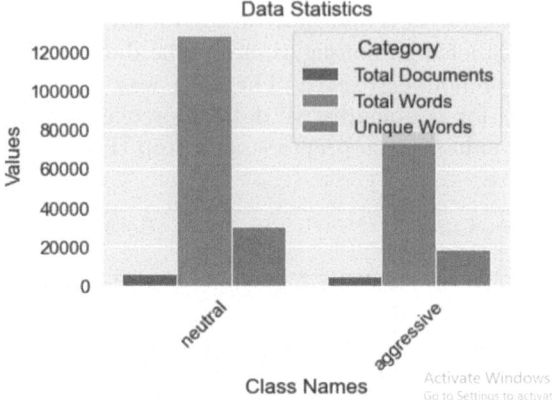

Fig. 2. The number of unique words detected

the aggressive and non-aggressive text. TF.IDF-Vectorizer transforms the strings into numbers. The length of each text and word is also taken into consideration as the frequency of words is used by TF.IDF. Word Embedding is used to produce numerical or vector representations of text data. Also, the Keras OOV tokenizer was used to represent all the words in the dataset as vectors and those vectors were used as input for the embedding layer of the Deep Learning algorithm CNN (Fig. 3).

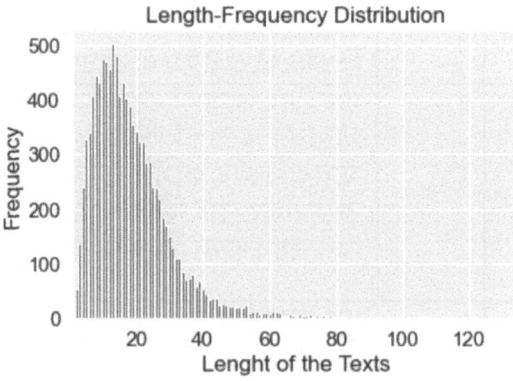

Fig. 3. The Frequency of words and texts

D. *Training and Testing Data Split*
After the preprocessing and feature selection step was completed, the authors split the 15 thousand data into 70% and 30% for training and testing respectively from the data set. After the splitting step was done, the data set contained 10,500

texts for training and 4500 for testing. Training sets of data were used to train the machine learning and deep learning models testing sets were used to test the same models. During the process, the authors randomly shuffled the data to get an evenly distributed dataset.

E. *Machine Learning Algorithms*
In recent research papers, it is shown that the classification models work well with categorically labeled data. Here in this research paper, the authors used both classification and Regression models which include Linear Regression, Linear SVM (Support Vector Machine), Kernel SVM, Random forest, Decision Tree, Multinomial Naive Bayes, and KNN.

Linear Regression: Linear Regression is a supervised machine learning algorithm that is used in regression problems. It predicts the continuous dependent variable with the help of independent variables from the data set. And to determine the best fit that precisely predicts the accurate value for the continuous dependent variable.

SVM: Support Vector Machine or SVM is used to classify data points which helps find a hyperplane in a K-dimensional space and K refer to a number of features. Among many possible features, the SVM chooses the feature with the maximum margin. Here in this research, the authors used both Linear and Kernel SVM.

Random Forrest: Random forest is a Supervised Machine Learning Algorithm that works very well with Classification and Regression problems. Based on different samples and inputs, creates a sample weight for a problem and then takes the highest valued weight for the classification problem and takes the average weight for regression. However, it works well with the classification problem.

Decision Tree: A supervised learning method used for both classification and regression. Analyses the data features and predicts the value of the target variable by implementing decision rules.

Multinomial Naïve Bayes: The multinomial Naïve Bayes algorithm is based on Bayes theorem and calculates the likelihood of occurrence of an event based on prior knowledge of the event's conditions using the Bayes theorem. The highest probability is the output. Probability of $P(A|x)$ where A is the of the possible outcomes given instance x. The equation is given below.

$$P(A|x) = P(x|A) * P(A)/P(x)$$

In this research, Multinomial Naïve Bayes is used.

KNN: The k-nearest neighbors (KNN) algorithm is also a supervised learning algorithm that relies on labeled input data to learn a function. And later based on that learned function produces a feature-appropriate output considering new unlabeled data.

F. Deep Learning Algorithm

Here, in this research paper neural network is used. The authors used the CNN (Convolutional Neural Network)model. CNN model works well with classification-type problems.

CNN: A Convolutional Neural Network or CNN is a neural network designed mainly for processing structured arrays of image and video data. It is one of the most used Neural networks in computer vision. Here is the structure of CNN that we used.

Embedding Layer: In this layer all the words in the data set turned into vectors and were used as the input of the embedding layer. This is the first layer of the neural network where a matrix is generated which represents the vectors and words.

Max-Pooling Layer: This layer is used to shrink down the spatial size of a convoluted feature. So, basically, dimension reduction is done in this layer and padded to reduce the noise from actual meaningful data.

Flatten Layer: This layer is used to convert the data into a 1-D array. It creates a flattened output of its previous layer. And a single long feature vector is created for the dense layer to perform classification through various models.

Dense Layer: This layer takes the values of all the previous layers and groups them into a dense feature to perform classification.

LSTM: LSTM is a special type of RNN(Recurrent Neural Network) that handles the long-term dependency problem by processing the previous and current data. There are three parts of the LSTM cell. These are the Forget gate which forgets irrelevant and excessive information, the input gate which adds or updates new information into the network and lastly, the output gate which then passes the updated information onto the network. LSTM network has a hidden state which is known as Short-term memory and a cell state which is known as Long-term memory.

Forget Gate: It is the first step of LSTM Network, where it decides which information to keep and which to forget from the previous instant.

Input Gate: In this step the network decides which relevant information is necessary to keep and add or update that information to the network.

Output Gate: Here based on the new updated information the network has to predict a relevant result for this instant. For better prediction, we used CNN-LSTM hybrid network.

4   Result Analysis

Here in this section, the detailed results are discussed. Over 15 thousand prepossessed data were given for training and testing purposes. And they are measured

Model	Accuracy	Precision	Recall	F1-Score
Linear Regression	88.31	88.31	88.31	88.31
Decision Tree	82.01	82.01	82.01	82.01
Random Forrest	86.71	86.71	86.71	86.71
Multinomial Naïve Bayes	90.46	90.46	90.46	90.46
KNN	79.95	79.95	79.95	79.95
Linear SVM	86.03	86.03	86.03	86.03
Kernel SVM	87.67	87.67	87.67	87.67
CNN+LSTM	85.15	90.28	86.85	84.55

by four metrics they are Accuracy, Precision, Recall, and F1-Score. Here are the following results and analyses.

Table: Results (Accuracy, Precision, Recall, F1-score)

The table shows that the highest accuracy from the machine learning algorithms was found from multinomial naïve Bayes about 90.46%. Which is about 2.15% higher than the second-best result of 88.31% from Linear Regression. Then the 3rd highest accuracy was gained from kernel SVM of about 87.67% which is 2.79% less than what we got in MNB. In the 4th position, we get Random Forrest resulting in 86.71% accuracy. It is around 3.75% less than the result of MNB. Next, we got 86.03% testing accuracy on the data set from Linear SVM. From the Decision Tree, we got 82.01% accuracy and from KNN we found 79.95% accuracy on the data set. In the case of the Deep Learning algorithm, for better prediction used CNN LSTM hybrid network and found an accuracy of 85.15%. The precision of 90.28% and Recall score of 86.85% and the F1-score is 84.55% (Figs. 4 and 5).

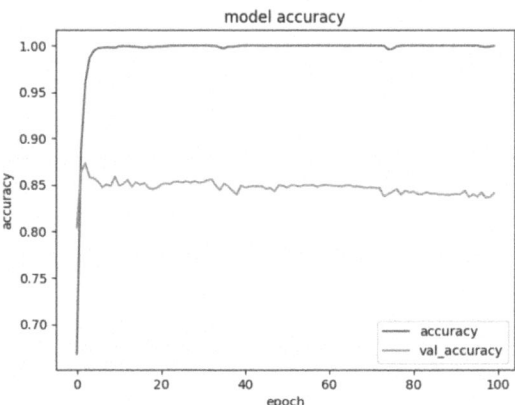

Fig. 4. CNN model accuracy

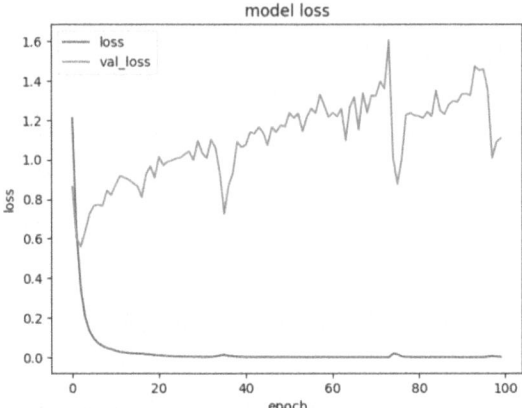

Fig. 5. CNN model loss

The model loss is calculated on training and validation and it is based on how the model is performing in these two sets. The sum of errors from each iteration in training or validation sets. This model Loss curve shows us how a model behaves after each iteration of optimization. A model accuracy curve shows the measurement of the algorithm's performance. The accuracy of a model is usually determined after the parameters are calculated. It curve shows the measurement of how accurate the model prediction is compared to the true positive data.

5   Conclusion

This approach explores the identification of aggressive and non-aggressive Bangla text in social media posts and comments through the application of machine learning and deep learning algorithms. Various classification and regression models such as Linear Regression, Linear SVM (Support Vector Machine), Kernel SVM, Random Forest, Decision Tree, Multinomial Naïve Bayes, and KNN were employed. The highest accuracy, 90.46%, was achieved with the Multinomial Naïve Bayes model. Linear Regression followed with an accuracy of 88.31%, and the CNN model yielded 85.15% accuracy. These findings suggest that classification algorithms perform effectively in Natural Language Processing (NLP) classification tasks.

6   Future Work

Our goal was to discern aggressive and non-aggressive Bangla text within social media comments and posts by applying machine learning and deep learning models. A range of machine learning models were employed, each yielding distinct outcomes. Furthermore, the CNN deep learning model was applied, resulting in an accuracy of 85.15%. Regarding future endeavors, we have proposed ideas to augment and fine-tune the obtained results. Exploring other Deep Learning algorithms, such as GRU, Attention-based models, RNN, etc., could contribute to further improvements. Incorporating more user-specific posts and comments has the potential to enhance results. Additionally, utilizing a larger dataset for training and testing purposes could be beneficial.

References

1. Editors of Encyclopedia Britannica. Bengali Language (2022). https://www.britannica.com/topic/Bengalilanguage
2. Abdhullah-AlMamun, Akhter, S.: Social media bullying detection using machine learning on Bangla text. In: 10th International Conference on Electrical and Computer Engineering (ICECE), pp. 385–388 (2018)
3. Ahmed, Md.T., Rahman, M., Nur, S., Islam, A.A., Das, D.: Deployment of machine learning and deep learning algorithms in detecting cyberbullying in Bangla and Romanized Bangla text: a comparative study. In: 2021 International Conference on Advances in Electrical, Computing, Communication and Sustainable Technologies (ICAECT), pp. 1–10 (2021)
4. Emon, E.A., Rahman, S., Banarjee, J., Das, A.K., Mittra, T.: A deep learning approach to detect abusive Bengali text. In: 7th International Conference on Smart Computing & Communications (ICSCC), pp. 1–5 (2019)
5. Islam, T., Ahmed, N., Latif, S.: An evolutionary approach to comparative analysis of detecting Bangla abusive text. Bull. Electr. Eng. Inform. **4**, 2163–2169 (2021)
6. Tuhin, R.A., Paul, B.K., Nawrine, F., Akter, M., Das, A.K.: An automated system of sentiment analysis from Bangla text using supervised learning techniques. In: 2019 IEEE 4th International Conference on Computer and Communication Systems (ICCCS), pp. 360–364 (2019)
7. Awal, Md.A., Rahman, Md.S., Rabbi, J.: Detecting abusive comments in discussion threads using Naïve Bayes. In: 2018 International Conference on Innovations in Science, Engineering and Technology (ICISET), pp. 163–167 (2018)
8. Jahan, M., Ahamed, I., Bishwas, Md.R., Shatabda, S.: Abusive comments detection in Bangla-English codemixed and transliterated text. In: 2nd International Conference on Innovation in Engineering and Technology (ICIET), pp. 1–6 (2019)
9. The DataReportal. https://datareportal.com/essentialfacebook-stats
10. Editorial. https://www.oberlo.com/statistics/whatage-group-uses-social-media-the-most
11. Dave Chaffey (2022). https://www.smartinsights.com/social-media-marketing/social-mediastrategy/new-global-social-media-research/

A Comparative Evaluation of Machine Learning Techniques for Detecting Malicious Network Traffic

Prince Tayal[1] ⓘ, Rohan Kumar[1] ⓘ, and Hemlata[2](✉) ⓘ

[1] CSE(AIML), Manipal University Jaipur, Jaipur, India
princetayal07@gmail.com, rohankumar@icloud.com
[2] Department of CSE-AIML, Manipal University Jaipur, Jaipur, India
hparmar601@gmail.com

Abstract. The Malicious network activities such as malware attacks, unauthorized intrusions, and data theft have become a rising threat as organizations increasingly rely on interconnected systems. Detecting these security incidents rapidly and accurately is critical, but also poses challenges due to the constantly evolving techniques used by hackers. Machine learning has surfaced as a promising avenue for developing intelligent systems capable of analyzing massive amounts of network data, identifying anomalies, and detecting malicious patterns in real-time. This research presents a thorough comparative evaluation of the key machine learning algorithms applied for detecting various types of malicious network traffic. The models tested include random forests, regular neural networks, and logistic regression. Using labeled datasets containing benign and attack traffic flows, the algorithms are trained, and their performance evaluated across parameters like detection accuracy, confusion matrix, F1-scores, and receiver operating characteristics. The results demonstrate that ensemble methods such as random forests achieve exceptionally high accuracy, exceeding 99% in some tests, by combining diverse decision trees. With careful tuning of hyperparameters, neural networks also display comparable performance owing to their automated feature extraction capabilities. The analysis provides useful insights into the strengths, weaknesses, and applicability of each technique. These findings will enable security experts to select the most appropriate machine learning solutions tailored to their network environments and requirements for developing robust intrusion detection systems capable of identifying threats in real time.

Keywords: Machine learning · Network security · Intrusion detection · Malware detection · Logistic regression · Random forests · Regular Neural networks · Ensemble methods · Comparative evaluation

A. K. Bairwa et al. (Eds.): ICCAIML 2024, CCIS 2184, pp. 184–204, 2025.
https://doi.org/10.1007/978-3-031-71481-8_15

1   Introduction

In recent years, cyberattacks have been growing at an exponential rate with the expansion of the Internet and interconnected systems. Malware, unauthorized intrusions, data breaches – these threats pose a serious risk to individuals, businesses, and critical infrastructure. A major challenge is being able to detect these malicious activities in real-time as they occur on networks. This is extremely difficult given how rapidly hacking techniques evolve to avoid detection. Traditional security tools relying on pre-defined signatures cannot catch novel threats they have not yet encountered. There is a need for more intelligent, automated solutions that can analyze massive volumes of network data, spot anomalies, and identify potential attacks.

Machine learning has come up to be an exciting prospect to tackle these cybersecurity challenges. By applying statistical techniques and algorithms to uncover patterns in data, machine learning can develop systems capable of modelling normal network behavior, detecting deviations, recognizing new attack signatures, and preventing intrusions in real-time. Unlike rigid rule-based systems, machine learning models can be trained on large datasets to continuously improve accuracy as more data comes in. Various supervised, unsupervised and hybrid machine learning algorithms have already been applied for network anomaly detection and malware analysis with promising results.

This paper aims to provide a comparative evaluation of the major machine learning techniques used for detecting malicious network traffic. The key algorithms examined include random forests, regular neural networks, and logistic regression. Using publicly available datasets labelled as benign and malicious, these techniques are tested on performance parameters like accuracy, confusion matrix, F1-scores, and receiver operating characteristics. In addition to monitoring detection capabilities, factors such as interpretability, computational efficiency and ease of implementation are analyzed. The objective is to thoroughly understand the benefits and drawbacks of every algorithm. This knowledge will serve as a guidepost for identifying optimal solutions that meet the specific needs of real-world network intrusion detection systems built using machine learning methodologies.

2   Literature Review

2.1   Literature Overview

First research paper has proposed using artificial neural networks for misuse detection to identify network intrusions without requiring full attack specifications. Preliminary analyses indicate neural networks show promise in improving detection of novel attacks by recognizing partial indicators and complex patterns not easily encoded into explicit rules. Our research aims to further evaluate the potential of neural networks to enhance misuse detection against unpredictable threats [1].

Previous research has looked into integrating soft computing approaches such as artificial neural networks (ANNs), support vector machines (SVMs), and multivariate adaptive regression splines (MARS) inside ensemble models for network intrusion detection. These hybrid approaches aim to leverage the strengths of individual techniques to improve classification accuracy. In a recent investigation, researchers discovered that combining these techniques yielded improved results in identifying cyber-attacks compared to employing each method individually. Our research aims to delve deeper into the realm of ensemble models, incorporating a mix of soft computing and hard computing techniques. The goal is to create intrusion detection systems that are not only more effective but also resilient in the face of constantly evolving cyber threats [2].

In a study conducted by Apruzzese and colleagues in 2019, they delved into the realm of cybersecurity, specifically focusing on the analysis of machine learning techniques for detecting intrusions, malware, and spam. While machine learning shows promise, the research concluded rigorous evaluation is still needed to assess effectiveness on real systems and address limitations before these techniques can be reliably deployed for automated threat detection. Our research provides further comparative assessment of machine learning algorithms tailored for malicious network traffic detection [3].

Mukkamala and his team conducted research back in 2002, exploring the utilization of support vector machines (SVMs) and neural networks to detect network intrusions. They based their investigation on KDD benchmark datasets. They demonstrated both techniques can build efficient and highly accurate classifiers for detecting attacks. Further experiments found using just 13 most significant features delivered comparable accuracy. The study provides early benchmarking of SVM and neural network performance for misuse detection using standardized datasets. Our work provides further comparative assessment on modern malicious traffic [4].

Research by Toosi and Kahani (2007) proposed an intrusion detection approach combining soft computing techniques like neuro-fuzzy classifiers and genetic algorithms. Their model uses neuro-fuzzy networks for initial attack classification, then a fuzzy inference engine for final decision-making optimized by a genetic algorithm. The experiments involved the assessment of the hybrid technique using the KDD Cup 99 benchmark dataset. Study demonstrates the potential of integrating multiple soft computing methods to improve anomaly detection accuracy. Our work provides further analysis using modern machine learning algorithms tailored for malicious traffic detection [5].

In a survey conducted by Tsai and collaborators in 2009, they examined intrusion detection systems that employed machine learning techniques, focusing on publications from the years 2000 to 2007. It examined single, hybrid approaches, and ensemble classifiers applied on benchmark datasets. The review compared classifier designs and experimental setups to assess achievements and limitations in developing ML-based intrusion detection. It provides useful insights into the state of machine learning for network security prior to the rise of deep learning. Our work provides an updated comparative benchmarking focused on detecting modern malicious traffic [6].

Bhuyan and colleagues conducted a survey in 2014, offering a structured overview of methods and systems for network anomaly detection based on computational techniques. It summarized and compared a wide range of anomaly detection approaches, including machine learning algorithms. The paper also discussed available tools and datasets for research. It highlighted open challenges and future directions in advancing network intrusion detection. Our work provides an updated comparative benchmarking focused specifically on machine learning techniques using recent datasets [7].

Research by Sharafaldin (2018), addressed the significant issue of DDoS attacks in network security. It comprehensively reviewed existing DDoS datasets and introduced the CICDDoS2019 dataset to overcome previous limitations. Leveraging this dataset, a novel detection approach based on network flow features was proposed. The paper also presented essential feature sets for detecting various DDoS attack types, along with their respective weights [8].

In a study undertaken by Javaid A. in 2016, the primary focus was on crafting a Network Intrusion Detection System designed to assist network admins in pinpointing vulnerabilities within their respective companies. The study aimed to address challenges related to flexibility and efficiency in detecting unforeseen and unpredictable attacks. To accomplish this goal, a proposed approach centered around deep learning was introduced, incorporating Self-taught Learning (STL), and applied to the NSL-KDD benchmark dataset for the purpose of network intrusion detection. The paper showcased the performance results of this approach and conducted a comparative analysis with prior research, taking into account parameters such as accuracy, confusion matrix, and F1 score [9].

2.2 Requirements

Drawing insights from the literature review, the following key requirements are defined for an effective machine learning based network intrusion detection system:

- Attaining high detection accuracy across diverse types of malware and network attacks.
- Low false positive rate to minimize false alarms
- High true positive rate to detect most threats
- Fast detection capability for real-time attack prevention
- Robustness against evasion techniques used by adversaries
- The capability to identify zero-day and unknown threats.
- Capability to work with imbalanced datasets
- Incremental learning potential to adapt as new data becomes available
- Interpretability for cybersecurity experts to analyze decisions
- Computational efficiency for high-speed networks
- Easy integration into existing security infrastructure

2.3 Outcome of Literature Review

The literature review emphasizes the application of a broad range of classical machine learning techniques for the detection of malicious network traffic. Ensemble methods, such as random forest and gradient boosting, often exhibit remarkably high detection accuracy when applied to contemporary datasets [8]. However, deep learning models are emerging as promising techniques due to their automated feature learning capabilities [3]. Tree ensembles, SVM, KNN, discriminant analysis, and Bayesian networks are the most mature techniques [10].

Most studies have relied on standard datasets like NSLKDD and UNSW-NB15 for benchmarking. However, newer datasets that capture modern attack techniques would be valuable for more robust evaluation [11]. There is a need for comparisons on very large real-world network data [12]. Assessment of computational performance and limitations is also required for practical deployment [13]. While accuracy metrics are widely reported, other factors like model interpretability also need to be considered based on the application context [14].

Overall, the review indicates machine learning has excellent potential for developing intelligent malware detection systems. However, more research is required to determine optimal algorithms and best practices tailored to different network environments and security needs [10]. Our experiments will provide further insights in this direction.

2.4 Problem Statement

Derived from the literature review, the central problem we seek to address in this study can be succinctly summarized as follows:

Determining the most appropriate machine learning algorithms and techniques for accurately detecting various types of malicious network traffic in real-world environments, given the constraints of computational efficiency, evolving cyber threats, class imbalance, and the need for interpretable models.

2.5 Research Objectives

The objectives of our research are:

- A comparative assessment of the fundamental categories of machine learning algorithms used in network intrusion detection, considering detection accuracy, computational efficiency, and other pertinent metrics.
- To analyze the performance trade-offs, advantages, and limitations of supervised, unsupervised, and hybrid models for detecting malicious network traffic.
- To identify the factors and best practices that allow maximizing the malware detection capability for different algorithms and network environments.
- To provide data-driven recommendations on selecting suitable algorithms for developing real-world intrusion detection systems (Fig. 1).

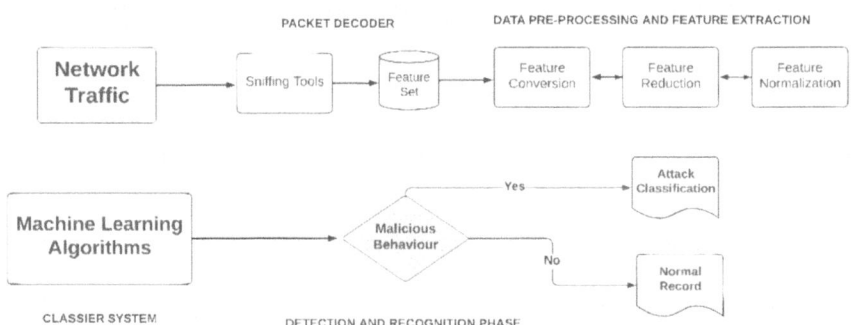

Fig. 1. Flowchart for depicting how classification works

3   Methodology

The methodology employed in this study is fundamental to understanding the research process behind evaluating machine learning algorithms for identifying malware in network traffic. The following part delves into the intricacies of data collection and preprocessing, feature extraction, machine learning algorithm selection, and the criteria for evaluating their performance.

3.1   Data Collection and Preprocessing

Effective malware detection relies on high-quality data. In this study, network traffic data is collected from various sources, including network logs, packet captures, and flow records [15]. These raw data sets are often voluminous, containing both legitimate and potentially malicious traffic. To prepare the data for analysis, rigorous preprocessing is essential.

Data preprocessing involves several key steps:

Data Cleansing: It involves the identification and correction of errors or inconsistencies in a dataset, including tasks such as addressing missing values, handling outliers, and rectifying inaccuracies. This step holds utmost importance in guaranteeing the quality and dependability of the data for any future evaluation or modeling.

Data Reduction: Considering the abundance of raw data, it becomes imperative to diminish its dimensionality. For this purpose, methods such as feature selection and PCA can be used, with the goal of preserving the most relevant attributes while reducing computational complexity [15].

Normalization: Data normalization is a process that standardizes the data to a common scale, preventing certain features from exerting disproportionate influence on the analysis due to their larger values. This ensures a more balanced consideration of all features in the dataset. Techniques such as Z-score normalization are often used [15].

Balancing The Dataset: The dataset may exhibit class imbalance, where benign traffic significantly outnumbers malicious instances. To correct imbalances in class distribution within a dataset, various techniques such as oversampling, undersampling, or generating synthetic samples (such as using Synthetic Minority Over-sampling Technique) can be used. These approaches aim to improve the learning process for machine learning models by creating a more equitable representation of different classes [15].

Feature Scaling: Ensuring proper scaling of features is crucial for enhancing the performance of certain machine learning models. Some of the scaling techniques are MinMax scaling and Z-score standardization. These techniques help maintain consistent scales across features, contributing to improved algorithmic performance [15].

Data partitioning: It is critical to divide the dataset into training, validation, and test sets to train and assess machine learning algorithms. This separation allows for effective model development, tuning, and unbiased assessment of performance.

3.2 Feature Extraction

Feature extraction is a crucial step in the process, preparing the data for machine learning. In the context of malware detection, features are characteristics derived from network traffic data that enable the algorithms to make predictions [16]. These features can be classified into multiple types, like statistical, behavioral, and structural.

Statistical Features: These encompass attributes such as packet counts, byte counts, and traffic rate statistics, which provide insights into the traffic's basic characteristics [15].

Behavioral Features These features capture the behavior of network traffic, including patterns of communication, network protocol usage, and communication entropy. They enable the identification of anomalous behaviors that might indicate malware presence [15].

Structural Features: Structural features represent the relationships within network traffic data, such as IP address pairs, port pairs, or subgraph structures in communication patterns [15].

The choice of feature extraction techniques is influenced by the specific objectives of the malware detection task and the characteristics of the dataset. Feature engineering is an iterative process involving the selection, extraction, and transformation of features to optimize model performance.

3.3 Machine Learning Algorithm Selection

Several algorithms are considered for malware detection in network traffic, each with its strengths and weaknesses.
Commonly evaluated algorithms include:

- **Random Forest:** This ensemble learning method is acknowledged for its proficiency in managing high dimensional data and conducting feature importance analysis [17].
- **Support Vector Machine (SVM):** This method is used to find extensive application in binary classification tasks and possess the capability to handle both linear and non-linear data [18].
- **Deep Learning:** Deep neural networks, particularly Regular Neural Networks (RNNs), have gained prominence in feature learning from network traffic data [19].
- **Naïve Bayes:** A simple and probabilistic approach that can effectively model relationships between features and classes [20].
- **Extreme Gradient Boosting:** Algorithms such as XGBoost are renowned for their effectiveness in managing imbalanced datasets, leading to an enhancement in classification accuracy [21].

The selection process considers factors such as the nature of the dataset, algorithm efficiency, computational resources, and the task's complexity.

3.4 Evaluation Metrics

Evaluating the performance of machine learning models is vital in assessing their efficacy in malware detection. The selection of appropriate evaluation metrics is dependent on the specific objectives and features of the specific assignment at hand. Commonly used parameters include:

- **Accuracy:** It is defined as the proportion of correctly predicted instances in relation to the total number of instances, and it offers a comprehensive evaluation about the model's efficacy across all classes or categories in the dataset [20].

$$Accuracy = \frac{TP + TN}{TP + FP + FN + TN} \tag{1}$$

- **F1-Score:** It is the harmonic average of precision and recall, which offers an equitable analysis of precision and recall, providing a more thorough evaluation [20].

$$F1Score = \frac{2 \times Precision \times Recall}{Precision + Recall} \tag{2}$$

- **Receiver Operating Characteristic (ROC) Curves:** It depicts a model's performance across different thresholds, assisting in the selection of an optimal balance between true positives and false positives [21].

These evaluation metrics guide the selection of the best performing algorithm and support further analysis of model robustness and generalization to real-world scenarios.

4 Dataset

The dataset is sourced from the GitHub repository of Prateek Lalwani, a security blogger. You can find the repository by clicking here at https://github.com/prk54/malware-detection-machinelearning-approach/blob/master/data.csv.

Dataset contains 41,323 binary files (including exe, DLL files) which are legitimate in nature means they are malware free. Also, the dataset contains 96,724 malware samples which are taken from virusshare.com. So, we can now conclude there are in total approximately 1,38,047 samples in the entire dataset (Figs. 2 and 3).

```
Index(['Name', 'md5', 'Machine', 'SizeOfOptionalHeader', 'Characteristics',
       'MajorLinkerVersion', 'MinorLinkerVersion', 'SizeOfCode',
       'SizeOfInitializedData', 'SizeOfUninitializedData',
       'AddressOfEntryPoint', 'BaseOfCode', 'BaseOfData', 'ImageBase',
       'SectionAlignment', 'FileAlignment', 'MajorOperatingSystemVersion',
       'MinorOperatingSystemVersion', 'MajorImageVersion', 'MinorImageVersion',
       'MajorSubsystemVersion', 'MinorSubsystemVersion', 'SizeOfImage',
       'SizeOfHeaders', 'CheckSum', 'Subsystem', 'DllCharacteristics',
       'SizeOfStackReserve', 'SizeOfStackCommit', 'SizeOfHeapReserve',
       'SizeOfHeapCommit', 'LoaderFlags', 'NumberOfRvaAndSizes', 'SectionsNb',
       'SectionsMeanEntropy', 'SectionsMinEntropy', 'SectionsMaxEntropy',
       'SectionsMeanRawsize', 'SectionsMinRawsize', 'SectionMaxRawsize',
       'SectionsMeanVirtualsize', 'SectionsMinVirtualsize',
       'SectionMaxVirtualsize', 'ImportsNbDLL', 'ImportsNb',
       'ImportsNbOrdinal', 'ExportNb', 'ResourcesNb', 'ResourcesMeanEntropy',
       'ResourcesMinEntropy', 'ResourcesMaxEntropy', 'ResourcesMeanSize',
       'ResourcesMinSize', 'ResourcesMaxSize', 'LoadConfigurationSize',
       'VersionInformationSize', 'legitimate'],
      dtype='object')
```

Fig. 2. A list of all the columns from the dataset

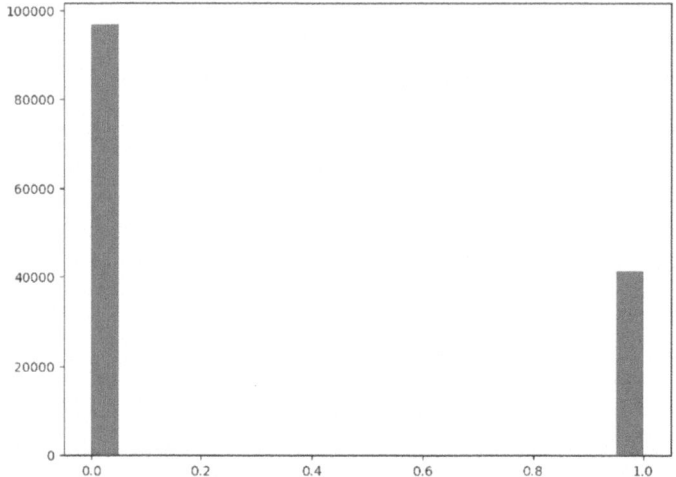

Fig. 3. A bar graph depicting legitimate and malware files from the dataset

5 Feature Engineering

This section delves into the critical aspects of feature selection and extraction techniques employed in the research, shedding light on how they contribute to enhancing the performance of machine learning models for malware detection within network traffic.

5.1 Feature Selection and Extraction Techniques

Feature engineering as a crucial stage in crafting machine learning models intended for malware detection. It involves the careful curation of relevant attributes (features) derived from unprocessed network traffic data. The goal is to provide the model with meaningful information while reducing dimensionality and computational complexity.

Feature Selection Techniques: Several techniques are used in feature selection, each with its advantages and applications. These include-

- *Filter Methods:* It analyzes feature significance by utilizing statistical metrics like Information Gain, Chi-Square, or Correlation Coefficient [22]. Features are ranked based on these metrics, and a subset is selected.
- *Wrapper Methods:* Wrapper methods employ machine learning models to assess feature subsets. By assessing the model's performance with different combinations of features, these methods enable the identification of the most optimal feature set [23].
- *Embedded Methods:* These methods integrate the process of feature selection into the model training phase. Examples of such techniques include L1 regularization (LASSO) or decision tree feature importance are examples [24].

Feature Extraction Techniques: Feature extraction refers to the process of converting raw data into a condensed and more representative format. Regarding the identification of malicious software in traffic over networks, the following techniques are notable-

- *Principal Component Analysis (PCA):* It is a technique to reduce dimensionality of information by portraying it onto a lower-dimensional subspace. The goal of this transformation is to keep as much variance in the original data as possible. This can be particularly useful in handling high-dimensional network traffic data [25].
- *Wavelet Transform:* The wavelet transform decomposes data into different scales, capturing both low and high frequency information. This can be valuable in extracting time frequency domain features from network traffic [26].

5.2 Feature Importance Analysis

Feature importance analysis is instrumental in understanding the contribution of individual features to the predictive power of machine learning models. This analysis provides insights into which features have the most significant influence on the model's decisions.

- **Random Forest Feature Importance:** Random Forest algorithms assign importance scores to features based on their contribution to reducing impurity in decision tree nodes [28].
- **Gradient Boosting Feature Importance:** Gradient Boosting models calculate feature importance by assessing how often a feature is used in decision tree splits and the improvement in accuracy it brings [29].
- **Deep Learning Feature Analysis:** In deep learning models, the importance of features can be explored through techniques such as gradient-based methods and activation maximization [30].

6 Machine Learning Algorithms

This section provides detailed descriptions of the selected machine learning algorithms used in the research for malware detection in network traffic. These algorithms have been chosen for their effectiveness in various aspects of this task.

6.1 Random Forest Classifier

The Random Forest Classifier is an ensemble-based machine learning methodology that has been tailored for classification tasks. It harnesses the strengths of decision trees by amalgamating several trees to enhance predictive accuracy and foster model resilience.

The Random Forest Classifier works by combining a set of decision trees. Each tree in the ensemble is trained on a slightly different subset of data and produces its own prediction. In classification scenarios, such as distinguishing whether an email is spam or not, each tree within the forest contributes its own prediction, effectively casting a "vote." The final decision hinges on the majority "votes" from the trees, resulting in a collective decision that typically demonstrates heightened accuracy and reliability compared to a single decision tree's output.

Some measures which were taken to ensure high accuracy are-

Feature Scaling (Standard Scaler): The data undergoes standardization through employment of the Standard Scaler. The goal of this procedure is to rescale the features to attain 0 as the value of average and 1 as the value of standard deviation. Standardization serves as a typical preprocessing step, ensuring uniform scaling across all features. This practice significantly aids the model in achieving improved performance and heightened accuracy.

Class Imbalance Handling (RandomOverSampler): An imbalance within the classes can result in bias within the model's predictions. For this, we use the RandomOverSampler from the imbalanced-learn library. It oversamples the minority class (malware) to create a more balanced dataset. This ensures that the model has enough examples of both classes to learn from.

Feature Selection (SelectFromModel): After addressing class imbalance, the code applies feature selection using the SelectFromModel method. It selects the most relevant features based on the feature importance calculated by a Random Forest Classifier (CLF). The model retains the features that have the most significant impact on its performance, while less important features are removed. This can help in reducing noise and improving accuracy.

Hyperparameter Tuning (max_depth): In the Random Forest model (CLF), the max_depth hyperparameter is adjusted to 10. The parameter max_depth regulates the highest depth achievable by each individual decision tree within the ensemble. By setting a limit on the tree depth, you can prevent overfitting and make the model generalize better.

These changes collectively aim to improve the model's accuracy by ensuring that the data is well-preprocessed, the class imbalance is addressed, irrelevant features are removed, and overfitting is mitigated through hyperparameter tuning. Each step contributes to enhancing the model's performance for malware detection (Figs. 4, 5 and 6).

```
Accuracy on the train dataset: 100.00%
Accuracy on the test dataset: 99.54%
F1 score on the test dataset: 99.24
```

Fig. 4. F1 Score and Accuracy of model on test and train dataset

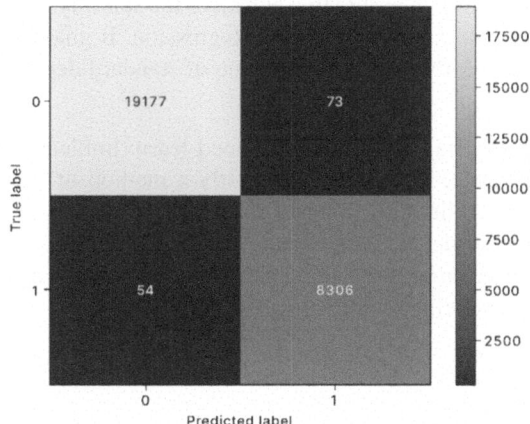

Fig. 5. Confusion Matrix for Random Forest Classifier

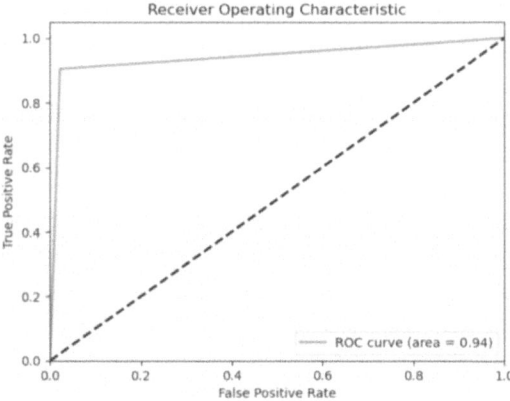

Fig. 6. ROC Curve for Random Forest Classifier

6.2 Logistic Regression

Logistic regression serves as a statistical technique employed for binary classification tasks. It establishes a model to depict the correlation between a binary dependent variable (such as yes/no or 1/0) and one or multiple independent variables, providing the probability of an event occurring. It is widely used in various fields, including medicine, marketing, and social sciences, for predictive modeling and decision making.

Some measures which were taken to ensure high accuracy are-

Standardization (Feature Scaling): The StandardScaler from scikit-learn is used to scale the features. While not a form of regularization in the traditional sense, standardization stands as a preprocessing measure that enhances the stability and efficacy of various machine learning models, including Logistic Regression. It guarantees that every feature maintains an average of 0 and 1 as the value of standard deviation, enabling more efficient convergence.

RandomOverSampler: It is from the imbalanced-learn (imblearn) library and is used to address class imbalance. Though not explicitly a method of feature selection, this method includes oversampling the minority class to address class imbalance within the dataset. As a result, the model has a more balanced representation of the various classes, allowing for improved learning and understanding of the minority class, which may lead to more relevant features being identified by the model.

Hyperparameter Tuning: The code specifies hyperparameters for the Logistic Regression model, including-

- **max_iter:** The maximum number of iterations for the solver to converge.
- **C:** The regularization parameter. In this code, $C = 1.0$ is used, indicating relatively weak regularization.
- **penalty:** The type of regularization, where penalty = 'l2' specifies L2 regularization. L2 regularization encourages small coefficients and helps prevent overfitting.

These hyperparameters are adjusted to control the regularization strength and convergence behavior of the Logistic Regression model (Figs. 7, 8 and 9).

```
Accuracy on the train dataset: 97.86%
Accuracy on the test dataset: 97.85%
F1 score on the test dataset: 97.85
```

Fig. 7. F1 Score and Accuracy of model on test and train dataset

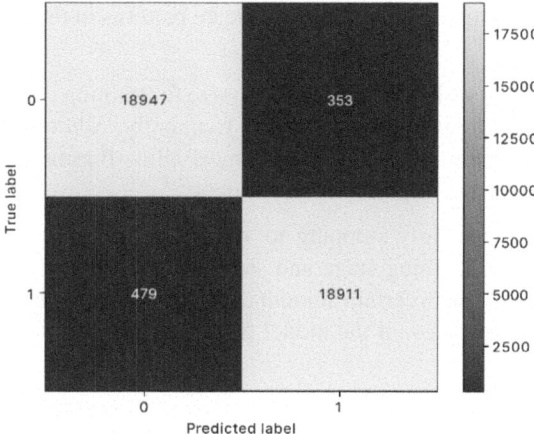

Fig. 8. Confusion Matrix for Logistic Regression

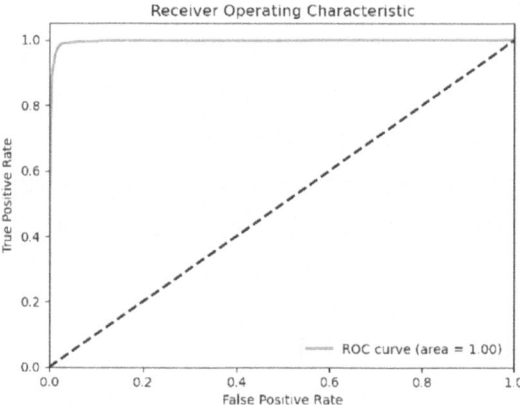

Fig. 9. ROC Curve for Logistic Regression

6.3 Regular Neural Network (RNN)

Deep Learning, particularly in the form of Regular Neural Networks (RNNs), has brought transformative advancements to the realm of image and pattern recognition. RNNs are versatile, multilayered architectures designed to autonomously extract intricate features from data, rendering them ideal for tasks such as image analysis. In the context of identifying malicious software, RNNs can be harnessed for the analysis of network traffic data, treating it as sequential data. Research has demonstrated the efficacy of RNNs in discerning nuanced and complex patterns associated with malware.
Some measures which were taken to ensure high accuracy are-

Neural Network Architecture: The neural network architecture is modified to have more layers and neurons (e.g., 64, 32, 16) compared to a simpler network. This increased complexity allows the model to learn more intricate patterns in the data, which can lead to improved accuracy.

Batch Size and Number of Epochs: The batch size for training is adjusted to 64, which can help the model update its parameters more frequently. Additionally, the number of training epochs is increased to 50. These changes allow the model to undergo more iterations and may improve its ability to fit the data.

Early Stopping: We use early stopping to avoid overfitting. It monitors the validation loss throughout the learning stage and stops the procedure if there is no obvious advancement in the loss after a certain amount of consecutive epochs. This precautionary technique is performed to prevent the model from being overly tailored to the training data.

Activation Function: Within the concealed layers, we utilize the ReLU activation function. It is also known for its effectiveness in deep neural networks and can facilitate the training process.

Binary Classification Threshold: In binary classification, a threshold of 0.5 is established. When the predicted value is equal to or more than to 0.5, it is considered as 1 (malware), whereas values below 0.5 are classified as 0 (legitimate data) (Figs. 10, 11 and 12).

```
Accuracy on the train dataset: 95.48%
Accuracy on the test dataset: 95.57%
F1 score on the test dataset: 92.51
```

Fig. 10. F1 Score and Accuracy of model on test and train dataset

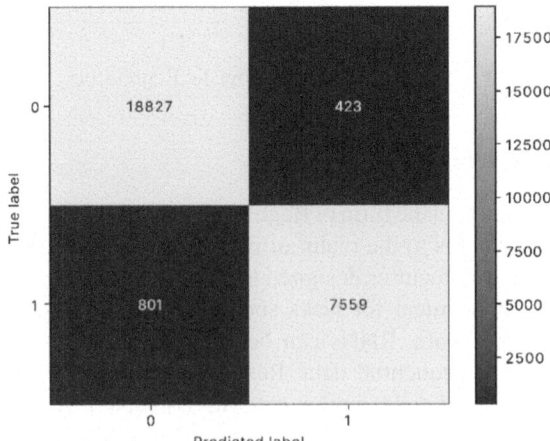

Fig. 11. Confusion Matrix for Regular Neural Network

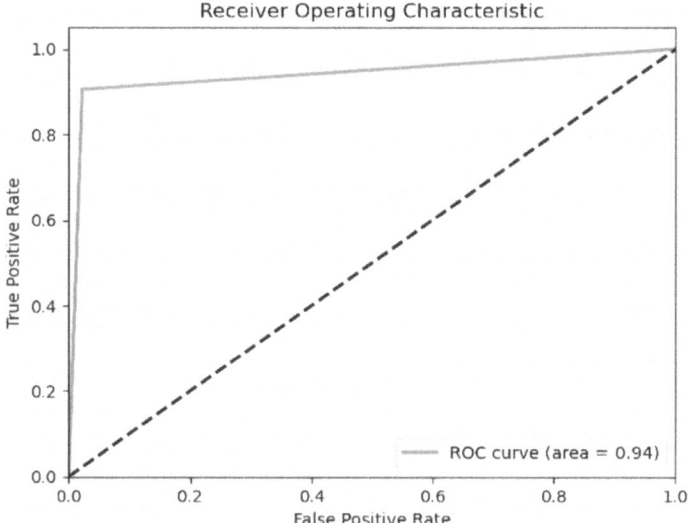

Fig. 12. ROC Curve for Regular Neural Network

7   Experimental Setup

This part outlines the essential parameters and configurations utilized in the research to ensure the validity and reproducibility of the experiments, including data splits, hyperparameter tuning, and cross-validation methods.

7.1   Testing and Training Data Splits

Within machine learning experiments, breaking down the dataset in groups for both training and testing is a pivotal step. In this study, we adopt a carefully planned strategy for dividing the data to evaluate how well the chosen machine learning algorithms perform. The dataset is partitioned into three primary subsets:

1. **Training Set:** This specific subset is used for training the models. It forms the basis for learning the relationships between features and labels. Typically, a significant portion of the dataset is allocated to the training set to ensure that the models capture the underlying patterns and characteristics of the data [15].
2. **Validation Set:** The validation set plays a crucial role in hyperparameter tuning. It is used to fine-tune the algorithms and optimize their configurations. Different combinations of hyperparameters are tested to identify the ones that yield the best results. This helps in preventing overfitting and achieving a balance between model complexity and generalization [30].
3. **Test Set:** The test set is set aside specifically to assess how well the algorithms perform. It simulates real-world scenarios where the models encounter unseen data. The test set's results provide an objective measure of how well the algorithms can generalize to new, potentially malicious, network traffic [20].

The division of the dataset into these subsets is typically performed randomly, with care taken to ensure that the class distribution is maintained to prevent bias.

7.2 Hyperparameter Tuning

The adjustment of hyperparameters is a vital component of the experimental setup since it directly affects the efficacy of the algorithm. Each method is comprised of specific hyperparameters that govern and influence its behavior. The procedure for finding the optimal hyperparameters involves systematic experimentation and evaluation.

7.3 Cross-Validation Methods

Cross-validation methods are integral to the research design, enabling the robust evaluation of the algorithms and their generalization to different data subsets [31]. There are several prevalent techniques such as K-fold Cross-Validation, Stratified Cross-Validation, and Leave-One-Out Cross-Validation. The selection of a cross-validation method relies on factors such as the dataset size, the algorithm used, and the research goals.

8   Conclusion

This research carried an in-depth relative analysis of significant machine learning algorithms employed to detect different forms of malicious network traffic. The algorithms tested include random forests, regular neural networks, logistic regression. Using labelled datasets of benign and attack traffic, the models were trained and their performance analyzed on parameters like ROC curves, F1-scores, accuracy, and confusion matrix (Figs. 13 and 14) (Table 1).

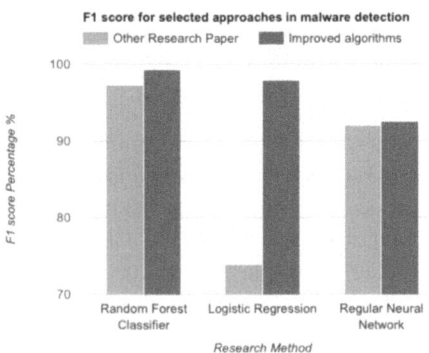

Fig. 13. Bar graph comparing F1 score of our proposed model and the average of other's model

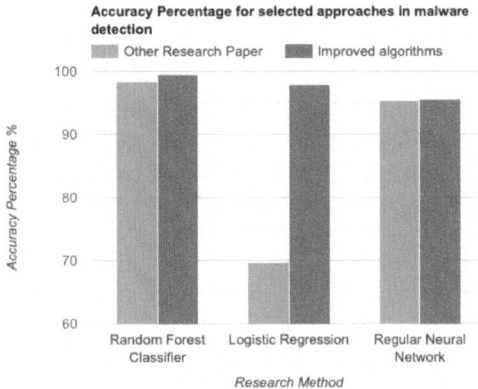

Fig. 14. Bar graph comparing accuracy of our proposed model and the average of other's model

Table 1. Comparison Table with Accuracy and F1 Score.

Technique	Machine learning algorithms			
	Average of results from other existing papers		Result from our proposed model	
	Accuracy (on test dataset)	F1 Score (on test data set)	Accuracy (on test dataset)	F1 Score on test data set)
Random Forest Classifier	98.38%	97.30%	99.54%	99.24%
Logistic Regression	69.72%	73.89%	97.85%	97.85%
Regular Neural Network	95.38%	91.97%	95.57%	92.51%

The results demonstrate that ensemble methods like random forests achieve very high accuracy, exceeding 99% in some tests, owing to their strategy of combining diverse decision trees. With careful tuning of hyperparameters, neural networks also display comparable performance due to their automated feature learning capabilities. The analysis indicates each technique has strengths – tree ensembles excel at handling high dimensional data, SVMs perform well on binary classification, while regular neural networks can uncover deep patterns.

Hence, we can conclude from the accuracy of different classifiers, for the given dataset the order which is followed is:

Random Forest Classifier > Logistic Regression > Regular Neural Network.

The study provides useful insights into selecting the most suitable ML solutions tailored to specific network environments and security requirements. Random forest is ideal for networks with abundant data. Deep learning (RNN) shows promise for feature extraction from raw traffic data. The research highlights the importance of continuous model validation, testing on recent datasets, and striking a balance between accuracy and interpretability.

In summary, the experiments highlight the substantial potential of machine learning in crafting smart, real world intelligent intrusion detection systems. With careful feature engineering and model optimization, ML can enable robust and scalable cybersecurity surveillance against evolving threats.

9   Future Prospects

While this study provides a comprehensive benchmarking of algorithms, further research can build on these findings in several directions-

- Evaluating deep learning architectures like CNNs and LSTMs for spatial and temporal pattern recognition in network traffic data.
- Testing the algorithms on very large real-world network datasets to assess computational feasibility.
- Enhancing models with active learning and incremental learning capabilities to adapt to new unlabeled data.
- Developing ensemble stacking models that combine multiple algorithms to improve attack detection accuracy.
- Analyzing model explain ability and visualization techniques to increase trust and transparency.
- Comparing cloud-based implementations for scaling to big data workloads in high-speed networks.
- Expanding the algorithms to detect zero-day and polymorphic attacks using behavior-based anomaly detection.
- Assessing model robustness against adversarial evasion attacks and developing mitigation strategies.
- Incorporating domain knowledge from cybersecurity experts into the machine learning pipeline.

Exploring these research paths could facilitate the connection between academic machine learning research and the practical implementation of intelligent intrusion detection systems in the real world. With rigorous benchmarking and innovation, machine learning is poised to transform network security defenses against increasingly sophisticated threats.

References

1. Cannady, J.: Artificial neural networks for misuse detection. Natl. Inform. Syst. Secur. Conf. **26**(2), 36881 (1998)
2. Mukkamala, S., Sung, A.H., Abraham, A.: Intrusion detection using an ensemble of intelligent paradigms. J. Netw. Comput. Appl. **28**(2), 167–182 (2005)
3. Apruzzese, G., Colajanni, M., Ferretti, L., Mirto, M.: On the effectiveness of machine and deep learning for cyber security. In: 2018 10th International Conference on Cyber Conflict (CyCon), pp. 371–390 (2019)

4. Mukkamala, S., Janoski, G., Sung, A.: Intrusion detection using neural networks and support vector machines. In: Proceedings of the 2002 International Joint Conference on Neural Networks, vol. 2, 17021707 (2002)
5. Toosi, A.N., Kahani, M.: A new approach to intrusion detection based on an evolutionary soft computing model using neuro-fuzzy classifiers. Comput. Commun. **30**(10), 2201–2212 (2007)
6. Tsai, C.F., Hsu, Y.F., Lin, C.Y., Lin, W.Y.: Intrusion detection by machine learning: a review. Expert Syst. Appl. **36**(10), 11994–12000 (2009)
7. Bhuyan, M.H., Bhattacharyya, D.K., Kalita, J.K.: Network anomaly detection: methods, systems and tools. IEEE Commun. Surv. Tutorials **16**(1), 303–336 (2014)
8. Sharafaldin, I., Lashkari, A.H., Hakak, S., Ghorbani, A.A.: Developing realistic distributed denial of service (DDoS) attack dataset and taxonomy. In: 2019 International Carnahan Conference on Security Technology (ICCST), pp. 1–8 (2018)
9. Javaid, A., Niyaz, Q., Sun, W., Alam, M.: A deep learning approach for network intrusion detection system. In: Proceedings of the 9th EAI International Conference on Bio-inspired Information and Communications Technologies, pp. 21–26 (2016)
10. Buczak, A.L., Guven, E.: A survey of data mining and machine learning methods for cyber security intrusion detection. IEEE Commun. Surv. Tutorials **18**(2), 1153–1176 (2016)
11. Ring, M., Wunderlich, S., Grüdl, D., Landes, D., Hotho, A.: Flow based benchmark data sets for intrusion detection. In: Proceedings of the 16th European Conference on Cyber Warfare and Security, pp. 361–369 (2019).
12. Shiravi, A., Shiravi, H., Tavallaee, M., Ghorbani, A.A.: Toward developing a systematic approach to generate benchmark datasets for intrusion detection. Comput. Secur. **31**(3), 357–374 (2012)
13. Gao, N., Gao, L., Gao, Q., Wang, H.: An intrusion detection model based on deep belief networks. In: Scientific Programming (2019)
14. April, A., et al.: A comprehensive survey on intrusion detection systems. University of Amsterdam, Netherlands (2009)
15. Dainotti, A., Claffy, K.C., King, A.: Analysis of internet background radiation. In: Passive and Active Measurement (PAM), pp. 31–40 (2011)
16. Kolter, J.Z., Maloof, M.A.: Learning to detect malicious executables in the wild. J. Mach. Learn. Res. **7**(May), 2721–2744 (2006)
17. Breiman, L.: Random forests. . Mach. Learn. **45**(1), 5–32 (2001)
18. Schölkopf, B., Platt, J.C., Shawe-Taylor, J., Smola, A.J., Williamson, R.C.: Estimating the support of a high-dimensional distribution. Neural Comput. **13**(7), 1443–1471 (2001)
19. Saxe, J., Berlin, K.: Deep neural network based malware detection using two-dimensional binary program features. In 2015 10th International Conference on Malicious and Unwanted Software (MALWARE), pp. 11–20. IEEE (2015)
20. Kang, Y., Tan, J., Jiang, X., Wang, Y., Jia, C.: Malware classification with deep convolutional neural networks. In 2017 10th International Conference on Malicious and Unwanted Software (MALWARE), pp. 110–119. IEEE (2017)
21. Chen, T., Guestrin, C.: XGBoost: A scalable tree boosting system. In: Proceedings of the 22nd ACM SIGKDD International Conference on Knowledge Discovery and Data Mining, pp. 785–794 (2016)
22. Huang, J., Li, K., Tan, C.L.: Advances in feature selection: a review. Data Min. Knowl. Disc. **13**(1), 13–54 (2006)
23. Kohavi, R., John, G.H.: Wrappers for feature subset selection. Artif. Intell. **97**(1–2), 273–324 (1997)
24. Friedman, J., Hastie, T., Tibshirani, R.: The elements of statistical learning, vol. 1. Springer series in statistics (2001)

25. Jolliffe, I.T.: Principal component analysis. Wiley Online Library (2002)
26. Vetterli, M., Kovacevic, J.: Wavelets and Subband Coding. Prentice-Hall (1995)
27. Breiman, L.: Random forests. Mach. Learn. **45**(1), 5–32 (2001)
28. Friedman, J.H.: Greedy function approximation: a gradient boosting machine. Ann. Stat. **29**(5), 1189–1232 (2001)
29. Simonyan, K., Vedaldi, A., Zisserman, A.: Deep inside convolutional networks: Visualizing image classification models and saliency maps. arXiv preprint arXiv:1312.6034 (2013)
30. Bailey, M., Jorgensen, L.: The effects of data breach disclosure laws. IEEE Secur. Priv. **3**(3), 40–45 (2011)
31. Zuech, R., Bishop, M.: A survey of Malware Detection Techniques. University of California, Department of Computer Science (2013)

Enhancing Cyclone Intensity Prediction Through Deep Learning Analysis of Imagery Datasets

Jyoti Dinkar Bhosale[1]($\boxtimes$) (ID), Suraj S. Damre[2], Ujwala V. Suryawanshi[3], and Rajkumar B. Pawar[4]

[1] VDF Group of Institution, Latur, India
bhosale.jyoti3@gmail.com
[2] D.Y. Patil College of Engineering Akurdi, Pune, India
[3] Computer Science Department, Rajarshi Shahu Mahavidyalaya (Autonomous), Latur, India
[4] Gharda Institute of Technology, Lavel, Ratnagiri, India

Abstract. Cyclones pose a significant threat, causing widespread devastation and loss of life. Early prediction of cyclone intensity plays a crucial role in mitigating their impact. In recent years, deep learning has emerged as a promising technique for image analysis. This paper introduces a deep learning-based method for estimating cyclone strength using image datasets. By leveraging convolutional neural networks (CNNs), the proposed approach extracts essential information from satellite imagery to forecast cyclone intensity. The model is trained on a comprehensive historical dataset sourced from the National Hurricane Center's HURDAT2 database and validated on new cyclone data. Evaluation metrics such as mean absolute error, mean squared error, and root mean squared error demonstrate the effectiveness of the CNN model in accurately estimating cyclone intensity. Training the CNN model on the historical dataset employs supervised learning, where labeled examples consisting of satellite data and corresponding cyclone intensities are utilized. Through this process, the model discerns patterns and correlations within the satellite data, enabling it to make precise predictions for unseen cyclone data.

Keywords: Convolutional Neural Network · Cyclone · Deep Learning · Intensity · Meteorology

1 Introduction

1.1 A Subsection Sample

Cyclones are large-scale weather phenomena that often originate over warm tropical oceans and can cause coastal areas to experience severe winds, heavy rain, storm surges, and flooding. These storms, commonly known as tropical cyclones or hurricanes, are formed over warm ocean waters when the conditions are just right. The combination of warm water, low pressure, and favorable atmospheric conditions creates a powerful system that can cause widespread destruction [1].

© The Author(s), under exclusive license to Springer Nature Switzerland AG 2025
A. K. Bairwa et al. (Eds.): ICCAIML 2024, CCIS 2184, pp. 205–217, 2025.
https://doi.org/10.1007/978-3-031-71481-8_16

Each category on the Saffir-Simpson scale is determined by the sustained wind speeds of the cyclone. For example, Category 1 cyclones have sustained winds of 74 to 95 miles per hour, while Category 5 cyclones have sustained winds of 157 miles per hour or greater. The scale also takes into account the potential for damage caused by the storm, with higher categories indicating more severe and potentially catastrophic conditions. As a civil engineer and as the director of the National Hurricane Center, Robert Simpson developed this scale in the 1970s [2–4]. It is designed to give an estimate of the potential damage and flooding a hurricane could cause based on its wind speed.

The way a CNN works is by progressively down sampling thereby gradually increasing the number of filters, so that the neural network can learn from the different features in the image. By using convolutional filters, the CNN is able to learn from the raw pixel data and extract features without the need for manual feature engineering. CNNs have been employed in a range of cyclone-related applications, including detecting and tracking storm movement using satellite imagery, predicting storm severity using multiple meteorological factors, and generating realistic simulations of storm behavior [5]. These apps have the potential to increase our ability to forecast and prepare for extreme weather occurrences, potentially saving lives and reducing property damage. The proposed strategy has the potential to improve cyclone early warning systems and lower the risk of damage and loss of life caused by these natural disasters [9–12]. The accurate and reliable prediction of cyclone intensity can aid in the timely and effective evacuation of people and protection of property. The proposed strategy helps to detect the behavior of cyclones and their influence on the environment, which can aid in the development of cyclone mitigation strategies [17]. By detecting the behavior of cyclones, such as their intensity, trajectory, and speed, scientists can gain valuable insights into the patterns and characteristics of these destructive weather phenomena. This information can then be used to develop more effective cyclone mitigation strategies, including early warning systems, evacuation plans, and infrastructure improvements in vulnerable areas.

2   Related Work

Using geometric features from cyclone images coupled with a multilayer perceptron (MLP) machine learning algorithm, the paper [1] proposes a new method for detecting tropical cyclone intensities. The proposed involves extracting geometric features from cyclone images, such as the area, perimeter, and eccentricity. These features were then used as inputs to an MLP machine learning algorithm, which was trained to predict the intensity of the cyclone. The method also demonstrated better performance than existing methods in terms of accuracy of 84%, speed and scalability. The findings of this research could help in improving the forecasting of cyclones in the region, and thus contributing to disaster management.

The classifiers in paper [2] were evaluated on a dataset of tropical cyclone band images collected from low-Earth orbiting satellites. The results were compared with existing state-of-the-art models and showed that the proposed framework outperformed them in terms of accuracy and Kappa coefficient. Additionally, the experiment was also conducted using different datasets and classification models, framework was robust and effective across different datasets and classifiers. The results in paper [3] also indicate model robust in terms of data variations and noise, and can effectively capture the cyclone intensity distribution over a large geographical area. The model is also highly scalable, and can be easily extended to other related applications. This process allows the network to learn from the data more effectively over time, and the hybrid similarity measurement allows the network to distinguish between features in the data that may not be easily recognizable to the human eye. This improvement in classification performance was particularly evident when small training sets were used, as the proposed method was able to outperform several existing methods. Overall, the proposed method achieved promising results, demonstrating its potential in improving cyclone intensity estimation from satellite images.

Further research is needed to refine and optimize the model and assess its accuracy in different areas.The validation process revealed that the DAV technique is more accurate than the Dvorak technique for storms with intensities greater than or equal to category 2. This is because the DAV technique takes into account additional factors, such as the size and shape of the storm, which the Dvorak technique does not. This means that the DAV technique is better able to predict the strength of a storm more accurately. This is especially important for forecasting potential damage from severe storms. Ultimately, this improved accuracy can help authorities make better plans to protect citizens and properties from the destructive effects of severe storms. It can also be used to inform people about potential risks posed by storms. Thus, the DAV technique provides an invaluable tool for reducing the impact of destructive storms, not only in terms of minimizing potential damage but also informing people of the risks they may face [4].

This novel model underwent testing on a dataset comprising over 500,000 images, achieving an accuracy of up to 0.25 m/s. Notably, it demonstrated proficiency in identifying various storm types, including tropical cyclones, monsoons, and mid-latitude cyclones. Additionally, the model displayed a capacity to identify new storms with an accuracy of up to 0.5 m/s. Offering a real-time solution, this approach enables swift and precise wind speed estimations through a situational awareness portal. Specifically designed to tackle the challenges of wind speed estimation from satellite imagery, the model's comparison with Dvorak T-number images facilitated accurate wind speed assessments across diverse conditions. Moreover, its ability to swiftly and efficiently detect new storms and provide real-time wind speed estimates through the situational awareness portal was confirmed [5].

The method exhibited high reliability, with a Root-Mean-Square Error (RMSE) of 3.38 m/s and a Mean Absolute Error (MAE) of 2.64 m/s, underscoring its efficacy in predicting wind speed for tropical cyclones. Evaluation against observations of tropical cyclone intensity by the U.S. National Hurricane Center and measurement of wind speed from aircraft reconnaissance data yielded a remarkable 97% accuracy rate. This indicates the method's proficiency in accurately estimating the intensity of most tropical cyclones. Results indicated that the proposed method could achieve a robust estimation of maximum wind speed with a root mean square error of 5.68 m/s and an approximate accuracy rate of 94% [6].

A technique has been devised to estimate tropical cyclone (TC) intensity automatically using data from the Special Sensor Microwave Imager (SSM/I). A recent study [7] evaluates the efficacy of this method by gathering 1,040 sample images from 142

TCs across the North Pacific, Atlantic, and Indian Oceans. Among these, 942 images serve as training samples, and a feature-selection algorithm is employed to identify the most optimal feature combination. The K-nearest-neighbor algorithm then compares the selected features of each testing sample to those of the training samples, identifying the K closest neighbors. In this study, the algorithm is applied to 98 testing samples using 15 selected features to estimate TC intensity. Utilizing the best-track intensity as the ground truth, the algorithm computes a root-mean-square error (RMSE) of 19.8 kt, signifying the accuracy of the estimation, which improves to 18.1 kt. Furthermore, a comparison is made between the RMSE generated using the best-track intensity and reconnaissance data for two recent (1999) Atlantic hurricanes. This comparison underscores the effectiveness of the additional feature in enhancing research accuracy and the superior reliability of best-track intensity data in predicting hurricane intensity over reconnaissance data.

The results highlight that best-track intensity data can be up to 25% more accurate than reconnaissance data, emphasizing its importance as the primary source for hurricane intensity information. In paper [8], a new technique was developed to calculate the intensity of tropical cyclones (TCs) by using data from the TRMM Microwave Imager (TMI) channels of 85,37,21,19,10 GHz from 1999 to 2003.

The connection between the TMI brightness temperature (TB) parameters and the TC maximum wind speed was investigated by computing the TBs in various circles of different radii at different TMI frequencies. The lower frequency channels of 10 or 19 GHz were found to provide higher correlation due to their direct connection to sea surface wind speed. Multiple regression equations were used with a few parameters to create the TC intensity estimation method, and after selecting 3 parameters and computing the regression coefficients, 10 equations with the lowest root mean square error (RMSE) were selected. The method was verified using independent data by comparing its results with the TC best track data. This comparison revealed that the method had an RMSE of about 8 m/s in the whole basin and about 6 m/s in the northwestern Pacific, indicating a relatively accurate estimation of TC intensity change.

In paper [9], a dependable and strong deep convolutional neural network (CNN) was proposed for estimating tropical cyclone intensity using satellite images. Regularization techniques were employed to effectively extract complex features from hurricane images, resulting in superior accuracy and lower root-mean-square error compared to 'state-of-the-art' methods.

In paper [10], A CNN model was optimized by fine-tuning the network architecture, adjusting hyperparameters such as learning rate and batch size, and optimizing the loss function. Through these iterative steps, the model's performance was significantly improved, achieving a lower root mean square error (RMSE) and outperforming the previous model by 35%. This highlights the effectiveness of deep learning approaches in accurately estimating tropical cyclone intensity from satellite images. In the paper [11], The performance of the TCICENet model was compared with several existing models in the field of tropical cyclone intensity prediction. The results showed that the TCICENet model outperformed other models in terms of both root mean square error and mean absolute error, indicating its superiority in accurately predicting the intensity of tropical cyclones. Furthermore, the model is able to accurately predict the intensity of different categories of cyclones.

According to the paper [12], TC intensity and wind radius can be estimated from infrared imagery over the northern Atlantic Ocean using deep learning and prior knowledge of tropical cyclones (TC). Results show that deep learning improves the accuracy of TC intensity and wind radius estimates. Additionally, the paper demonstrates the potential of using deep learning for TC forecasting in the northern Atlantic Ocean. The DeepTCNet also had a consistent performance across all regions of the world, with no significant differences between different basins. The results demonstrate the effectiveness of the DeepTCNet model in providing accurate estimates of TC intensity and wind radii.

Researchers developed a model that uses convolutional neural networks (CNNs) to estimate tropical cyclone intensity. The model was optimized using residual learning and attention mechanisms to improve its feature extraction ability. The model was tested on a dataset of more than 10,000 cyclones to demonstrate its accuracy. Results showed that the model was able to accurately predict cyclone intensity with a high degree of accuracy. The research team concluded that the model could be a useful tool for forecasting cyclone intensity. The model was able to pick up on subtle patterns in the data that are not immediately apparent to the human eye.

This was achieved by using residual learning, which allows the model to learn from the difference between the estimated output and the actual output. Additionally, the attention mechanism allows the model to focus on the most important features in the data, which further improves the model's accuracy [13].

To estimate tropical cyclone intensity from geostationary satellite images, the paper [14] developed STE-TC, a spatiotemporal encoding module. By integrating temporal information from TCs into satellite images, the module enhanced the spatial features. The dense connectivity strategy was applied to a ConvMixer network to create a model for estimating TC intensity called DenseConvMixer. STE-TC module and DenseConvMixer network were evaluated through experiments and visualizations. CNN-based models could be improved by using the STE-TC module. The results of the experiments showed that the STE-TC module improved CNN-based models and the DenseConvMixer network achieved better performance than existing methods. The model was successfully applied to satellite images and provided more accurate TC intensity estimation.

The paper [15] proposed a model called TFG-Net that uses a graph convolutional neural network for the estimation of tropical cyclone intensity with great detail. It was composed of three parts, for producing general wind speed information. After testing on the GridSat dataset, the TFG-Net obtained an RMSE metric of 11.12 knots, outperforming the traditional method by 33.33% and deep learning method by 2.54%.

TCNNs are deep networks formulated using tensor algebra, as described in the paper [16]. They are capable of implementing convolutional and pooling operations with the help of tensor decomposition. Furthermore, TCNNs can exploit the structure of convolutional networks, enabling faster and more efficient computation. This makes TCNNs ideal for use in computer vision tasks, such as image classification, object detection, and segmentation. TCNNs have been shown to provide significant improvements in accuracy and speed compared to traditional convolutional networks. In addition, TCNNs can be used to reduce the size and complexity of convolutional networks, making them more suitable for deployment on mobile devices and embedded systems. This makes TCNNs attractive for use in a wide range of applications, including autonomous driving, medical imaging, and robotics. TCNNs can also be used to improve generalization performance. Furthermore, they can be used to implement deep architectures with fewer layers than traditional convolutional networks, making them more suitable for applications where the model size and computational complexity are key constraints.

In paper [17], a study was conducted in order to examine how the Advanced Dvorak Technique (ADT), using geostationary meteorological satellite data, could be further improved to produce a more accurate interpretation of the intensity of tropical cyclones (TCs) around the world. The results of the study showed that the ADT had a higher accuracy than the existing methods of TC intensity estimation. This improvement was attributed to the higher resolution data obtained from the geostationary satellite. The findings of the study contribute to the development of more reliable TC intensity estimates.

An infrared and water vapor image-based model of tropical cyclone intensity was presented in the paper [19]. RMSE and MAE were respectively 5.13 and 4.03 m/s, respectively, for the 60×60 pixel input image. The same model was applied to a larger resolution image with a size of 128×128 pixel. The results improved significantly, with RMSE and MAE values of 3.76 and 3.23 m/s, respectively. This indicates the importance of higher resolution images for accurate cyclone intensity predictions. The model also demonstrated excellent temporal stability, with RMSE and MAE values of 4.13 and 3.17 m/s, respectively, when applied to the same image with a one-hour time lag. This indicates the robustness of the model for cyclone intensity predictions. Additionally, the model was able to generate more accurate predictions when applied to higher resolution images, thus confirming the importance of image resolution for cyclone intensity predictions. Furthermore, the temporal stability of the model was also excellent, as it was able to generate very similar predictions when applied to the same image with a one-hour time lag. This confirms the robustness of the model for cyclone intensity predictions.

For estimating tropical cyclone intensity with satellite remote sensing, a convolutional neural network (CNN) hybrid model is proposed in [20]. Three different intensity regression models were used to analyze the TCs by their intensity. Small samples were fitted more quickly with piecewise thinking. Unlabeled TC samples were classified with a classification network, and the results were used to determine the regression network that should be used to estimate them. The model was validated with both synthetic and real-world data, and was found to perform better than other existing models. The proposed model is capable of estimating the intensity of TCs with a high degree of accuracy. The results of this study demonstrate that piecewise thinking and deep learning are effective methods for estimating the intensity of TCs. This model can be used to improve our understanding of TCs and their potential effects on climate.

3   Methodology

The research proposed in the study will involve the following steps:

1. Obtaining the necessary data
2. Preparing the data for analysis
3. Training the model
4. Verifying the accuracy of the results.

Data Gathering. The National Centers for Environmental Information's HURSAT data project makes use of cyclone images. These images were obtained from a database that contains satellite images of cyclones in NetCDF file format. One noteworthy feature of this database is that the center of each hurricane corresponds to the center of its corresponding image. Furthermore, the National Hurricane Center's HURDAT2 database supports the analysis by providing numerical data on cyclone images with components such as time, latitude, longitude, wind speed, and so on in the form of best track data. This extensive database contains data on all reported cyclones in the Atlantic and Pacific basins. It includes the wind speeds of these cyclones recorded at 6-hour intervals. Data scraping is used to obtain cyclone imaging data from the HURSAT data project for specified years. It scrapes a website to acquire files including satellite photos for each storm. The algorithm restricts the data to cyclones in the Atlantic and Pacific basins that have wind speed and position records in the National Hurricane Center's Best Track dataset. NetCDF files containing images from satellites of each cyclone are extracted and saved locally from the files generated above. To guarantee data integrity, redundant images of the same cyclone taken at the same time were removed.

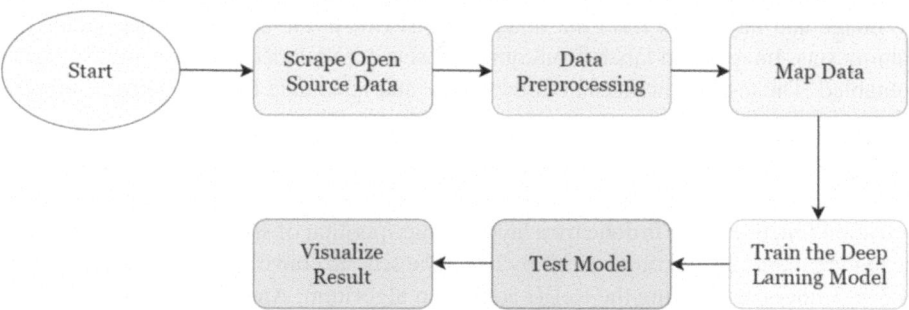

Fig. 1. Proposed Architecture

Overall, this data collection process ensured the acquisition of high-quality cyclone satellite imaging data, which aligned with the research goal of assessing and understanding the dynamics of these extreme weather phenomena. After that the wind speed is divided into various categories which are as as follows:

- *If the wind speed is 33 km per hour or less, it is categorized as a 'Tropical Depression'.*
- *If the wind speed is between 34 and 64 (inclusive), it is categorized as 'T. Storm'.*

- *For wind speeds between 65 and 83 (inclusive), the function assigns the label'Category 1'.*
- *If the wind speed falls between 84 and 95 (inclusive), it is labeled as'Category 2'.*
- *For wind speeds ranging from 96 to 113 (inclusive), the category assigned is'Category 3'.*
- *Wind speeds falling between 114 and 134 (inclusive) are classified as'Category 4'.*
- *For wind speeds greater than 134, the function assigns the label'Category 5'*

Data Preprocessing. A critical component of analyzing and preparing cyclone data for future examination is reading and processing the dataset to guarantee its suitability for analysis. To do this, data is first loaded from files, which include image and label databases containing critical information on cyclones. The dataset is then divided into subsets to aid model evaluation, with k-fold cross-validation being used in particular.

An option for image augmentation is provided to enhance the dataset. This entails creating new images and labels for each subset using image augmentation techniques. Different augmentation procedures are used based on various parameters given by the labels. If the label is between 50 and 75 (indicating a tropical storm), the code generates two new images by flipping the original image horizontally and vertically. The modified photos, together with their labels, are then put to the appropriate listings. Similarly, if the label is between 75 and 100, rotation is used to generate six new images. The modified photos and labels are attached to the respective lists once more. Furthermore, if the label is equal to or greater than 100, the code generates 12 new images by rotating them. These augmented photographs are added to the lists, along with their labels.

Augmentation increases the amount of training data available and has the potential to improve the performance of subsequent models.

Following that, the subsets are joined to form sets for model iterations. Concatenating the image and label data from the other subsets, except the current subset, yields the training sets. Images and labels from other subsets are included if image augmentation is enabled. The test set includes both the image and label data from the current subset.

Model Training. The model is designed to predict a continuous value for grayscale images. The network takes a grayscale image with a resolution of 50×50 pixels as input. The first convolutional layer consists of 32 3×3 filters, each using the ReLU activation function. This is done by a layer of max-pooling of size 2×2, which reduces the feature maps' spatial dimensions by half. The second convolutional layer consists of 64 3×3 filters, each using the ReLU activation algorithm. Another max-pooling layer of the same size is then placed.

There are 64 3×3 filters and ReLU activation function are used in the final convolutional layer. This layer's output is flattened and sent to a fully connected neural network. To prevent overfitting, the fully linked layers include a dropout rate of 0.5. This is followed by two dense layers, both of which make use of ReLU activation functions. The output layer contains a single neuron without an activation function, allowing the model to predict a continuous value for each input image.

The rmsprop optimizer is used for model optimization, with the mean squared error loss function. During training, this optimizer changes the network's weights to minimize loss. During training, the model calculates two metrics to evaluate its performance on the training set. These measures aid in spotting potential overfitting or underfitting issues. Finally, the CNN model constructed using the Keras toolkit is wellsuited for predicting continuous values in grayscale photos. The architecture, rmsprop optimization, and evaluation metrics all contribute to good training and performance evaluation on the training set.

Validation of Results. The dataset was split into two sections, training and validation. The model was then trained for a predetermined number of epochs and batch size. After each epoch, the model's performance on the validation set was assessed to see if it was overfitting. An Early Stopping callback was implemented to stop training if the validation error didn't improve after a certain number of epochs, in order to stop overfitting and guarantee that the model could be applied to new data. Based on the best validation error, the best weights are restored. Following training, the model's performance by epoch can be displayed. The training and testing loss, as well as the mean absolute error (MAE) for each epoch metric, are saved in a data frame for subsequent analysis, together with the relevant epoch numbers. The measurements are then plotted on a graph with Seaborn's lmplot function, which fits exponential decay curves.

The epoch number is represented on the x-axis, and the metric values are represented on the y-axis. There are two rows in the graph: one for loss and one for MAE. Using distinct colors, each row distinguishes between training and testing data. The graph is displayed, providing information about the model's performance during the training phase.

4   Algorithm Used

Convolutional Neural Network (CNN). The Convolutional Neural Network, or CNN, is a deep learning model which has been designed to process data that is structured in a grid-like form, such as images [21]. This model is based on the organization of the visual brain in animals and attempts to learn hierarchical features, ranging from basic to complex patterns, in an autonomous and adaptive way. The CNN consists of three types of layers: convolutional, pooling, and fully connected. The convolution and pooling layers are in charge of extracting meaningful features from the input data. The convolution layer is made up of mathematical processes, specifically convolution, which entails applying a kernel (a tiny grid of optimizable parameters) to each point in the input image grid (2D array of pixel values). This allows CNNs to process images more effectively because features can appear everywhere in the image.

Convolutional Neural Networks (CNNs) are powerful models used for grid-like data, such as photographs, that draw from the concepts of the animal visual brain. Features are collected and transmitted through additional layers, increasing in complexity in a hierarchical manner. The fully connected layer then translates these features into a final output, often used for classification. To train a CNN, the model's parameters are adjusted to reduce the difference between the predicted and reality, typically done with backpropagation and gradient descent algorithms. In summary, CNNs are able to extract and learn increasingly complicated characteristics with the use of convolution, pooling, and fully connected layers.

5   Result and Interfaces

The MAE and RMSE calculates the average size of errors by first squaring the differences between predictions and actual values, then taking the square root of the resulting average. An RMSE value that is lower than the MAE value indicates that the model is better at predicting the data. This is shown in Table 1. The column in Table 1 lists the precise fold or subset of the data set that was utilized in the study's validation. The purpose of reporting both MAE and RMSE is to provide different perspectives on the performance of the prediction model. The MAE over all folds for our cyclone intensity prediction model was 9.96 knots. In our situation, an MAE of 9.96 knots shows that, on average, the predictions of our model were 9.96 knots off from the actual cyclone strength estimates.

The RMSE for our study was found to be 13.65 knots. A score of 13.65 knots indicates that, on average, our model's predictions were off by about 13.65 knots. The RMSE provides an indication of the typical forecast error (Fig. 2).

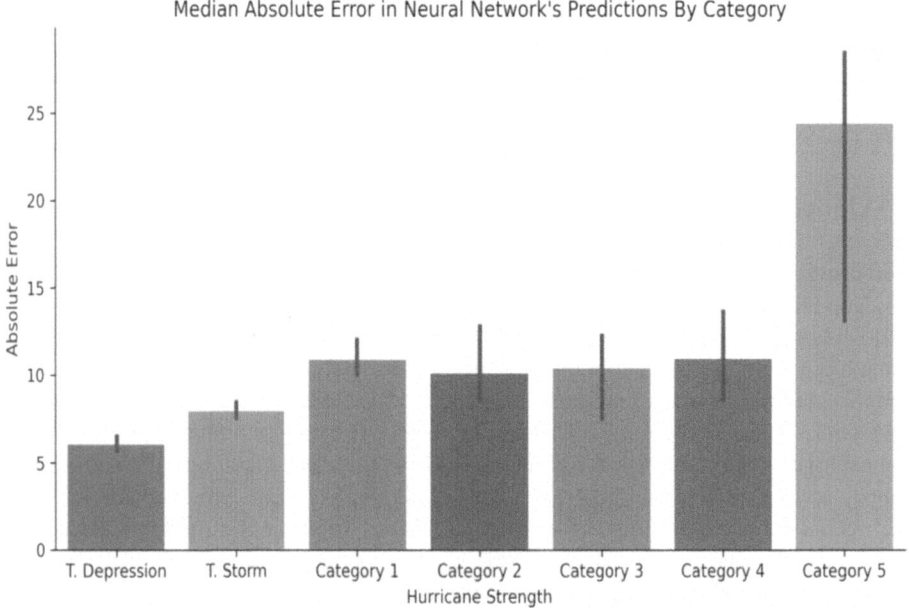

Fig. 2. Mean Absolute Error in Neural Network's Predictions by Category

Figure 1 visualizes the mean absolute error in the Hurricane's strength which are divided into seven categories. Our model performed best on the cyclones that were categorized into the Depression type with a MAE of and performed unfavorable on the Category 5 type cyclone with a MEA. The following Table 1 gives a summary of validation folds of MAE and RMSE.

Table 1. Contrasting the Mean Absolute Error (MAE) and Root Mean Squared Error (RMSE) across various folds.

Validation Fold	MAE (Mean Absolute Error)	RMSE (Root Mean Squared Error)
I	12.8365 knots	16.2777 knots
II	10.4333 knots	13.2222 knots
III	10.4146 knots	13.2352 knots
IV	9.9171 knots	12.6129 knots
V	9.4262 knots	12.4232 knots
All Folds	9.96 knots	13.65 knots

6 Conclusion and Future Scope

Concluding this research paper; it has been recommended that utilizing deep learning methods can accurately estimate the severity level of hurricanes employing satellite image sets with subsequent analysis returning impressive results - due primarily to Convolutional Neural Networks' use within storm modelling applications. The effective extraction of valuable data from satellite imaging permits the early detection of intensifying hurricane strengths that can lead to reduced adverse impacts on communities. Through comprehensive measuring techniques, we were able to test and verify this technique's effectiveness.

These findings reinforce and emphasize the significance of deep learning methods in disaster management preparation. Considering the potential future work, First, more sophisticated deep learning architectures that can capture temporal relationships and improve the model's capacity to analyse sequential satellite images, such as recurrent neural networks (RNNs) or attention mechanisms, can be incorporated into the proposed CNN model. Utilizing additional data sources, such as oceanic and atmospheric data, can also deliver supplementary information for a more precise estimation of cyclone intensity. Additionally, the study can be expanded to incorporate areas and basins than the Atlantic and Pacific, increasing the model's generalizability. Last but not least, efforts can be made to combine real-time data streams and create an operational system for ongoing cyclone intensity monitoring and prediction, enabling prompt response and proactive mitigation techniques.

References

1. Kar, C., Kumar, A., Banerjee, S.: Tropical cyclone intensity detection by geometric features of cyclone images and multilayer perceptron. SN Appl. Sci. **1**, 1099–1099 (2019). https://doi.org/10.1007/s42452-019-1134-8
2. Chen, Z., Yu, X., Chen, G., Zhou, J.: Cyclone intensity estimation using multispectral imagery from the FY-4 satellite. In: 2018 International Conference on Audio, Language and Image Processing (ICALIP), pp. 46–51 (2018). https://doi.org/10.1109/ICALIP.2018. 8455603

3. Chen, G., Chen, Z., Zhou, F., Yu, X., Zhang, H., Zhu, L.: A Semisupervised Deep Learning Framework for Tropical Cyclone Intensity Estimation. In: 10th International Workshop on the Analysis of Multitemporal Remote Sensing Images (MultiTemp), pp. 1–4 (2019). https://doi.org/10.1109/Multi-Temp.2019.8866970

4. Ritchie, E.A., Wood, K.M., Rodríguez-Herrera, O.G., Piñeros, M.F., Tyo, J.S.: Satellite-Derived Tropical Cyclone Intensity in the North Pacific Ocean Using the Deviation-Angle Variance Technique. Wea. Forecasting **29**, 505–516 (2014). https://doi.org/10.1175/WAF-D-13-00133.1

5. Maskey, M., et al.: Deepti: deep learning-based tropical cyclone intensity estimation system. JJ Miller IEEE Journal of Selected Topics in Applied Earth Observations and Remote Sensing **13**, 2020–2020. https://doi.org/10.1109/JSTARS.2020.3011907,2020

6. Piñeros, M.F., Ritchie, E.A., Tyo, J.S.: Estimating tropical cyclone intensity from infrared image data. Wea. Forecasting **26**, 690–698 (2011). https://doi.org/10.1175/WAF-D-10-050 62.1

7. Bankert, R.L., Tag, P.M.: An Automated Method to Estimate Tropical Cyclone Intensity Using SSM/I Imagery. J. Appl. Meteor. Climatol **41**, 461–472 (2002). https://doi.org/10.1175/1520-0450(2002)041<0461:AAMTET>2.0.CO;2

8. Shunsuke, H., Tetsuo, N.: Estimation of Tropical Cyclone's Intensity Using TRMM/TMI Brightness Temperature Data. J. Meteorolog. Soc. Japan. Ser. II **85**(4), 437–454 (2007). https://doi.org/10.2151/jmsj.85.437

9. Pradhan, R., Aygun, R.S., Maskey, M., Ramachandran, R., Cecil, D.J.: Tropical cyclone intensity estimation using a deep convolutional neural network. IEEE Transactions on Image Processing **27**, 692–702 (2018). https://doi.org/10.1109/TIP.2017.2766358

10. Lee, J., Im, J., Cha, D.-H., Park, H., Sim, S.: Tropical Cyclone Intensity Estimation Using Multi-Dimensional Convolutional Neural Networks from Geostationary Satellite Data (2019). https://doi.org/10.3390/rs12010108

11. Zhang, J., Wang, X.-J., Ma, L.-M., Lu, X.-Q.: Tropical cyclone intensity classification and estimation using infrared satellite images with deep learning. IEEE Journal of Selected Topics in Applied Earth Observations and Remote Sensing **14**, 2070–2086 (2021). https://doi.org/10.1109/JSTARS.2021.3050767

12. Zhuo, J., Tan, Z.: Physics-augmented deep learning to improve tropical cyclone intensity and size estimation from satellite imagery. Mon. Wea. Rev. 2097–2113 (2021). https://doi.org/10.1175/MWR-D-20-0333.1

13. Tan, J., Yang, Q., Hu, J., Huang, Q., Chen, S.: Tropical Cyclone Intensity Estimation Using Himawari-8 Satellite Cloud Products and Deep Learning (2022). https://doi.org/10.3390/rs14040812

14. Zhang, Z., et al.: A neural network with spatiotemporal encoding module for tropical cyclone intensity estimation from infrared satellite image (2022). https://doi.org/10.1016/j.knosys.2022.110005

15. Xu, G., et al.: TFG-Net: Tropical Cyclone Intensity Estimation from a Fine-grained perspective with the Graph convolution neural network. Eng. Appl. Artif. Intel. **118** (2023). https://doi.org/10.1016/j.engappai.2022.105673

16. Chen, Z., Yu, X.: A novel tensor network for tropical cyclone intensity estimation. IEEE Trans. Geosci. Remote Sens. **59**(4), 3226–3243 (2021). https://doi.org/10.1109/TGRS.2020.3017709

17. Olander, T.L., Velden, C.S.: The Advanced Dvorak Technique (ADT) for Estimating Tropical Cyclone Intensity: Update and New Capabilities, pp. 905–922 (2019). https://doi.org/10.1175/WAF-D-19-0007.1

18. Wang, G., et al.: Tropical cyclone intensity estimation from geostationary satellite imagery using deep convolutional neural networks. IEEE Trans. Geosci. Remote Sens. **60**, 1–16 (2022). https://doi.org/10.1109/TGRS.2021.3066299

19. Zhang, R., Liu, Q., Hang, R.: Tropical cyclone intensity estimation using two-branch convolutional neural network from infrared and water vapor images. IEEE Trans. Geosci. Remote Sens. **58**, 586–597 (2020). https://doi.org/10.1109/TGRS.2019.2938204
20. Tian, W., Huang, W., Yi, L., Wu, L., Wang, C.: A CNN-based hybrid model for tropical cyclone intensity estimation in meteorological industry. IEEE Access **8**, 59 158–59 168 (2020). https://doi.org/10.1109/ACCESS.2020.2982772
21. Yamashita, R., Nishio, M., Do, R.K.G.: Convolutional neural networks: an overview and application in radiology. Insights Imaging **9**, 611–629 (2018). https://doi.org/10.1007/s13244-018-0639-9

Revolutionizing Road Safety: Machine Learning Approaches for Predicting Road Accident Severity

Meenakshi Malik[1]([✉]) [iD], Rainu Nandal[2], and Rita Chhikara[3]

[1] Department of Computer Science and Engineering, BML Munjal University, SOET, Gurgaon, India
`meenakshimalik16@gmail.com`
[2] Department of Computer Science and Engineering, U.I.E.T, Maharshi Dayanand University, Rohtak, India
`rainunandal.uiet@mdurohtak.ac.in`
[3] Department of CSE, The NorthCap University, Gurgaon, India
`ritachhikara@ncuindia.edu`

Abstract. Millions of people die every year from road accidents, which are becoming more commonplace worldwide. They cost society heavily in terms of money and the economy. Road accident prediction has primarily been researched as a classification problem in the literature, meaning that the goal is to forecast whether or not a traffic mishap will occur in future deprived of delving into the intricate interactions between the many components that contribute to these incidents. Fewer studies have examined a subcategory of models of collaborative machine learning modals than majority of research conducted to date on the significance of subsidizing aspects for road accidents and respective severity. Thus, using the road accident dataset as a basis, we assessed a number of ML model in this study for forecasting the traffic accidents' severity. Additionally, we have examined the anticipated outcomes and utilized an explainable machine learning (XML) approach to assess the significance of contributing factors to traffic accidents. This work has taken into consideration various ensembles of ML models, like Random Forest (RF), Extra Tree, Extreme Gradient Boosting (XGBoost), to forecast road accidents with varying injury severity. Extra tree is the top classifier, according to the comparison results, with 81.06% accuracy, 81.06% recall, 75.91% precision, and 77.34% of F1-Score models.

Keywords: Machine Learning (ML) · Accident severity · XGB · Random Forest

1 Introduction

In today's rapidly evolving society, a substantial surge has been noticed in the automobiles volume, leading to a significant rise in traffic accidents with considerable life and economic consequences (Micheale [1]). The WHO's reports that annually, traffic mishaps claim the lives of millions of individuals, while nonfatal accidents also impact a staggering number of people (Bahiru et al. [2]). It is evident that traffic mishaps are emerging as a prominent global cause of fatalities and injuries. Consequently, the prevention and prediction of traffic accidents have become focal points in both areas; traffic sciences and intellectual vehicle research. The traffic mishaps severity as a crucial metric for assessing their harm is influenced by various factors. Numerous studies on traffic accidents have referenced diverse algorithms and factors. Lu et al. [3] conducted an analysis considering the location of a car, road safety rating, surface quality, visibility, car condition, and driver condition, resulting in the establishment of a prediction accuracy model of 86.67%. Additionally, Ren et al. [4] utilized a methodology of deep learning to develop a model of short-term traffic accident risk prediction based on factors such as air pollution, weather, traffic accidents, and traffic flow. Ren et al. [5] proposed spatio-temporal correlation of vehicle accidents in city accidents risk predictions.

Huang et al. [6] introduced a temporal aggregation neural network layer, that autonomously catches relation scores from the temporal dimensions to forecast traffic accident occurrences. Murphey et al. [7] very efficiently diagnosed driver behaviour with the help of data mining methods. In order to prevent and predict traffic mishaps plenty of algorithms have been applied, with the random forest algorithm and it is gaining popularity in recent years. This algorithm, widely employed in diverse fields like medicines (Iwendi et al. [8]), meteorology area (Ding et al. [9]), and statistics area (Schonlau and Zou [10]). It has shown good output in addressing traffic accidents analysis too.

Yan and Shen [11] utilized the random forest algorithm and Bayesian optimization to explore the impact of influencial factors on traffic accidents' severity. Zhao et al. [12] projected an accident risk prediction algorithm which was based on a random forest and deep convolutional neural network. Chen and Chen [13] evaluated prediction performances using accuracy, specificity and sensitivity, identifying the best effective prediction models and input variables with the higher positive impact on these metrics.

Koma et al. [14] applied the random forest algorithm to identify cognitive driver distraction, focusing on various types of eye movement. Wang et al. [15] examined various factors such as time period, grade of road, lanes, closeness to infrastructure, and mishap-prone areas as pointers influencing traffic. The experimental outcomes demonstrated the efficacy in avoiding congested roads and determining high-speed routes. Traffic accidents typically stem from the interplay of human factors, vehicle dynamics, road conditions, and the environment. Leveraging the random forest algorithms' commendable capabilities, our paper highlighted assessment of impact of various accident reasons on the traffic mishaps severity. Subsequently, the model was employed for predicting the traffic accidents' severity and conclusive prediction results highlight the superior performance of the proposed traffic accident elements in forecasting the traffic accidents' severity.

2 Related Work

Our paper concentrates on forecasting accident severity through the examination of historical accident data. To achieve similar objectives, researchers globally have explored diverse methodologies across various domains and data science (data analysis and machine learning model construction) emerged as one of the most prevalent domains.

For prediction and classification, statistical analysis and ML algorithms were applied. Systematically, multiple stages have been executed, including data collection and preparation for ML model construction. The culmination involved result comparison to derive a precise model tailored to the identified problem. In [16], the authors integrated "Convolutional Neural Network and Random Forest" coining the approach as RFCNN.

To predict Traffic-Related Accident (TRA) severity, the researchers employed Fully Connected Neural Network (FCNN) on a dataset comprising accident records spanning between 2016–2020 in the United States. Subsequently, they successfully constructed a highly accurate model, contributing to improved decision-making processes and road management. In [17] authors adopted Artificial Neural Network (ANN) for predicting traffic mishaps harm severity, utilizing a dataset encompassing records of accidents. Initially comprising forty-eight attributes, counting target variable—a categorical variable with 4 classes: minor, moderate, severe, and death. The dataset underwent preprocessing, resulting in a reduction to sixteen features. The ANN classifier attained an accuracy rate of 0.74. [18] authors analysed traffic mishaps in Bangladesh, employing Machine Learning (ML) techniques on an incident dataset to assess accident severity.

Moreover, the researchers conducted a comprehensive comparative analysis, evaluating the outcomes obtained from four machine learning (ML) algorithms through two distinct experiments. In the initial experiment, the focus was on categorizing accident severity into four classes: "Fatal, Grievous, Simple Injury, Motor Collision." Subsequently, in the following experiment, the target variable underwent a transformation into "Fatal/Grievous," with Grievous representing the amalgamation of the last three classes. In their study, the authors [19] devised a robust framework based on six diverse machine learning algorithms: Decision Tree, Naïve Bayes, Random Forest, Logistic Regression, AdaBoost, and Bagging. The primary objective of this framework was to predict crash severity, effectively distinguishing between fatal, serious, or minor incidents. The intended implementation aimed to make meaningful contributions to road safety and traffic control by providing valuable insights for road authorities.

3   Methodology for Road Traffic Accident Predictions

This section will delve into the classifiers and dataset employed for Highway Severity Forecast. The projected method of this research work, illustrated in Fig. 1, outlines the data and workflow. The classifiers play a critical part during prediction of road accidents' severity, and their selection is pivotal in the effectiveness of the predictive models. Additionally, the dataset used in this study serves as the foundation for training and evaluating the machine learning algorithms. The clear depiction in Fig. 1 provides a visual representation of the structured approach adopted in this research, guiding the subsequent discussions on the outcomes and insights gained from the analysis.

3.1   Random Forest (RF)

Data mining technology encompasses various techniques such as association, prediction, classification, clustering, sequential pattern mining, and more [21]. Our work employs a random forest algorithm for prediction of traffic accidents' severity. The random forest algorithm, projected by Breiman [22], is a composite classifier that incorporates multiple decision trees. Known for its strengths in handling multidimensional data, the random forest stands out as one of the leading classification algorithms in current practice.

Historically, researchers have introduced various ensemble techniques, with bagging and boosting being prominent examples. Random Forest (RF), initially proposed by Breiman [23], is a notable instance. RF algorithms, involving the construction of N trees, iterate through four steps. Step 1 includes training the data using a bootstrap dataset, a subset of the original dataset. In Step 2, trees are generated, followed by Step 3 where attributes are randomly shortlisted. The final step, Step 4, determines the ultimate prediction through majority voting based on the results of the selected trees [24].

3.2 XGBoost Model

The XGBoost algorithm, is nothing but an artificial intelligence-integrated ML algorithm introduced by Chen et al. [25], offers notable advantages such as rapid parallelism, fault tolerance, controlled complexity and robust generalization capabilities. This algorithm iteratively synthesizes multiple decision tree models having low accuracy into a high-precision strong learner. It employs second-order Taylor expansion, and a regularization term is incorporated into the loss function to manage the model's complexity and diminish overfitting possibility.

3.3 Extra Tree Classifier

The Extra Tree Classifier (ETC) distinguishes itself by employing a random subset of features for splitting tree nodes, setting it apart from the Random Forest (RF) methodology. While RF builds trees using the complete dataset and randomly selects cut points for node splits, ETC takes a unique approach by incorporating a multi-linear approximation rather than RF's piecewise constant strategy. This heightened level of randomization in ETC not only underscores its superior performance compared to RF but also brings the added advantage of minimizing correlation among errors made by base learners. A study documented in [26] showcases ETC's remarkable accuracy, surpassing that of RF.

3.4 Evaluation Matrix

It can be calculated as the ratio of correct predictions to the total number of predictions.

$$Accuracy = \frac{No.\ of\ correct\ predictions}{Total\ no\ of\ input\ samples} \tag{1}$$

$$Accuracy = \frac{TP + TN}{TP + TN + FP + FN} \tag{2}$$

In this scenario, the symbols denoting true negative, true positive, false positive, and false negative are TN, TP, FN, and FP, respectively. As elucidated in [27], their interpretations are as follows.

- True Positives: Instances where our prediction is positive, and the actual output is also positive.

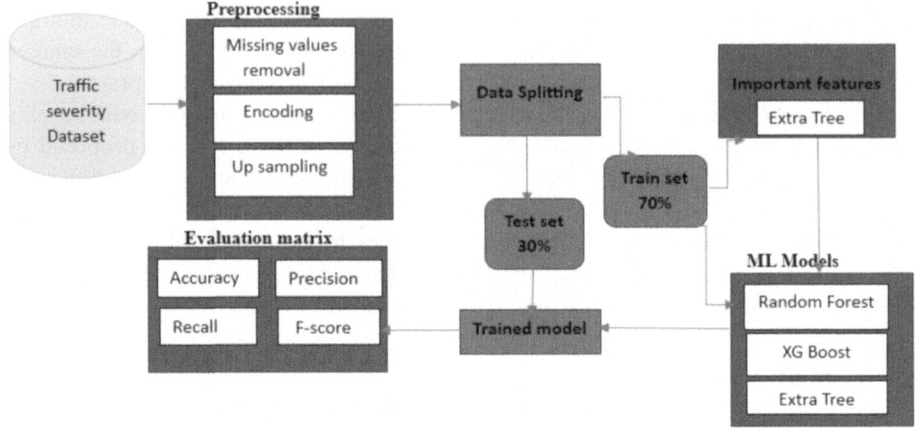

Fig. 1. Architecture of the proposed models

- True Negatives: Instances where our prediction is negative, and the actual output is also negative.
- False Positives: Instances where we predicted positive, but the actual output is negative.
- False Negatives: Instances where we predicted negative, but the actual output is positive.

3.5 F1 Score

The F1 score, as a machine learning evaluation metric, gauges a model's accuracy by incorporating both precision and recall scores.

$$F1 = 2 * \frac{1}{\frac{1}{precision} + \frac{1}{recall}} \tag{3}$$

3.6 Precision

Precision indicates the frequency with which an ML model is accurate in predicting the target class. It is calculated as the number of true positive predictions divided by the sum of true positive and false positive predictions

$$Precision = \frac{TP}{TP + FP} \tag{4}$$

3.7 Recall

Recall assesses the ratio of correctly identified actual positive labels by the model.

$$Recall = \frac{TP}{TP + FN} \tag{5}$$

4   Experiments and Results

In this segment, we showcase outcomes obtained from implementing ensemble models—RF, XGB and ETC—on the road accident dataset [28–30]. Figure 2 illustrates the significant variables identified by the Extra Trees classifier. In our study, out of the 32 original feature set variables, Extra Trees identified 10 crucial feature sets. Table 1 provides a comparative analysis of accuracy, precision, recall, and F1 score by juxtaposing the outcome of the ensemble models.

Table 1. Comparison results of all models.

Model	Accuracy	Precision	Recall	F
Extreme gradient boosting	79.68	76.17	79.68	77.64
Random Forest	79.7	76.39	79.7	77.69
Extra Tree	81.06	75.91	81.06	77.34

XGB   **Random Forest**

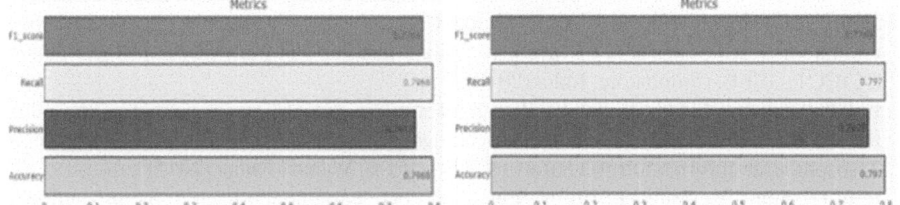

Extree

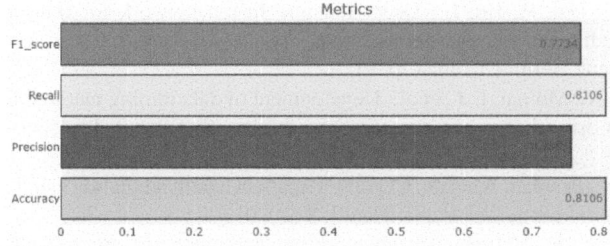

Fig. 2. Performance of the models

5   Conclusion

In conclusion, the escalating global incidence of road accidents, leading to millions of annual fatalities, poses substantial economic and commercial burdens on humanity. Prior researches predominantly framed prediction of road accident as a classification challenge, focusing on foreseeing whether an accident might occur without delving into the intricate interplay of contributing factors. While some studies have explored the significance of these factors concerning accident occurrence and severity, only a limited number have investigated a subset of ensemble machine learning (ML) models.

This paper contributes to the field by evaluating various ML models to predict road accident severity using a dedicated dataset. The study employs ensemble ML models like Random Forest, Extra Tree and Extreme Gradient Boosting to predict accidents of varying severity. The findings indicate that Extra Tree emerges as the most effective classifier, boasting an 81.06% accuracy, 75.91% precision, 81.06% recall, and a 77.34% F1-Score. Furthermore, the paper employs explainable ML techniques to assess the significance of road mishaps causative elements, enhancing the interpretability of the predictive models. This research underscores the potential of ensemble ML models, particularly Extra Tree, in enhancing the accuracy of road accident severity predictions and provides valuable insights into the underlying factors influencing these incidents.

References

1. Micheale, K.G.: Road trafc accident: human security perspective. Int. J. Peace and Develop. Stud. **8**(2), 15–24 (2017)
2. Bahiru, T.K., Singh, D.K., Tessfaw, E.A.: Comparative study on data mining classifcation algorithms for predicting road trafc accident severity. In: Proceedings of the 2018 Second International Conference on Inventive Communication and Computational Technologies (ICICCT). IEEE, Coimbatore, India (2018)
3. Lu, T., Dunyao, Z.H.U., Lixin, Y., Pan, Z.: Te trafc accident hotspot prediction: based on the logistic regression method. In: Proceedings of the 2015 International Conference on Transportation Information and Safety (ICTIS). IEEE, Wuhan, China (2015)
4. Ren, H., Song, Y., Wang, J., Hu, Y., Lei, J.: A deep learning approach to the prediction of short-term trafc accident risk (2017). https://arxiv.org/abs/1710.09543
5. Ren, H., Song, Y., Wang, J., Hu, Y., Lei, J.: A deep learning approach to the citywide trafc accident risk prediction. In: Proceedings of the 2018 21st International Conference on Intelligent Transportation Systems (ITSC). IEEE, Maui, HI, USA (2018)
6. Huang, C., Zhang, C., Dai, P., Bo, L.: Deep dynamic fusion network for trafc accident forecasting. In: Proceedings of the 28th ACM International Conference on Information and Knowledge Management. Beijing, China (2019)
7. Murphey, Y.L., Wang, K., Molnar, L.J., et al.: Development of data mining methodologies to advance knowledge of driver behaviors in naturalistic driving. SAE Int. J. Transport. Safety **8**(2), 77–94 (2020)
8. Iwendi, C., Bashir, A.K., Peshkar, A., et al.: COVID-19 patient health prediction using boosted random forest algorithm. Frontiers in Public Health **8** (2020)
9. Ding, J., Dai, Q., Fan, W., et al.: Impacts of meteorology and precursor emission change on O3 variation in Tianjin, China from 2015 to 2021. J. Environ. Sci. **126**, 506–516 (2023)
10. Schonlau, M., Zou, R.Y.: Te random forest algorithm for statistical learning. STATA Journal **20**(1), 3–29 (2020)
11. Yan, M., Shen, Y.: Trafc accident severity prediction based on random forest. Sustainability **14**(3) (2022)
12. Zhao, H., Li, X., Cheng, H., Zhang, J., Wang, Q., Zhu, H.: Deep learning-based prediction of trafc accidents risk for Internet of vehicles. China Communications **19**(2), 214–224 (2022)

13. Chen, M.-M., Chen, M.-C.: Modeling road accident severity with comparisons of logistic regression, decision tree and random forest. Information **11**(5) (2020)
14. Koma, H., Harada, T., Yoshizawa, A., Iwasaki, H.: Detecting cognitive distraction using random forest by considering eye movement type. Int. J. Cognit. Info. Natu. Intell. **11**(1), 16–28 (2017)
15. Wang, L., Wu, J., Li, R., et al.: A weight assignment algorithm for incomplete trafc information road based on fuzzy random forest method. Symmetry **13**(9) (2021)
16. Manzoor, M., et al.: RFCNN: traffic accident severity prediction based on decision level fusion of machine and deep learning model. IEEE Access **9**, 128359–128371 (2021). https://doi.org/10.1109/ACCESS.2021.3112546
17. Alkheder, S., Taamneh, M., Taamneh, S.: Severity prediction of traffic accident using an artificial neural network: traffic accident severity prediction using artificial neural network. J. Forecasting **36** (2016). https://doi.org/10.1002/for.2425
18. Labib, Md.F., Rifat, A.S., Hossain, Md.M., Das, A.K., Nawrine, F.: Road accident analysis and prediction of accident severity by using machine learning in Bangladesh. In: 2019 7th International Conference on Smart Computing & Communications (ICSCC), pp. 1–5. Sarawak, Malaysia, Malaysia (2019). https://doi.org/10.1109/ICSCC.2019.8843640
19. Malik, S., El Sayed, H., Khan, M.A., Khan, M.J.: Road accident severity prediction — a comparative analysis of machine learning algorithms. In: 2021 IEEE Global Conference on Artificial Intelligence and Internet of Things (GCAIoT), pp. 69–74 (2021). https://doi.org/10.1109/GCAIoT53516.2021.9693055
20. Sharma, B., Katiyar, V.K., Kumar, K.: Traffic accident prediction model using support vector machines with gaussian kernel. In: Proceedings of Fifth International Conference on Soft Computing for Problem Solving, pp. 1–10. Singapore (2016). https://doi.org/10.1007/978-981-10-0451-3_1
21. Hussain, S.: Survey on current trends and techniques of data mining research. London J. Res. Comp. Sci. Technol. **17**(1) (2017)
22. Breiman, L.: Random forests. Mach. Learn. **45**(1), 5–32 (2001)
23. Breiman, L.: Random forests. Machine Learning **45**(1), 5–32 (2001)
24. Sekhar, C.R., Madhu, E., et al.: Mode choice analysis using random forrest decision trees. Transportation Research Procedia **17**, 644–652 (2016)
25. Chen, T., Guestrin, C.: Xgboost: a scalable tree boosting system. In: Proceedings of the 22nd Acm Sigkdd International Conference on Knowledge Discovery and Data Mining, pp. 785–794. San Francisco, CA, USA (2016)
26. Geurts, P., Ernst, D., Wehenkel, L.: Extremely randomized trees. Machine Learning **63**(1), 3–42 (2006)
27. Umer, M., Ashraf, I., Mehmood, A., Ullah, D.S., Choi, G.S.: Predicting numeric ratings for google apps using text features and ensemble learning. ETRI J. (2020). https://doi.org/10.4218/etrij.2019-0443
28. Bedane, T.T.: Road Traffic Accident Dataset of Addis Ababa City. Addis Ababa Science and Technology University, Addis Ababa (2020)
29. Jaroli, P., Singla, C., Bhardwaj, V., Mohapatra, S.K.: Deep learning model based novel semantic analysis. In: 2022 2nd International Conference on Advance Computing and Innovative Technologies in Engineering (ICACITE), pp. 1454–1458. IEEE (2022)
30. Bhardwaj, V., Rahul, K.V., Kumar, M., Lamba, V.: Analysis and prediction of stock market movements using machine learning. In: 2022 4th International Conference on Inventive Research in Computing Applications (ICIRCA), pp. 946–950. IEEE (2022)

Survey and Analysis of Machine Learning Methods for Parkinson's Disease Diagnosis

Poonam Yadav[1], Meenu Vijarania[1], Meenakshi Malik[2(✉)], and Ritu[1]

[1] K.R. Mangalam University, Gurugram, India
[2] BML Munjal University, Gurugram, India
meenakshimalik16@gmail.com

Abstract. Parkinson's disease (PD) is a debilitating state that hinders quality of life. It results from the demise of dopamine-generating cells across the region of substantia nigra of the central nervous system (CNS), that has an impact on the body. Individuals with Parkinson's disease experience difficulty speaking, writing, and walking. To distinguish between individuals with Parkinson's disease and those in good health, various machine learning-based approaches are employed. A thorough assessment of methods based on machine learning for Parkinson disease prediction is presented in this work. With an emphasis on the application of research findings in real-world clinical settings, this review critically evaluates the usefulness of machine learning models regarding the practical identification of Parkinson's disease. In this work, literature analysis was carried out on published papers in past years utilizing the PubMed, University of California, Irvine (UCI) repository, and Xplore databases on Institute of Electrical and Electronics Engineers (IEEE) offer an extensive summary among the techniques for machine learning and data modalities used in differential diagnosis of Parkinson's disease. We have proposed methodology also based on our study.

Keywords: Parkinson's disease · machine learning · datasets

1 Introduction

Parkinson's disease (PD) is a neurological disorder which diminishes the capacity to direct motion. Several different motor and non-motor symptoms are present, and it normally progresses gradually over time. Resting tremors, rigidity of the muscles, and bradykinesia (slowness of movement) are the hallmarks of Parkinson's disease. Along with these neurological symptoms and non-motor signs and symptoms like sadness, altered cognition, and sleep disruptions are frequently present.

Sixty percent of those over 50 have Parkinson's disease. Speech and movement issues are common in patients with Parkinson's (PWP) disease [1]. In Parkinson's disease (PD) patients, dysarthria is also noted. It is typified by fragility, immobility, and inability to synchronize within the speech-motor system, which impacts breathing, phonetic transcription, expression, and prosody [2].

Parkinson's disease is generally brought on by a confluence of environmental triggers, including exposure to toxins, oxidative stress, and mitochondrial malfunction,

A. K. Bairwa et al. (Eds.): ICCAIML 2024, CCIS 2184, pp. 226–238, 2025.
https://doi.org/10.1007/978-3-031-71481-8_18

as well as hereditary predisposition. This degeneration of dopamine-producing brain cells is what ultimately causes Parkinson's disease. Indications of Parkinson's disease encompass an assortment of both motor and non-motor signs, providing a comprehensive picture of the diverse manifestations associated with the condition as shown in Fig. 1.

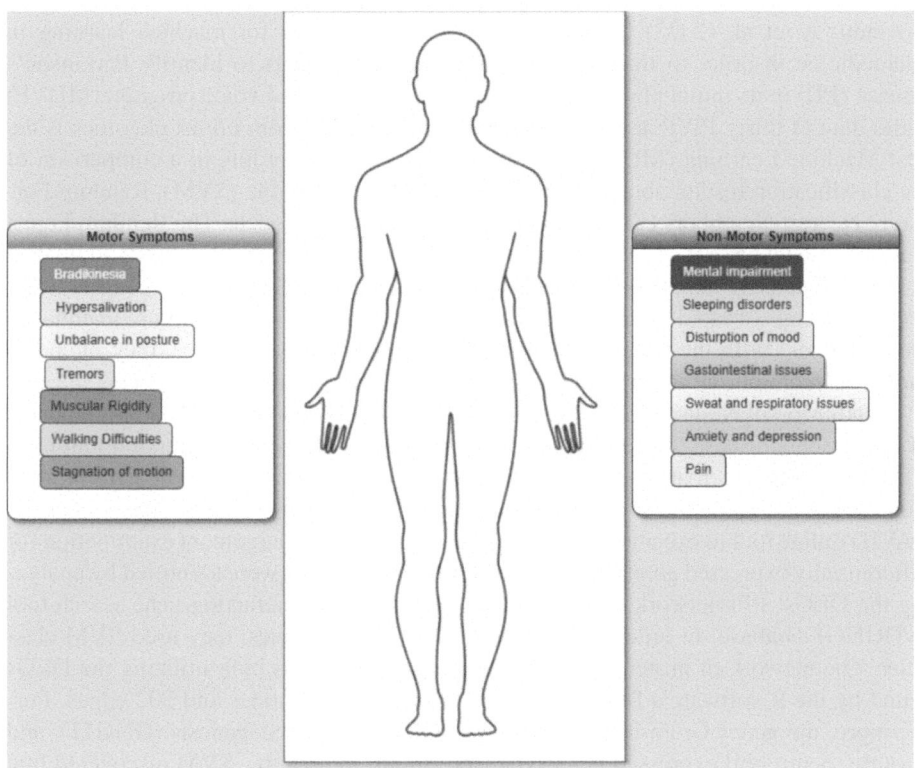

Fig. 1. Symptoms of Parkinson's disease (two types of signs- motor and non-motor)

Since Parkinson's disease cannot be cured, there is an urgent need for continued research as well as care for those who are affected and their families. While there are medications to control symptoms and enhance quality of life, finding a cure remains a major goal of medical research worldwide. The signs of Parkinson's disease are often managed and controlled by medication, but for the purpose of managing and controlling symptoms, additional therapies and treatments such as employment counselling, physical rehabilitation, and speech-language therapy may also be beneficial [3]. Approximately 60,000 new clinical diagnoses of Parkinson's disease (PD) are made each year, and the neurodegenerative illness affects one million people in the United States of America (US) [4].

Parkinson's disease can be effectively managed earlier thanks to early detection, which gives patients and healthcare professionals more power. It opens the door to a variety of treatments, interventions, and assistance programs that jointly enhance patients' functional capacities and well-being while also possibly lessening the social and financial burden brought on by the disease's later stages.

2 Literature Review

Govindu, A. et al. (2023) [5] have presented technologies for machine learning in telemedicine in order to train four machine learning models to identify Parkinson's disease (PD) in its initial phases. They used multi-dimensional voice program (MDVP) audio data of thirty PWP and healthy individuals. The Random Forest classifier is the best Machine Learning (ML) method for PD detection, according to a comparison of the classification results obtained from Support Vector Machine (SVM), Random Forest, K-Nearest Neighbors (KNN), and Logistic Regression models. The Random Forest classifier model has a 0.95 sensitivity and a 91.83% detection accuracy. Since audio data is insufficient as a biomarker for Parkinson's disease categorization, they suggest using both recordings of rapid eye movement (REM) sleep and audio in the future to enhance the outcomes. With the usage of audio recordings on a mobile device, they intend to further telecommuting in medicine classification of Parkinson's disease (PD).

Shadi Moradi et al. (2022) [6] purposed this study to find genes related to the onset of Parkinson's disease. For additional study, they used the Gene Expression Omnibus (GEO)-provided microarray dataset (GSE22491). To analyze and evaluate gene expression and find DEGs, the R software's Limma package was utilized. They used the DAVID online tool to execute the KEGG pathway and GO enlargement examination for differentially expressed genes (DEGs). In addition, hub genes were identified by analyzing the DEGs' PPI network with Cytoscape and the gene-interacting-gene search tool (STRING) database. In order to forecast the precision of genes, they used SVM classifier. The network of protein-protein interactions (PPIs) was built utilizing the DEGs found by the R software's Limma package, and it had 264 nodes and 502 edges. Furthermore, the genes Gram-negative bacteria (GNB5), gluconeogenesis (GNG11), and elastase, neutrophil expressed (ELANE) were found to be hubs. SVM discovered that an 88% prediction accuracy may be attained by combining three genes. The three hub genes GNB5, GNG11, and ELANE may be employed as PD biomarkers for detection, per the results of the study. Furthermore, As stated by authors, the highly accurate SVM conclusions may be used as PD biomarkers.

Mei J. et al. (2021) [7] examined materials published up until February 14, 2020, in a study of the literature conducted utilizing the IEEE Xplore and PubMed databases. Before being included, selected for relevant content, and presented in this review, a total of 209 papers were reviewed for their goals, sources of data, data types, machine learning approaches, and related outcomes. This study demonstrates a great deal of promise for using cutting-edge machine learning methods and biomarkers to clinical judgment, which will result in a more methodical and knowledgeable confirmation of Parkinson's disease.

T. J. Wroge et al. (2018) [4] suggested automated speech analysis to validate a physician's diagnosis because they rely on the patient's self-report of their diagnosis in this case. They investigate how well supervised classification algorithms similar to a deep neural network work for correctly diagnosing patients with the illness. When using a pathological post-mortem analysis as the reference point, their maximum accuracy was 85% produced by the machine learning models surpasses the non-experts' average precision of clinical diagnosis was 73.8%. as well movement disorder specialists' average precision (79.6% without follow-up, 83.9% after follow-up).

W. Wang et al. (2020) [8] suggested using a deep learning model to automatically distinguish between healthy individuals and those with Parkinson's disease (PD) based on premotor characteristics (such as olfactory loss, Disorder of sleep behavior (RBD), and rapid eye movement (REM). The suggested model achieved an accuracy of 96.45%, demonstrating high detecting capability. Their findings demonstrated that, when it came to differentiating between patients with Parkinson's illness as well as normal individuals, the deep learning model that was built performed better at detection than the twelve machine learning models that were taken into consideration. The writers claim that deep learning outperforms the others. This is as a result of the fact that they created deep learning using tiny PD data sets gathered from 584 people (401 early PD cases and 183 healthy cases). Nonetheless, as time goes on and data becomes larger and more complex, deep learning should be able to prove its mettle.

Zehra Karapinar Senturk (2020) [9] suggested diagnosing Parkinson's illness using machine learning. The author's method involves the stages of feature selection and classification. The feature selection task took into account the Recursive Feature Elimination and Feature Importance approaches. In the tests, Parkinson's patients were classified using Support vector machines, artificial neural networks, and regression trees. It was demonstrated that Recursive Feature Elimination with Support Vector Machines outperformed the other techniques. The lowest number of vocal features required to identify Parkinson's illness resulted in 93.84% accuracy.

Almeida, J. S. et al. (2019) [10] looks at how voice sounds are processed in order to identify Parkinson's disease. Their methodology assesses the application of four machine learning approaches and eighteen feature extraction strategies for the classification of data from continuous phonation and speech activities. The analysis of the classification performance was done using five metrics: Accuracy, Specificity, Sensitivity, and Equal Error Rate (EER) and Area Under Curve (AUC) values from curves representing characteristics of the receiver operation and detection error trade-off (DET). They evaluate this method against alternative methods utilizing identical data sets. They demonstrate that phonation tasks were superior to speech tasks in terms of disease detection efficiency. The accuracy, AUC, and EER of the acoustic cardioid (AC) channel with the best performance were 94.55%, 0.87, and 19.01%, respectively. We have obtained 92.94% accuracy, 0.92 AUC, and 14.15% EER when utilizing the smartphone (SP) channel.

Thakur, K. et al. (2023) [11] found that one of the disease's common symptoms is voice deterioration. In order to diagnose Parkinson's disease and assess performance, the authors of this research build and contrast cutting-edge machine learning techniques like Random Forest and Support Vector Machine, Decision Tree Classifier, and Extra Trees Classifier. Their approach, which outperformed all other models for Parkinson's disease prediction, suggests combining the Extra Trees Classifier with entropy-based feature selection. The greatest results are 94.34% accuracy, 0.9388 precision, 0.9684 F1-score, and 1 recall value.

Ranjan, N. M. et al. (2023) [12] suggested a system that uses wave and spiral drawings of both healthy and Parkinson's disease patients as input and two different forms of handwriting analysis. They employed the oriented gradient histogram for feature extraction. The method that was created employed a random forest classifier in conjunction with a machine learning algorithm to identify Parkinson's disease in patients. With spiral drawing, their model's accuracy was 86.67%, while with wave drawing, it was 83.30%.

Alalayah KM et al. (2023) [13] suggested innovative methods for enhancing the methods for PD early diagnosis through the assessment of particular aspects and ML algorithm hyperparameter tuning for PD diagnosis based on voice abnormalities. The recursive feature elimination (RFE) approach sorted features in light of their impact on the desired attribute, while the synthetic minority oversampling technique (SMOTE) balanced the dataset. To lower the dataset's dimensions, they utilized principal component analysis (PCA) and t-distributed stochastic neighbor embedding (t-SNE). Support vector machines (SVM), random forests (RF), decision trees (DT), K-nearest neighbors (KNN), as well as multilayer perception (MLP) are examples of classifiers were ultimately fed the final characteristics using PCA and t-SNE. Furthermore, MLP using the PCA algorithm produced results with an F1-score of 96.66%, 98% accuracy and 96.66% precision and recall of 96%.

J. Divya et al. (2023) [14] discovered that to ascertain if an individual has Parkinson's disease or not, it is straightforward for implementing a system of machine learning that evaluates the variations in their speech pattern. One of the possible outcomes of the author's algorithm's regular use of Extreme Gradient Boosting (XGBoost) classifiers and parameter extraction is the audio featured dataset, which can be found in the UCI dataset database. In terms of assessing the palladium patient's state of health, the algorithm has produced a result that is significantly more precise. XGBoost produced the maximum level of precision that could be attained.

Yuan, L. et al. (2023) [15] used 252 participants' speech sounds as the database. To enhance the accuracy of Parkinson's disease (PD) classification, this study employed machine learning algorithms with linguistic signal data as input. The generated classifiers were then merged. The testing results showed that these machine learning algorithms might achieve up to 95% diagnosis accuracy. Furthermore, a clinical experience-based feature extraction method for subject language signal analysis was presented.

T. Wasif et al. (2021) [16] used five distinct classifiers. The objective of this research is to optimize and improve the feature employed in the dataset. For optimization, they combined the wrapper and filter approaches into a hybrid feature selection method. Recursive Feature Elimination is the name of the technique. XGBoost achieved the greatest accuracy of 97.43% using nine out of 22 features in its classifier.

Vinora, A. et al. (2023) [17] provided a brief evaluation of the best available at the time and suggested prospective directions for future study. Since tremor classification and detection are essential to identify and treatment of PD patients, machine learning techniques can be applied in this situation. One of the most prevalent movement disorders in clinical practice, it is commonly classified based on behavioral or etiological characteristics. Determining and assessing gait patterns associated with Parkinson's disease (PD), such as gait start and freezing, which are indicative of PD symptoms, is another crucial difficulty. Since vocal impairment affects 90% of Parkinson's disease patients, speech data analysis is essential to distinguishing the condition from healthy persons.

Yadav, D. et al. (2022) [18] discovered that it impacts a person's speech in the early stages. Six machine learning models were compared using six measurements (recall, accuracy, precision, F1-score, false positive rate, and area under the Receiver Operating Characteristic (ROC) curve): Decision trees, Support Vector Machine, Random Forest, Logistic Regression, K Nearest Neighbors, as well as Gaussian Naïve Bayes.

Johri, A. et al. (2019) [19] presented the VGFR Spectrogram Detector and Voice Impairment Classifier, two models based on neural networks designed to aid medical professionals and patients in early disease diagnosis. Convolutional neural networks, or CNNs, have undergone a thorough empirical examination in order to forecast disease. Deep dense artificial neural networks, or ANNs, have been utilized for voice recordings and substantial image categorization of gait information converted into spectrogram pictures. While the accuracy of the Voice Impairment Classifier is 89.15% and the classification accuracy recorded by the Spectrogram Detector (VGFR) is 88.1%.

3 Decoding the Frameworks: An In-Depth Examination of Machine Learning Methodologies

3.1 Support Vector Machines

A supervised machine learning method called Support Vector Machine (SVM), is employed for both regression as well as classification. It's a technique that works with both non-linear and linear data [20]. Finding the optimal hyperplane in an N-dimensional space to partition data points into different feature space classes are the primary objective of the SVM method. As shown in Fig. 2, SVM chooses the extreme points and vectors to aid in the creation of the hyperplane. The hyperplane aims to maximize the margin between the closest points of various classes. Its dimension is determined by the quantity of attributes; if there are two input aspects, it is a line; if there are three input features, it is a two-dimensional (2-D) plane. Using features taken from clinical evaluations and voice recordings, among other medical data, supervised learning model (SVM) identify patients as positive or negative for Parkinson's disease.

3.2 Decision Trees

The decision tree (DT) is best supervised learning methods for classification and regression applications. A decision tree is a tree-structured classifier in which the outcomes are represented by each leaf node, the inside nodes reflect the characteristics of a dataset, and the branches indicate the decision rule. The "highest decision" characteristic, which forms the tree's root node at first, is the crucial component in the DT creation process [14]. It is constructed by continuously splitting the training data into subsets based on the attribute values, continuing this process until a halting condition is met, such as the minimal number of samples required to divide a node or the maximum tree depth. A measurement, such as Gini impurity or entropy, which assesses the level of impurity or unpredictability within the subsets, is employed by the Decision Tree approach to

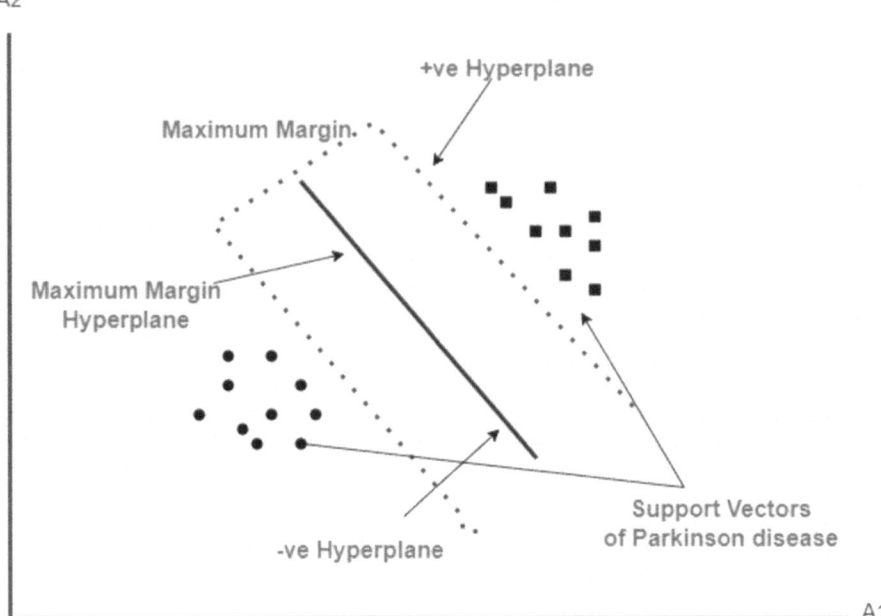

Fig. 2. Support Vector Machine (audio recordings used as test and training data)

ascertain the ideal attribute to divide the data in training. Formula to calculate entropy is:

$$\text{Entropy}(S) = -\text{Prob}(Y)\log_2\text{Prob}(Y) - \text{Prob}(N)\log_2\text{Prob}(N) \tag{1}$$

where;

S = Total no of samples

Prob(Y) = Probability of Y

Prob(N) = Probability of N

The goal is to determine which property best maximizes the information gain or the impurity decrease after the split. This model helps to categorize individuals into groups based on whether patient have Parkinson's disease or not.

3.3 Random Forest

The Random Forest relies on ensemble learning, aggregating the output of multiple classifiers and selecting a candidate based on a majority vote [21]. The random forest's output for classification tasks is the class that the majority among the trees chose. In regression tasks, the average prediction, or mean, is given back for each individual tree. It is used to find patterns suggestive of Parkinson's disease by analyzing a variety of datasets, including those obtained from speech and gait studies. Figure 3 shows the construction of multiple decision trees during training to create the class mode for classification or the mean prediction for regression, known for its robustness and high predictive accuracy.

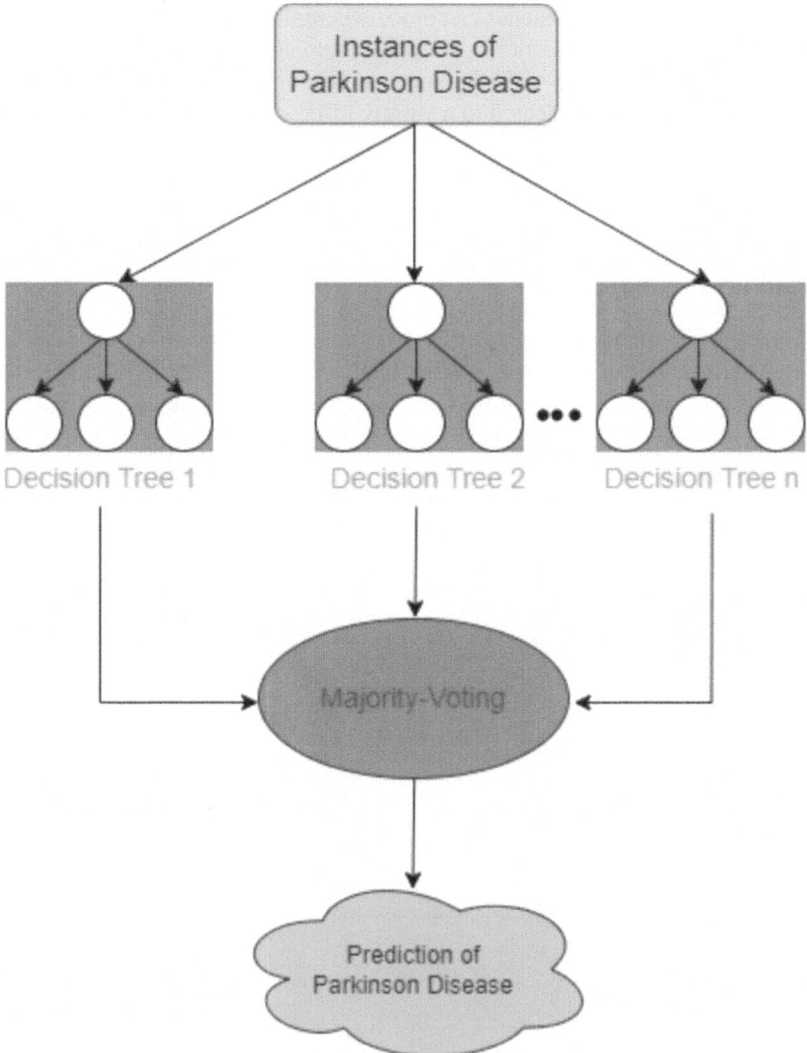

Fig. 3. Random Forest Algorithm used numerous decision trees for classification or regression

3.4 Neural Networks

The basis of deep learning [26] techniques is neural networks. Additionally, they are mentioned to as artificial neural networks (ANNs) or simulated neural networks (SNNs). One portion of machine learning is neural networks. Neural networks are made to resemble the composition and capabilities of the human brain. by modeling them accordingly. The effective node layers of the neural network are composed of an output layer, a single or several hidden levels, and an input layer. Every node in this network links to other nodes and has a threshold and weight assigned to it. Artificial neural systems and other in-depth education approaches can process complex data, including genetic information and brain imaging Functional Magnetic Resonance Imaging (fMRI), Magnetic Resonance Imaging (MRI) etc. to find minute patterns or signs linked to Parkinson's disease. Patient data may be subjected to a variety of procedures, including examination, segmentation, augmentation, scaling, normalization, sampling, aggregation, and sifting, to be able to produce precise predictions which, benefit the healthcare system and its interested parties [22].

3.5 Time Series Analysis

A machine learning method called time series forecasts target values only from known past target values. In the literature, this particular type of regression is referred to as auto-regressive modeling. Time series in Fig. 4 refers to a sequence of data points indexed in chronological order, typically at uniform intervals, allowing for the analysis of trends, patterns, and behaviors over time, often applied in various fields such as finance, economics, and environmental science. In order to monitor changes in motor performance over time, temporal data, such as sensor readings from wearable devices, can be analyzed using time series analysis techniques like autoregressive modelling or recurrent neural networks.

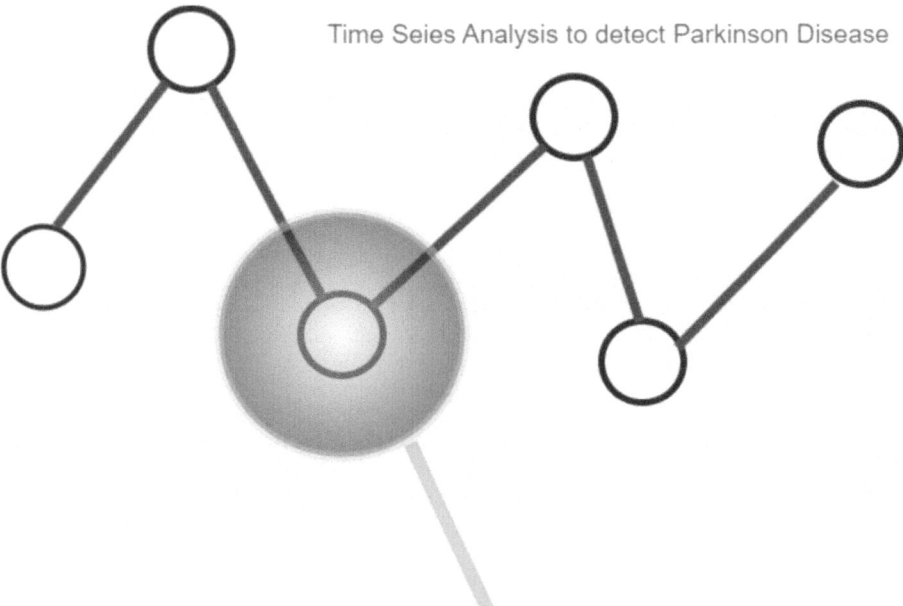

Fig. 4. Time series analysis for Parkinson's disease diagnosis

3.6 Dimensionality Reduction

It is a technique for transforming higher dimensional datasets into lower dimensional datasets while maintaining information similarity. This method is frequently applied in machine learning to solve regression and classification problems and produce a more well-fitting predictive model. It is the process of converting high-dimensional data while maintaining the necessary components of the source data into a lower-dimensional context. An unsupervised method for reducing dimensionality that might expedite training and testing is Principal Component Analysis (PCA) [23]. PCA can facilitate the identification of important elements in the diagnosis of Parkinson's illness.

3.7 Ensemble Learning

Multiple machine learning models are combined into a single model through ensemble learning. The model's performance is to be improved. While stacking seeks to increase prediction accuracy, bagging aims to lower variation., and boosting aims to lessen prejudice. The predictions from various models are counted independently. As a result, the forecast produced by the majority of models is taken as final. Higher accuracy is achieved by using ensemble approaches, which capitalize on the diversity and complementary features of individual models [24].

3.8 Cross-Validation

Cross validation is a machine learning technique that evaluates a model's performance on missing data. A number of folds, or subsets, are created from the given data; the model is trained on one-fold and validated on the other folds. This process here is carried out more than once, along with a new fold acting as the constant validation set. The results of each validation phase are averaged at the conclusion to get a more trustworthy assessment of the model's performance. To evaluate how well machine learning [27] models generalize to the data and make sure they are not overfitting, cross-validation techniques like the k-fold cross-validation method are crucial. The outcomes are validated using the cross-validation technique [25].

4 Proposed Methodology

Our proposed methodology is based on a thorough analysis of prior research on machine learning-based Parkinson's disease detection, with a focus on dimensionality reduction and feature refinement techniques. By incorporating well-known algorithms such as Gradient Boosting and Support Vector Machines, we hope to enhance the interpretability and the model's accuracy. Furthermore, motivated by effective group tactics, our methodology investigates the combination of complimentary models to establish a strong foundation for precise and timely identification of Parkinson's disease. The accompanying Fig. 5 provides a visual representation of the proposed methodology, highlighting the steps and connections that need to be adhered to in order to implement the analysis-based approach. The Parkinson's disease dataset used in this flowchart is sourced from

PubMed and the UCI repository. The dataset includes handwriting, neuroimaging, voice recording, gait, and Parkinson's disease. The data is preprocessed using feature extraction and feature selection techniques. SVM, Random Forest, and ANN models etc. were chosen, and three phases of training—training, testing, and validation—were conducted using various metrics. Optimizing a performance of the model without overfitting or generating an excessively high variance is referred to as tuning. In machine learning, this is achieved by choosing suitable "hyperparameters". Selecting the right set of hyperparameters can be computationally demanding, but it is essential for model accuracy. In contrast to other model parameters, the model does not learn hyperparameters on its own via training methods. These parameters must be manually set instead. There are numerous ways to choose the right hyperparameters. By employing all-encompassing approach, we hope to further the progress of precise and effective machine learning-based Parkinson's disease diagnosis. Below is the diagram of proposed methodology based on analysis.

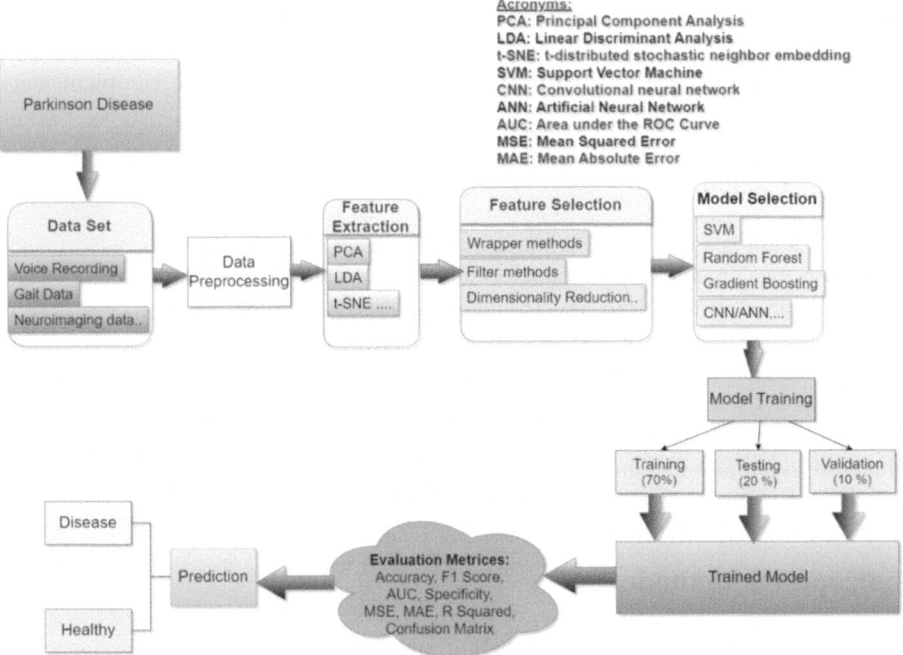

Fig. 5. Proposed methodology based on detailed review

5 Conclusion

Machine learning-based Parkinson disease prediction model was thoroughly reviewed in this research. A succinct overview of the several computational intelligence-based methods for Parkinson disease prediction is provided. In order to forecast Parkinson illnesses, a synthesis of the findings from numerous researchers is also provided in the literature. In our next research project, we hope to introduce a hybrid model that combines the advantages of several algorithms, providing a more thorough and reliable approach to early Parkinson's disease detection by utilizing the lessons learned from successful frameworks in the literature.

References

1. Prabhavathi, K., Patil, S.: Tremors and Bradykinesia. Techniques for Assessment of Parkinsonism for Diagnosis and Rehabilitation, 135–149 (2022)
2. Fraser, K.C., Meltzer, J.A., Rudzicz, F.: Linguistic features identify Alzheimer's disease in narrative speech. J. Alzheimers Dis. **49**(2), 407–422 (2016)
3. Heisters, D.: Parkinson's: symptoms, treatments and research. British Journal of Nursing **20**(9), 548–554 (2011)
4. Wroge, T.J., et al.: Parkinson's disease diagnosis using machine learning and voice. In: 2018 IEEE signal processing in medicine and biology symposium (SPMB), pp. 1–7. IEEE (2018)
5. Govindu, A., Palwe, S.: Early detection of Parkinson's disease using machine learning. Procedia Comp. Sci. **218**, 249–261 (2023)
6. Moradi, S., Tapak, L., Afshar, S.: Identification of novel noninvasive diagnostics biomarkers in the Parkinson's diseases and improving the disease classification using support vector machine. BioMed Research International 2022 (2022)
7. Mei, J., Desrosiers, C., Frasnelli, J.: Machine learning for the diagnosis of Parkinson's disease: a review of literature. Frontiers in Aging Neuroscience **13**, 633752 (2021)
8. Wang, W., Lee, J., Harrou, F., Sun, Y.: Early detection of Parkinson's disease using deep learning and machine learning. IEEE Access **8**, 147635–147646 (2020)
9. Senturk, Z.K.: Early diagnosis of Parkinson's disease using machine learning algorithms. Med. Hypotheses **138**, 109603 (2020)
10. Almeida, J.S., et al.: Detecting Parkinson's disease with sustained phonation and speech signals using machine learning techniques. Pattern Recogn. Lett. **125**, 55–62 (2019)
11. Thakur, K., Kapoor, D.S., Singh, K.J., Sharma, A., Malhotra, J.: Diagnosis of parkinson's disease using machine learning algorithms. In: Congress on Intelligent Systems, pp. 205–217. Springer Nature Singapore, Singapore (2022)
12. Ranjan, N.M., Mate, G., Bembde, M.: Detection of parkinson's disease using machine learning algorithms and handwriting analysis. J. Data Mining and Manage **8**(1), 21–29 (2023). e-ISSN: 2456-9437
13. Alalayah, K.M., Senan, E.M., Atlam, H.F., Ahmed, I.A., Shatnawi, H.S.A.: Automatic and early detection of parkinson's disease by analyzing acoustic signals using classification algorithms based on recursive feature elimination method. Diagnostics **13**(11), 1924 (2023)
14. Divya, J., et al.: Detection of parkinson disease using machine learning. In: 2023 International Conference on Inventive Computation Technologies (ICICT), pp. 53–57. IEEE (2023)
15. Yuan, L., Liu, Y., Feng, H.M.: Parkinson disease prediction using machine learning-based features from speech signal. Ser. Orient. Comp. Appl. 1–7 (2023)
16. Wasif, T., Hossain, M.I.U., Mahmud, A.: Parkinson disease prediction using feature selection technique in machine learning. In: 2021 12th International Conference on Computing Communication and Networking Technologies (ICCCNT), pp. 1–5. IEEE (2021)
17. Vinora, A., Ajitha, E., Sivakarthi, G.: Detecting parkinson's disease using machine learning. In: 2023 International Conference on Artificial Intelligence and Knowledge Discovery in Concurrent Engineering (ICECONF), pp. 1–6. IEEE (2023)

18. Yadav, D., Jain, I.: Comparative analysis of machine learning algorithms for parkinson's disease prediction. In: 2022 6th International Conference on Intelligent Computing and Control Systems (ICICCS), pp. 1334–1339. IEEE (2022)
19. Johri, A., Tripathi, A.: Parkinson disease detection using deep neural networks. In: 2019 Twelfth international conference on contemporary computing (IC3), pp. 1–4. IEEE (2019)
20. Bind, S., et al.: A survey of machine learning based approaches for Parkinson disease prediction. Int. J. Comput. Sci. Inf. Technol **6**(2), 1648–1655 (2015)
21. Ahmed, I.A., Senan, E.M., Shatnawi, H.S.A., Alkhraisha, Z.M., Al-Azzam, M.M.A.: Hybrid techniques for the diagnosis of acute lymphoblastic leukemia based on fusion of CNN features. Diagnostics **13**(6), 1026 (2023)
22. Alzubaidi, M.S., et al.: The role of neural network for the detection of Parkinson's disease: a scoping review. In: Healthcare. Vol. 9(6), p. 740. MDPI (2021)
23. Rao, D.V., Sucharitha, Y., Venkatesh, D., Mahamthy, K., Yasin, S.M.: Diagnosis of parkinson's disease using principal component analysis and machine learning algorithms with vocal features. In: 2022 International Conference on Sustainable Computing and Data Communication Systems (ICSCDS), pp. 200–206. IEEE (2022)
24. Ali, A.M., Salim, F., Saeed, F.: Parkinson's disease detection using filter feature selection and a genetic algorithm with ensemble learning. Diagnostics **13**(17), 2816 (2023)
25. Lamba, R., Gulati, T., Al-Dhlan, K.A., Jain, A.: A systematic approach to diagnose Parkinson's disease through kinematic features extracted from handwritten drawings. J. Reliab. Intell. Environ. 1–10 (2021)
26. Jaroli, P., Singla, C., Bhardwaj, V., Mohapatra, S.K.: Deep learning model based novel semantic analysis. In: 2022 2nd International Conference on Advance Computing and Innovative Technologies in Engineering (ICACITE), pp. 1454–1458. IEEE (2022)
27. Bhardwaj, V., Rahul, K.V., Kumar, M., Lamba, V.: Analysis and prediction of stock market movements using machine learning. In: 2022 4th International Conference on Inventive Research in Computing Applications (ICIRCA), pp. 946–950. IEEE (2022)

Accelerated GPU-Based Clustering, Classification and Regression Using RAPIDS: A Comparative Study

Mayuri Gupta[1]([✉]) [iD], Ashish Mishra[1] [iD], Ashutosh Mishra[2] [iD], and Jayesh Gangrade[3] [iD]

[1] Jaypee Institute of Information Technology, Noida, India
mayurigupta2010@gmail.com, ashishmishra81@gmail.com
[2] Thapar University, Patiala, Punjab, India
ashutosh.mishra@thapar.edu
[3] Manipal University Jaipur, Jaipur, India
jgangrade@gmail.com

Abstract. Machine learning has become increasingly popular as big data and data science applications have grown in demand over the previous decade. GPUs, on the other hand, are ideally suitable for Machine learning challenges due to their tremendous performance. NVIDIA has released the RAPIDS framework, which includes the cuML library, to take benefit from GPU performance for machine learning. A GPU-based framework such as RAPIDS provides fast results on large datasets, whereas a CPU-based framework takes more time. In comparison to the CPU, there is no library other than RAPIDS that provides fast computation speed on the GPU. The purpose of this paper is to compare RAPIDS and Sci-kit clustering, regression, and classification algorithms. As a result, K-means, DBSCAN, Linear regression, Lasso regression, Ridge regression, Random Forest, Nearest Neighbor, and KNearest Neighbor algorithms are used on GPUs and CPUs, and the results are determined by calculating the time taken for execution and speed up. The results indicate that the RAPIDS framework can completely harness the power of the GPU compared to the CPU-based Sci-kit library.

Keywords: Machine Learning · GPU · NVIDIA · K-Means · cuML · CPU · RAPIDS · Sci-kit

1 Introduction

A dataset has been developed from different sources that have spread at an unprecedented rate over the prior decade. The processing of huge quantities of data in a timely way to generate important features and drive decision-making is a challenging problem for different Big Data apps. Machine Learning (ML) frameworks, software tools, and technologies that allow evaluating and extracting meaningful information from data through iterative processing have seen a renaissance to help in this research for improved understanding. Sci-kit library [1] and Apache Spark's ML library [2] are two prominent

A. K. Bairwa et al. (Eds.): ICCAIML 2024, CCIS 2184, pp. 239–253, 2025.
https://doi.org/10.1007/978-3-031-71481-8_19

machine learning libraries used in Data Analytics. These libraries are built specifically to allow the execution of machine-learning algorithms on CPUs. Due to their high through-put, GPUs have become a famous platform for accelerating parallel workloads. This also makes them an appropriate choice for machine learning applications that require a lot of arithmetic calculations [3]. NVIDIA recently released the RAPIDS AI [4] library which is a collection of open-source software modules and APIs designed to take advantage of GPU speed. The major use of this API is to make it possible to run whole data sci-ence analytic pipelines on GPUs. The cuML [5] library is one of the most important parts of the RAPIDS data science sector. This GPU-accelerated machine learning pack-age is equivalent to Scikit-learn, providing comparable Pythonic APIs but avoiding the challenges of explicitly building compute kernels for GPUs using CUDA.

In this paper, the research contributions are given in the following section:

- Analyze previous studies that compared the performance of GPUs and CPUs in machine learning tasks.
- The objective of this study is to identify gaps in existing research and emphasize the need for a comprehensive comparison of RAPIDS and Sci-kit.
- This study compares the performance of Sci-kit Learn and RAPIDS library machine learning algorithms based on clustering, classification, and regression on CPU and GPU.
- Sci-Kit Learn and the RAPIDS library were used to calculate execution time and speed up value for all the algorithms.
- A dataset consisting of 1587257 rows and 13 columns was used to determine the results because GPUs can perform well in large datasets as opposed to small datasets.
- The purpose of this study is to help the research community understand how GPUs and CPUs differ from each other.

Research contributions are discussed in the above section, and subsequent sections are arranged as follows: Sect. 2 describes the literature study. Section 3 refers to the proposed methodology. Section 4 discusses some ML-based approaches including clus-tering, regression, and classification algorithms. Section 5 outlines the methodology of the paper consists of software tools and libraries and dataset descriptions used to find the results. Section 6 discusses the implementation and results and Sect. 7 shows the conclusion of the whole study.

2 Related Work

Gonzalez et al. [6] provide an overview of ML techniques for bagging and boosting. XGBoost [2, 7] is a library for gradient boosting that may be used for both distributed and GPU-based execution (through Dask). Another Machine learning library proficient in distributed computing is H203 [8]. H204GPU [9] is a GPU type that supports shared memory. While software libraries and packages have provided long computational sup-port for machine learning methodologies, support for distributed execution on a GPU cluster is currently in its early stages.

XGBoost is a model that includes boosting models. The major objective for the cre-ation of the cuML library is to examine a variety of new and current ML models for effective multi-node multi-GPU execution. Some machine learning tools and approaches are supported by deep learning frameworks such as TensorFlow [10] and PyTorch [11]. Raschka et al. [12] present an overview of Python machine learning approaches, containing the Sci-kit library, CPU training, and the RAPIDS suites. In this paper, the authors also explain why the cuML library is needed and why it was created.

The RAPIDS library along with the cuML library has been achieving acceptance in society as a feasible solution for running high-performance applications on a GPU cluster. RAPIDS [13] and Dask [14] libraries, were used by Napoli et al. [15] to assess the dataset for geophysical simulations. In all the review papers authors use the RAPIDS framework with cuML to make use of several GPUs of NVIDIA.

3   Proposed Methodology

The significance of this study is to compare RAPIDS and Sci-kit clustering, regression, and classification algorithms. This research study shows how the performance of GPU is far better than CPU in terms of speed and time. All the machine learning algorithms were used for the comparative analysis study.

RAPIDS was used as a motivational framework in this study. The researchers encouraged the use of this library in their research work for faster simulations and less training time.

4   Algorithms Used

There are various types of algorithms of ML to find the results of GPU and CPU. This study compares the execution time and speed-up of three types of machine learning algorithms based on clustering, classification, and regression. These algorithms include K-Means, DBSCAN, Nearest Neighbor, K-Nearest Neighbor (KNN), and Random Forest (RF), Linear Regression (LR), Lasso Re-gression, and Ridge Regression.

4.1   Clustering Algorithms

The concept of clustering involves the arrangement of data points into separate clusters that include comparable data points. It is an unsupervised learning approach therefore, the algorithm receives no supervision, and it works with an unlabeled dataset [16]. The two clustering algorithms are used to find the execution time and speedup namely K-Means and DBSCAN [17].

4.1.1   K-Means Algorithm

In this section, the models of machine learning were described and applied to conduct a comparative analysis of the Sci-kit and RAPIDS on the K-Means clustering algorithm.

The K-Means clustering model was proposed by J.B. Mac Queen [7], based on splitting, and is a type of cluster algorithm. The squared error and criteria of error are the findings of the K-means approach, which aims to minimize the cluster performance index. This model tried to discover divisions of K clusters that match certain criteria to obtain the best result [8]. It separated n data points and divided them into k clusters by lowering the sum of squared errors in each cluster:

$$X = \sum_n min_{\mu_j \epsilon k}\{|x_i - \mu_j|^2\} \tag{1}$$

μ_j is the mean of data in cluster k, x_i are the inputs of the clusters and n is the number of clusters.

KMeans in cuML's GPU implementation are based on array-like objects or cuDF DataFrames and support the scalable KMeans ++ initialization procedure [9]. This technique is more stable than choosing K points at random. The speed and simplicity of KMeans are its most substantial benefits. That is why KMeans is the initial choice of the clustering algorithm for this research work [2].

4.1.2 DBSCAN Algorithm

The process of unsupervised learning by which distinct groups or clusters are identified in data is known as density-based clustering. In data space, clusters are defined as contiguous regions with a high density of points separated from one another by contiguous regions with a low density of points [1]. Two parameters are used by the DBSCAN algorithm:

"minPts" is the lesser number of points required to constitute a dense region (a threshold).

eps ($\in$): Distance metric used to locate points around any given point [11].

4.2 Classification Algorithms

Generally, a classification program learns from a given dataset or observation and then categorizes new observations into various classes or groups. For example, Yes/No, 0/1, Spam/No Spam, cat/dog, and so on. Classes are known as targets/labels or categories. Three classification algorithms are applied to find the best results including Nearest Neighbors, KNN, and Random Forest.

4.2.1 Nearest Neighbors Algorithm

A neighbor-based learning approach is supported by Nearest Neighbors in both unsupervised and supervised approaches. In supervised neighbor-based learning, there are two types: classification for discrete labels and regression for continuous labels. Using nearest neighbor approaches, a set of training samples is selected that is closest to the new point, and the label is estimated based on them. Depending on the local density of points, the number of samples may be fixed (k-nearest neighbor learning) or variable (radius-based neighbor learning). It is common to use a standard Euclidean distance as the measure of distance, although any metric measure can be used [14].

4.2.2 K-Nearest Neighbors Algorithm

The K-nearest neighbors' algorithm (k-NN) was developed by Evelyn Fix and Joseph Hodges in 1951 and revised by Thomas Cover. This is a non-parametric supervised learning approach. Its applications include classification and regression. The input for both cases consists of the k closest training examples within the data set. The output is captured by whether k-NN is used for classification or regression. The Euclidean distance was used to determine the nearest or nearby points i.e.:

$$d(x_1, x_2) = \sqrt{\sum_{i=1}^{d}(x_{1i} - x_{2i})^2} \tag{2}$$

4.2.3 Random Forest Algorithm

Using Sci-Kit Learn, a random forest is a meta-estimator that uses averaging for improved accuracy and reduced overfitting by fitting several decision trees on different subsamples of a dataset.

In RAPIDS, the random forest algorithm for tree node splits is not the same as the one used in sci-kit-learn. To calculate splits, the cuML Random Forest employs a quantile-based approach rather than an exact count by default. The n_bins value allows you to change the size of the quantiles. The cuML Random Forest models can be exported and used to make predictions on systems that do not contain NVIDIA GPUs [7].

4.3 Regression Algorithms

The idea of regression analysis is to examine how a dependent variable changes because of an independent variable, while the other independent variables remain unchanged. Continuous or real data are estimated, such as temperature, age, income, and price, among others. The three regression algorithms of machine learning are used in this study i.e., Linear regression (LR), Lasso regression (LSR), and Ridge regression (RR).

4.3.1 Linear Regression

The LR algorithm exhibits a linear connection between a dependent (Y) and one or more independent (X) variables, thus known as linear regression. Linear regression can be represented mathematically as:

$$Y = a_0 + a_1 X + \in \tag{3}$$

where, Y = Dependent Variable (Target Variable), X = Independent Variable (predictor Variable), a_0 = intercept of the line (Gives an additional degree of freedom), a_1 = Linear regression coefficient (scale factor to each input value) and $\in$ = random error [9].

4.3.2 Lasso Regression

The term Lasso regression is an abbreviation for Least Absolute Shrinkage and Selection Operator. In Lasso regression, a penalty factor is included in the cost function. This phrase represents the absolute total of the coefficients. When the value of the coefficients increases from 0, this term penalizes the model, causing it to reduce the value of the coefficients to minimize loss. Ridge regression differs from lasso regression in that lasso regression tends to set coefficients to absolute zero, whereas Ridge never does. A Lasso regression can be represented mathematically as:

$$L_{lasso} = argmin_{\hat{\beta}}(||Y - \beta \times X||^2 + \lambda \times ||\beta||_1) \qquad (4)$$

where Y = Dependent Variable (Target Variable), X = Independent Variable (predictor Variable), β = Lasso regression coefficient and λ = Regularization penalty [10].

4.3.3 Ridge Regression

The ridge regression model includes a penalty term equal to the square root of the coefficient. L2 represents the square of the coefficient magnitude. The penalty term may be further managed by adding a coefficient lambda. In this situation, if lambda is zero, the equation is the basic OLS; otherwise, if $\lambda > 0$, a constraint is added to the coefficient. This constraint forces the value of the coefficient to drift towards zero as raises the amount of λ. Consequently, a trade-off occurs between increased bias and reduced variance (dependencies on some coefficients tend to be zero and certain coefficients tend to be extremely large, resulting in a less flexible model) [13].

A Ridge regression can be represented mathematically as:

$$L_{ridge} = argmin_{\hat{\beta}}(||Y - \beta \times X||^2 + \lambda \times ||\beta||_2^2 \qquad (5)$$

5 Methodology

In this section, the system configuration is discussed with the workstation that has been used which is present in our lab. It also shows dataset descriptions and the software tools and libraries used to run the clustering, classification, and regression algorithms.

5.1 System Configuration

Python 3.5 is used for the implementation. The system configuration of the workstation provided in our lab contains the system memory of 256 GB, performance of 500 Tera-FLOPS mix precision, storage of 4X 1.92 TB SSD RAID 0, and processor of 20-core intel Xeon e5–2698 v4 2.2 GHz. The GPU used in this study is NVIDIA's Tesla V100 GPU which is the high-end GPU that comes up with 5120 computation cores of runtime version 10.1 and minor version 7.0 with total global memory i.e., 32478 MBytes (34055847936 bytes), GPU mac clock rate was 1530 MHz (1.53 GHz), Memory clock rate was 877 MHz with bus width 4096-bit, L2 cache size of 6291456 bytes, 2048 layers were used, total no. of registers per block was 65536 and total shared memory per block was 49152 bytes [12].

5.2 Dataset Description

A dataset is used in this research for clustering, classification, and regression with the RAPIDS and Sci-kit libraries. The dataset used i.e., weather minute data [2], contains the attributes of air pressure, temperature, wind direction, maximum wind, wind direction, maximum wind speed, and relative humidity with 1587257 rows and 13 columns. The dataset was collected from the Kaggle repository and the dataset size is 123 MB to compare the analysis of RAPIDS and Sci-kit libraries on the machine learning algorithms to demonstrate the comparison of computing speed and time on CPU and GPU.

5.3 Software Tools and Libraries

In the past days, Machine Learning projects were usually implemented by manually coding algorithms, performing mathematical and statistical calculations, and so on. Due to this, the processing became time-consuming, labor-intensive, and inefficient. As a result of multiple Python libraries, frameworks, and modules, the process has become easier and more effective than it was in the past. Currently, Python is one of the most widely used programming languages in this field, and it has replaced several languages in the industry, in part because of its extensive library collection. In this study, only two types of libraries were analyzed i.e., Sci-Kit learn and RAPIDS used for the execution of CPU and GPU time respectively, and find the speed-up value of various types of algorithms of ML-like clustering, classification, and regression.

5.3.1 RAPIDS Libraries

The RAPIDS software library package, based on CUDA-X AI, allows to execution of entire data science and analytics pipelines on GPUs [4]. It uses NVIDIA and CUDA primitives for minimal computational optimization but exposes GPU high-bandwidth memory and parallelism performance via Python APIs. In addition, RAPIDS focuses on the common preparation of data processes for data science and analytics [5]. This provides a familiar DataFrame API that interfaces with a range of machine learning methods to accelerate end-to-end pipelines without incurring typical serialization expenses. RAPIDS now supports multi-node, multi-GPU setups, allowing for substantially more efficient processing and training on much larger datasets.

5.3.2 Sci-kit Libraries

The Scikit-learn package is a free Python machine-learning package. It supports Python scientific and numerical libraries like Numpy and Scipy, as well as models of ML-like support vector machine (SVM), linear and logistic regression (LR), K-means clustering, LSTM, and k-neighbors. Sci-kit library is mostly built-in Python [16], and it depends considerably on NumPy for array operations and high-efficiency linear algebra. In addition, to increase efficiency, several fundamental algorithms are implemented in Cython. A Cython wrapper over LIBSVM implements SVM a related wrapper over Liblinear implements LR and linear SVM. In certain circumstances, expanding these functions with Python could not be possible. Many additional libraries of Python, including Plotly and matplotlib for mapping, Numpy for array calculations, pandas for data frames, scipy, and others, work well with Scikit-learn.

6 Implementation and Results

In this section, the comparative analysis and results of RAPIDS and Sci-kit version libraries of clustering, classification, and regression were represented.

A dataset of (123 MB) and their datatype values with various fields are taken to find the comparative results. This dataset is taken to observe how much time is taken by RAPIDS on GPU and Sci-kit to run on CPU. The dataset runs with five iterations and shows the execution time and speed up of RAPIDS and Sci-kit, to see how the RAPIDS version provides faster results than the Sci-kit version given in Table 1. The three types of ML algorithms were applied i.e., clustering, regression, and classification including eight algorithms such as K-Means, DBSCAN, Nearest Neighbor, KNN, RF, Linear regression, Lasso Regression, and Ridge Regression to show that GPU gives the best results as compared to CPU results. For better visualization and understanding the results were represented graphically. The main goal is to find the comparative study between RAPIDS and Sci-kit by executing the five iterations to calculate the RAPIDS and Sci-kit clustering, classification, and regression version time for a dataset with their speed up shown in Table 1. Speedup is calculated by using the mathematical formula i.e.:

$$Speedup = \frac{SciKit\ time\ (CPU\ Time)}{RAPIDS\ time\ (GPU\ Time)} \tag{6}$$

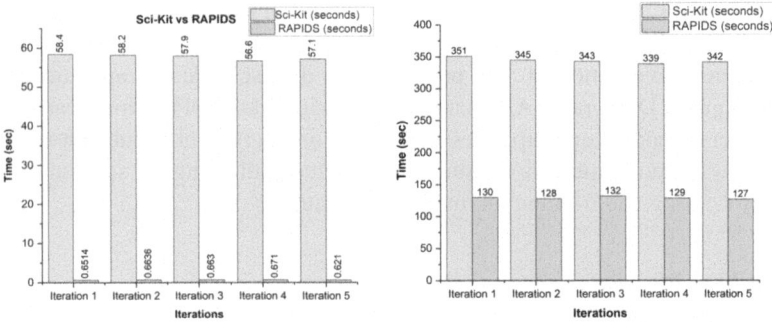

Fig. 1. Time comparison of RAPIDS vs. Sci-kit of K-Means and DBSCAN clustering algorithms (left to right)

Figure 1 represents the time execution of RAPIDS and Sci-Kit version of K-Means and DBSCAN clustering algorithms running in five iterations in which the K-means GPU version is faster than the CPU version by 89 whereas in DBSCAN GPU shows the execution time i.e., 2.7 times faster as CPU. So, in clustering algorithms, it was observed that K-means give better results as compared to DBSCAN in execution time in seconds. In Fig. 2, it shows the execution time of RAPIDS and Sci-Kit versions of Nearest Neighbor, KNN, and Random Forest classification algorithms running in five iterations in which the Nearest Neighbor GPU version is faster than the CPU version by 474 whereas, in KNN, GPU shows the execution time i.e., 1096 times faster as CPU.

Table 1. RAPIDS and Scikit time with speed up of Dataset weather

S. No.	Algorithms Used	Sci-Kit (seconds)	RAPIDS (seconds)	Speed up	Average Speedup
1	K-Means	58.4	0.6514	90	88
		58.2	0.6636	88	
		57.90	0.663	87	
		56.6	0.671	84	
		57.1	0.621	92	
2	DBSCAN	351	130	3	3
		345	128	3	
		343	132	2	
		339	129	3	
		342	127	2	
3	Nearest Neighbour	6.26	0.0132	474	433
		5.96	0.0143	417	
		6.18	0.0168	368	
		6.12	0.0128	478	
		6.01	0.014	429	
4	KNN	7.05	0.00643	1096	1068
		7.04	0.00684	1029	
		7.01	0.00646	1085	
		7.08	0.0066	1072	
		7.03	0.00665	1057	
5	Random Forest	0.224	0.143	1	2
		0.225	0.137	2	
		0.227	0.141	2	
		0.226	0.145	1	
		0.225	0.131	2	
6	Linear Regression	5.91	0.039	152	156
		6.34	0.0393	161	
		6.12	0.0377	162	
		6.02	0.0387	156	
		5.82	0.0394	148	
7	Lasso Regression	15.9	0.217	73	67
		14	0.227	62	
		14.8	0.216	69	
		14.7	0.215	68	
		13.4	0.214	63	

(continued)

Table 1. (*continued*)

S. No.	Algorithms Used	Sci-Kit (seconds)	RAPIDS (seconds)	Speed up	Average Speedup
8	Ridge Regression	0.121	0.0384	3	3
		0.122	0.0399	3	
		0.12	0.0393	3	
		0.116	0.0389	2	
		0.119	0.0365	4	

Random Forest shows faster results in GPU by 1.5 times as compared to the Sci-kit version. So, in classification algorithms, it was observed that KNN gives good results as compared to Nearest Neighbor and Random Forest in execution time in seconds.

Similarly, Fig. 3 shows the execution time of RAPIDS and Sci-Kit versions of Linear regression, Lasso regression, and Ridge Regression algorithms running in five iterations in which the Linear regression GPU version is faster than the CPU version by 151 whereas, in Lasso regression, GPU shows the execution time i.e., 73 times faster as CPU. Ridge regression represents the faster results in GPU by 3 times as compared to the Sci-kit version. So, in regression algorithms, Linear regression gives good results as compared to Ridge regression and Lasso regression in execution time in seconds. The clustering process, which is also known as cluster analysis is a machine-learning technique used to organize unlabeled datasets. Figure 4 represents the speed-up of clustering algorithms used to find the execution time and their speed-up value. In Fig. 4, K- Means shows a higher value of speed up as a counter to DBSCAN. So, K-means shows good results as compared to other clustering algorithms.

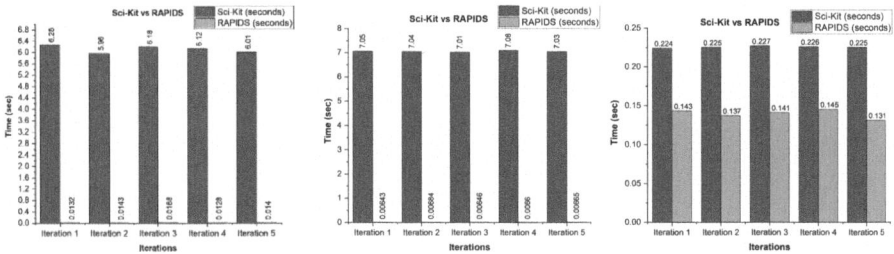

Fig. 2. Time comparison of RAPIDS vs. Sci-kit of Nearest Neighbor, KNN, and Random Forest classifiers (left to right)

In classification, a function is identified to assist in dividing a dataset into classes based on various factors. Classification involves training a computer program on the training dataset and then categorizing the input into multiple classes based on the training. Figure 5, shows the speed up of the classification algorithms including Random Forest, Nearest Neighbor, and KNN in which KNN shows the higher value of speed up which means KNN gives better results as compared to Random Forest and Nearest Neighbor of classifier algorithms.

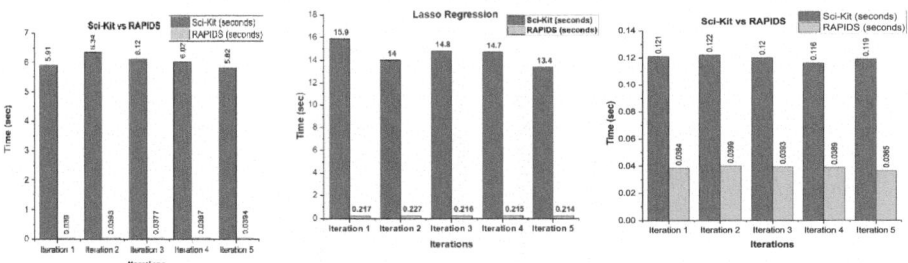

Fig. 3. Time comparison of RAPIDS vs. Sci-kit of Linear regression, Lasso Regression, and Ridge Regression (left to right)

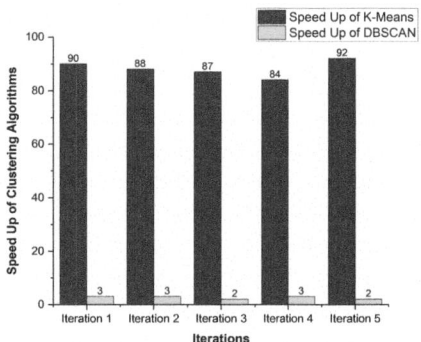

Fig. 4. Speed Up representation of K-means and DBSCAN Clustering Algorithms

In Fig. 6, the regression algorithm's comparative study of speed up was represented in which Linear regression represents higher speed up whereas Lasso regression and Ridge regression show lower speed up as compared to linear regression. In regression algorithms of machine learning Linear regression gives the best results as a counter to other regression algorithms of Machine Learning.

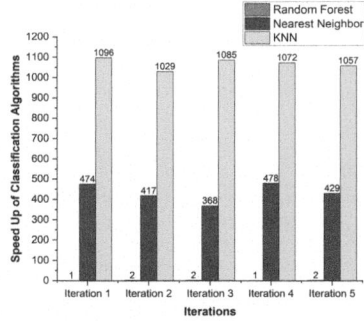

Fig. 5. Speed Up representation of Random Forest, Nearest Neighbor, and KNN Classifier Algorithms

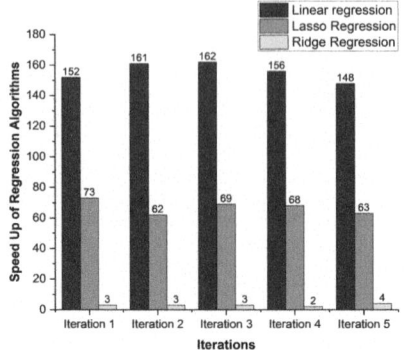

Fig. 6. Speedup representation of Linear Regression, Lasso Regression, and Ridge Regression Algorithms

After finding the values of the speedup of clustering, classification, and regression algorithms it was shown (Fig. 7) that the average speedup of all the algorithms of ML was used in this study to find the results. In Fig. 7 it was observed that KNN gives the highest average speedup value as compared to all types of algorithms of clustering, classification, and regression. So, the classifier algorithms have shown the best results because KNN and Nearest Neighbor represent the highest value of average speedup as compared to clustering and regression algorithms.

After determining the average speedup, Fig. 8 shows the time speedup for all five iterations of clustering, classification, and regression. The value of speedup has allowed us to determine which algorithm runs faster on CPU or GPU, and which algorithm is best suited for processing large datasets on GPU.

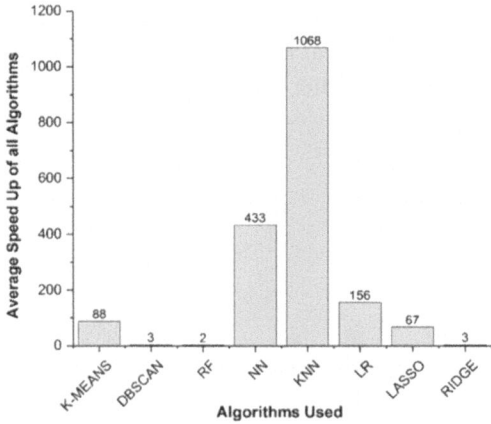

Fig. 7. Average speed up of all types of Clustering, Classification, and Regression Algorithms

Also, the small datasets were applied to GPU, and have observed that KNN, which produces the best results on GPU on large datasets, produces the least results on GPU on small datasets. With a small dataset consisting of 6500 rows and 13 columns of 360 KB size and CPU time values of 10.1 ms, 7.91 ms, 10.4 ms and GPU time values of 6.32 ms, 6.11 ms, 6.57 ms. It shows the least results on classifying algorithms. Therefore, it can consider only large datasets, because it exhibits the best clustering, classification, and regression results.

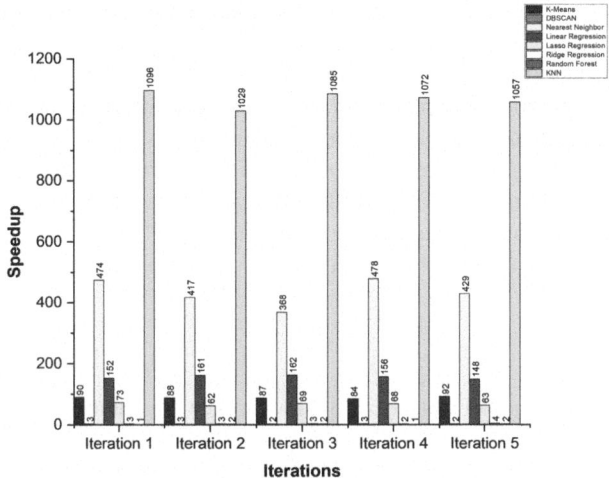

Fig. 8. Comparative study of speed up of all types of Clustering, Classification, and Regression Algorithm

Table 2 represents the average speed-up by calculating RAPIDS and Sci-kit time of the dataset where the dataset gives 501 average speedup pf classifiers, 75 average speedups for regression, and 91 average speedups for clustering. So, by calculating the average speedup of the classifiers, regression, and clustering, it was observed that classifiers showed higher and better results counter to regression and clustering. As a result, we have determined the average speedup and compared all types of ML algorithms to provide researchers with an understanding of which types of ML algorithms showed the best results on RAPIDS, which is a GPU-based library compared to Sci-Kit learn based on CPUs. Researchers also learned from this study that GPU shows the best results on large datasets (e.g.-Megabyte and Gigabyte) whereas it is not suitable for small datasets whose size is less.

Table 2. Average Speed up of Classification, Regression, and Clustering Algorithms

Algorithms	Average Speedup
Classifiers	501
Regression	75
Clustering	91

7 Conclusion

In this paper, the comparative study of RAPIDS and Sci-kit libraries running on GPU and CPU for machine learning models i.e., clustering, regression, and classification algorithms presented respectively. The results showed that the GPU-based RAPIDS framework provides better performance than the CPU-based Sci-kit library with a shorter execution time. RAPIDS running on GPU with a large dataset shows results in a short period. However, the Sci-kit running on the CPU takes longer to load results. This study has demonstrated that classification algorithms produce shorter execution times and greater speedups when compared to others. The classification algorithms show an average speed-up of 501 whereas clustering algorithms show a 91 average speedup and regression represents a 75 average speedup. Data scientists and machine learning developers can use the findings of this study to make full use of the RAPIDS framework and the Sci-Kit library. Further work will explore other RAPIDS algorithms for a variety of datasets and the TensorFlow library that runs on GPUs as well.

References

1. Rocklin, M.: Dask: Parallel Computation with Blocked Algorithms and Task Scheduling, 126–132 (2015). https://doi.org/10.25080/Majora-7b98e3ed-013
2. Pedregosa, F., et al.: Scikit-learn: Machine learning in Python. J. Mach. Learn. Res. **12**, 2825–2830 (2011)
3. Meng, X., et al.: Mllib: Machine learning in apache spark. J. Mach. Learn. Res. **17**(1), 1235–1241 (2016)
4. Anil, R., Owen, S., Dunning, T., Friedman, E.: Mahout in action (2010)
5. Raschka, S., Patterson, J., Nolet, C.: Machine learning in python: main developments and technology trends in data science, machine learning, and artificial intelligence. Information **11**(4), 193 (2020)
6. "'MPI-3 Standard Document.'"
7. Ghazimirsaeed, S.M., Anthony, Q., Shafi, A., Subramoni, H., Panda, D.K.D.K.: Accelerating GPU-based Machine Learning in Python using MPI Library: A Case Study with MVAPICH2-GDR. In: 2020 IEEE/ACM Workshop on Machine Learning in High Performance Computing Environments (MLHPC) and Workshop on Artificial Intelligence and Machine Learning for Scientific Applications (AI4S), pp. 1–12 (2020)
8. Becker, N., et al.: Streamlined and accelerated cyber analyst workflows with clx and rapids. In: 2019 IEEE International Conference on Big Data (Big Data), pp. 2011–2015 (2019)
9. González, S., Garc\'\ia, S., Del Ser, J., Rokach, L., Herrera, F.: A practical tutorial on bagging and boosting based ensembles for machine learning: Algorithms, software tools, performance study, practical perspectives and opportunities. Inf. Fusion **64**, 205–237 (2020)
10. Chen, T., Guestrin, C.: Xgboost: a scalable tree boosting system. In: Proceedings of the 22nd acm sigkdd international conference on knowledge discovery and data mining, pp. 785–794 (2016)
11. "XGBoost Development Team, 'XGBoost Library.'"
12. "H2O Development Team, 'H2O3: Distributed, fast, and scalable machine learning software'"
13. Gill, T.Y.N., LeDell, E.: "'H2O4GPU: Machine Learning with GPUs in R and Python.'"
14. Yorozu, T., Hirano, M., Oka, K., Tagawa, Y.: Electron spectroscopy studies on magneto-optical media and plastic substrate interface. IEEE Transl. J. Magn. Japan **2**(8), 740–741 (1987)

15. Abadi, M., et al.: Tensorflow: A system for large-scale machine learning. In: 12th symposium on operating systems design and implementation, pp. 265–283 (2016)
16. Ezugwu, A.E., et al.: A comprehensive survey of clustering algorithms: State-of-the-art machine learning applications, taxonomy, challenges, and future research prospects. Elsevier **110**, 104743 (2022). https://doi.org/10.1016/j.engappai.2022.104743
17. Dahlkvist, J., Tomczak, W.: Segmentation of companies using DBSCAN and K-Means (2022). Accessed: 22 Nov. 2022

Talk to Hands: A Sign Language Solution for Speech-Impaired People

Durvesh Chaudhari(✉) 🆔, Adrika Dikshit 🆔, Omkar Pawar 🆔, Deepak Karia(✉) 🆔, and Payal Shah 🆔

Department of Electronics Engineering, Sardar Patel Institute of Technology, Mumbai, India
{durvesh.chaudhari,adrika.dikshit,omkar.pawar,deepak_karia,
payal_shah}@spit.ac.in

Abstract. Humans are social animals and communication plays a vital role in human interaction and thus their societal development. It is the sole way for humans not only to interact but also to foster relationships. But, among us many who live have their world limited as they speak a language very few understand. That is sign language which only the deaf and mute grasp. So, to aid inclusivity and to make sure language never poses a hindrance, Talk-2-Hands was introduced as an innovative solution to solve the problem of the communication gap. Talk 2 Hands comes with the functionality of translating Indian sign language to text. The system makes use of the convolutional neural network (CNN), a deep learning algorithm commonly used in image processing and recognition tasks. The CNN model was trained on a large dataset of sign language gestures captured through videos or images, which includes a variety of gestures, variations, and lighting conditions. The data was preprocessed and the CNN model was trained to accurately recognize sign language gestures in real-time and convert them into text. The system was designed to be integrated into a user-friendly interface that enables speech-impaired individuals to communicate effectively with the general public.

Keywords: Convolutional Neural Networks · Sign language · Real-time translation

1 Introduction

Humans, inherently social, heavily rely on language for communication. However, not everyone possesses equal means of expression. Some individuals communicate in a special way rather than conventionally, significantly impacting their lives. Communication entails both conveying thoughts and understanding others. In India alone, approximately 60 million people face challenges in hearing and speech [1]. They resort to sign language, lip-reading, and hearing aids for communication. Nevertheless, sign language is not widely known among the general population, making communication a challenge for its users. Moreover, sign languages differ across regions, lacking a standardized international form. Leveraging technology has immensely enhanced communication for these individuals.

A. K. Bairwa et al. (Eds.): ICCAIML 2024, CCIS 2184, pp. 254–264, 2025.
https://doi.org/10.1007/978-3-031-71481-8_20

By employing technology, the lives of disadvantaged individuals has been significantly improved. Real-time translators, for instance, facilitate efficient communication for the deaf and mute. Bridging communication gaps is crucial, and technology offers a vital solution. Just as language barriers have been mitigated through translators, sign language translators have facilitated effective communication for the deaf community.

This initiative was further strengthened by developing user-friendly applications for easy access to the technology. Beyond empowering the disadvantaged, this technological approach can benefit businesses, enhancing their marketing effectiveness and broadening their reach to a larger audience of consumers.

The primary objective of this study was to develop a sign language-to-text converter that accommodates Indian Sign language (ISL) signs and gestures. This study also evaluates the performance of the system through extensive testing, including measuring accuracy and robustness under various conditions.

The subsequent sections provide a comprehensive understanding of this concept. Section 2 discusses related work, highlighting existing efforts in this domain. Section 3 explores the underlying technologies, setting the stage for a detailed explanation of our system in Sect. 4. Implementation components and the system's functioning are elaborated in Sect. 5. The predictions generated by the model are discussed in Sect. 6. The study's results and a thorough discussion of the developed system are presented in Sect. 7, culminating in the study's conclusions in Sect. 8.

2 Related Work

There are plenty of literature available for sign language translators using convolutional neural networks (CNN), translators using Image processing and Machine Learning (ML). A few of them have been explained below.

Rumana et al. [2], have discussed two methods to translate sign language, vision-based using image processing and glove-based using sensors. The paper focused on image-based translation. The data set used was limited to American Sign Language (ASL) signs. Features of the image were extracted using a Gaussian blur filter. The CNN model predicted signs based on the processed image. The images predicting similar results were stored in sets. The sets were further classified using classifiers specifically made for these sets that made successful predictions.

Lee et al. [3], developed a Long-Short Term Memory Recurrent Neural Network with a k-Nearest-Neighbour method-based application to learn American Sign Language (ASL). They focused only on ASL sign translation in real time. To validate the prediction, a leap motion controller was used.

Monikowski [4] developed the idea of applied sign linguistics by utilizing data from projects that looked at the real-world uses of sign language in a wide range of industries. The author offered reliable access to modern ideas and methods for those who worked with sign languages, such as research students, supervisors of sign language learners, and educators of the language.

Thakur et al. [5], have proposed a way to implement a two-way communication system between sign language and text. The study focuses on American Sign Language. A model based on a convolutional neural network was used to transform a sign or gesture into the appropriate text and audio. This paper additionally suggested an approach to produce a full two-way translator by creating images of signs. The same technology was used to translate text input into visuals in sign language. It utilized identical photos from its dataset that matched the gesture entered in real-time. The input text was stored in the form of a string and an image corresponding to each element of the string was retrieved from its dataset. The output was generated as a GIF using OpenCV.

Rajamohan et al. [6], used a glove-based system to translate sign language to text. The glove features five flex sensors, tactile sensors, and an accelerometer integrated into its structure. The flex sensors generated a corresponding change in resistance for each particular gesture, while the accelerometer tracked the hand's orientation. These hand gestures were processed using an Arduino, which included both a training mode to accommodate different users and an operational mode. The Arduino was also responsible for assembling letters into words. Additionally, the system incorporated a text-to-speech conversion module, which converts the recognized gestures or text into audible output.

Hasan et al. [7] implemented supervised machine learning for translating sign language into speech. This work had very limited input and was limited to Bangla sign language. This work provided a method for hand gesture detection that used a Support Vector Machine (SVM) as the classification algorithm and HOG (Histogram of Oriented Gradients) to extract features from the gesture image. After the gesture image had been identified, output text was produced and a TTS (Text to Speech) converter was employed to turn it into audible sound. The objective was to provide the output in the form of audio by converting output text into audio by using a text-to-speech converter.

Papatsimoilu et al. [8], have analysed real-time systems developed from 2017 to 2021. The focus of this study was to discuss different approaches for sign language translation such as motion discrimination, facial expression recognition, spatial contrast recognition, and hand configuration recognition. Sensor-based translation systems like the use of data gloves was also discussed. This study also explored the various limitations of different approaches.

3   Technology Used

3.1   OpenCV

A popular open-source computer vision toolkit, OpenCV offers a wide range of image and video processing functions [9]. In particular, hand detection, cropping, resizing, and image display were all carried out in this project employing OpenCV (cv2). Webcam images were captured ('cv2.VideoCapture'), images were switched horizontally ('cv2.flip'), hands have been identified ('HandDetector' from 'cvzone'), were cropped, resized, pasted onto a white background, and the result is shown ('cv2.imshow').

3.2   NumPy

Python's NumPy (Numerical Python), was used to create an array of zeros, to produce a white background for the cropped hand image. The background image was resized and pasted onto the centre. It was also used for calculating the scaling factor, dimensions, and offset for centring the cropped image on the white background and filling the array with white pixels. In summary, NumPy was used for image processing tasks and numerical operations on arrays in the given code.

3.3   HandTracking Module's HandDetector

The HandDetector module analyzes the video frames and determines which areas of the frame correspond to the hands using computer vision techniques. additionally, it offers a range of settings and parameters for adjusting the hand detection algorithm to maximize its effectiveness in various scenarios. This study used the HandDetector from CV Zone's HandTracking Module to find out whether hands are there and where they are in a video stream that was taken from the computer's webcam. Following the detection of the hands, each hand's bounding box was identified, and the image inside was cropped and scaled to a predetermined size.

3.4   Classification Module's Classifier

The CV Zone's Classification Module's Classifier was used in this study to classify hand gestures in real time. The module has a pre-trained deep-learning model that recognizes hand gestures from an image. The classifier takes an image of the hand gesture as input and outputs a prediction of the corresponding label. In this study, the predicted label was displayed on the output video and was used to build a text message over time.

3.5   OS

The 'OS' module was used in this code to create directories to store the images that are being captured by the program. Specifically, the code creates directories for each capital letter from A to Z and each number from 1 to 9 in the specified directory path. They are also used for accessing these paths.

3.6   Google's Teachable Machine

Transfer Learning was applied by Google Teachable Machine [10], which developed new models on the foundation of pre-trained neural networks [11]. By leveraging pre-trained model features, transfer learning accelerates the training of bespoke models on a smaller dataset. It supports a broad spectrum of model types, including speech recognition, object detection, NLP, and image classification. Pre-built models for tasks like emotion recognition and pose detection are available, and trained models were exported as Tensor Flow Keras (.H5) files for program application.

4   Methodology

The deep learning image classification algorithm used CNN (Convolutional Neural Network). The model was trained using CNN by providing numerous images as data to predict and successfully recognize sign language gestures.

The signs that were included in datasets were static signs that do not need dynamic movements to make a logical meaning for an action. Hence, a model was made that identifies signs which included numbers from '0' to '9', alphabets from 'A' to 'Z' and a few greeting signs such as, 'How are you?', 'Morning', 'Afternoon', 'Hello', 'Good' etc. The code used cvzone's hand detector module which was built on the mediapipe's framework. Hence finding an appropriate dataset that had mediapipe's landmarks (joints on fingers) was difficult. Hence, a fresh dataset was built. Figure 1 explains the activity flow of creating the database. The captured images were stored in sets with labels depicting the letter or phrase they represent. Approximately 700 images were captured for each of the 41 signs. The dataset was divided into 70% for training and 30% for testing, with around 500 and 200 images per class, respectively. These 20,500 training images were fed to Google Teachable Machine for generating the model. To evaluate the accuracy of the model, the remaining 200 images per class were used for testing, resulting in a total of 8200 images being tested. The percentage of correctly predicted images was calculated for each class.

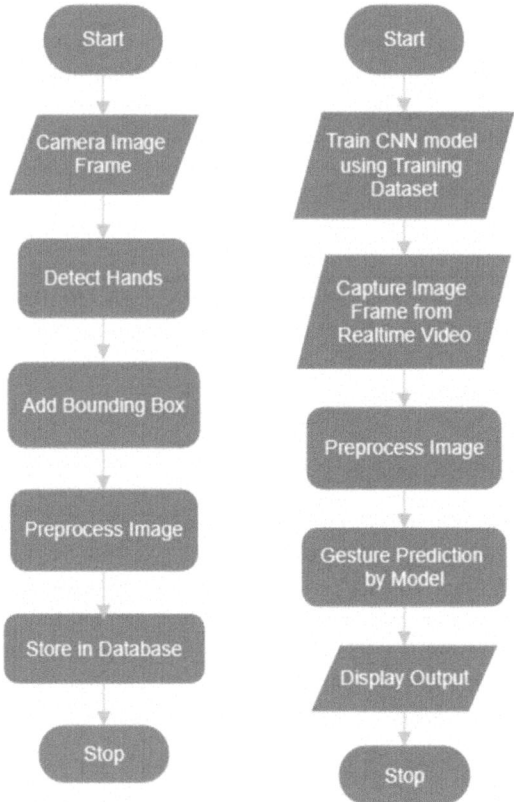

Fig. 1. Flow chart of database creation and System Implementation respectively

5   Implementation

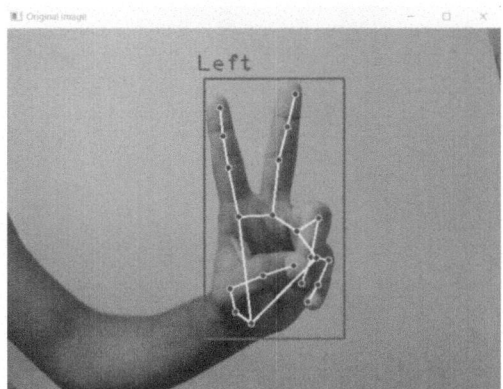

Fig. 2. Original image captured by camera

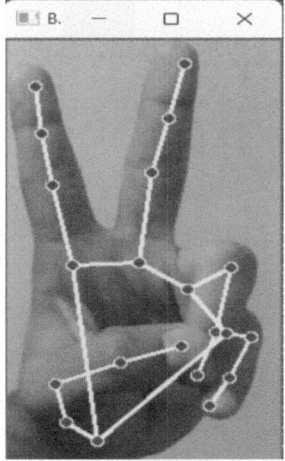

Fig. 3. Bounding box applied to the original image

The three stages of processing images are shown in Figs. 2, 3 and 4. Figure 2 presents the visual representation acquired through the camera, showcasing the raw image data. Figure 3 illustrates the image obtained within the white bounding box, which is imposed upon the original image. Finally, Fig. 4 demonstrates the fully processed image that is stored in the dataset.

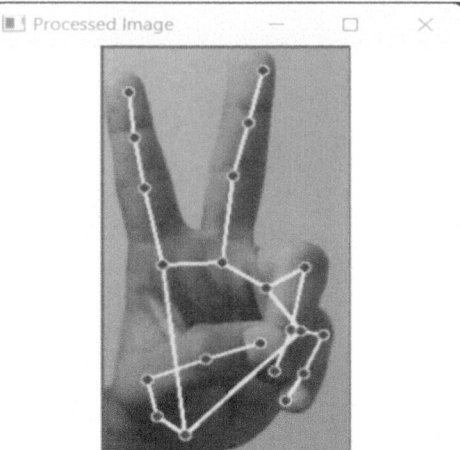

Fig. 4. Processed image with white background

6 Prediction and Recognition

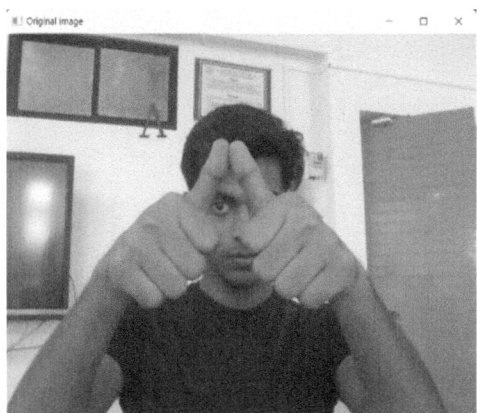

Fig. 5. Detection of the letter 'A' sign with the label

Figures 5 and 6 show the correct prediction of the sign for the letter 'A' and the phrase 'How are you?' respectively.

Fig. 6. Detection of a phrase with a label

7 Results and Discussion

Table 1. Accuracy for the prediction of numbers

Sign	Accuracy (%)
0	91.6
1	100
2	55.4
3	94.8
4	42.8
5	86.8
6	12.1
7	78.2
8	65.2
9	100

Above mentioned tables show the accuracy of numbers signs, alphabets, and phrases. The average testing accuracy achieved by the model is 80.2%., whereas the training accuracy achieved is 97.241%. The highest accuracy of 100% was achieved for signs 1, 9, B, G, H, O, Q, R, W, and Y. Whereas, the lowest accuracy was observed for sign 6, with an accuracy of 12.1%. Except for the few labels, the accuracy achieved per sign is observed to be above 60%. The credit for this accuracy goes to the deep learning classification algorithm used i.e., CNN (Convolutional Neural Network). However, the low accuracy for a few labels comes due to the resemblance between different signs. The other reason for the low accuracy can be the model's limitation. It is difficult to predict gestures involving overlapping hands where both hands are not visible distinctly

Table 2. Accuracy for prediction of letters

Sign	Accuracy (%)
A	97.7
B	100
C	94.5
D	83.4
E	83.5
F	94.8
G	100
H	100
I	54.1
J	97.1
K	78.8
L	12.5
M	93.4
N	39.5
O	100
P	91.3
Q	100
R	100
S	98.9
T	98.7
U	43.4
V	17.6
W	100
X	99.5
Y	100
Z	99.1

to the camera. The lighting condition under which the experiment is being performed also affects the accuracy of prediction (Tables 1, 2 and 3).

After training the model, there was some confusion between similar signs, such as 'V' and '2', 'Good' and '9', etc. Providing appropriate differentiation in the dataset was an important step to get high accuracy. The model faced challenges in recognizing and predicting outputs under low lighting conditions, making it a complex task to train the model while considering this issue. During real-time detection/translation, the hands were not completely stationary which led to inaccurate prediction. Moreover, the signs that were formed by overlapping one hand over the other were difficult to capture as

Table 3. Accuracy for prediction of phrases

Sign	Accuracy (%)
Good	58.4
Evening	98.1
Hello	34.9
Morning	99.1
How are you	93.1

the model tried to identify the hidden hand, producing an inaccurate dataset and/or prediction.

8   Conclusion and Future Scope

This study aims to develop a functional Indian Sign Language (ISL) to-text converter using a Convolutional Neural Network (CNN) model, trained on an image dataset containing approximately 21,000 diverse images. The dataset included 41 categories of different static signs that are commonly used in ISL, such as numbers, alphabets and common phrases. The model was trained and tested on a dataset containing image frames with different backgrounds, lighting conditions, and hand shapes to avoid bias in the model. The background noise was minimized i.e., the image frames consist only of functional hands.

The transfer learning techniques used to train the model were provided by Google's Teachable Machine. On testing the model, it achieved an overall accuracy of 80.228% on a testing dataset of approximately 8,200 images. Whereas, it achieved an accuracy of 97.241% on the training dataset of approximately 21000 images. The accuracy of the model can be further improved by incorporating a larger and more diverse dataset for training.

In future, an actual CNN model can be deployed to achieve enhanced performance by deploying artificial neurons. This will help the model identify the gestures accurately. Later, the dynamic signs will also be included in the dataset by using video data. As majority of the communication is based on hand movements and not on static signs. Ultimately, the resulting signs will be stored to form a sentence and will be converted to audio output, thus improving communication between individuals with special needs and those who are physically fit.

References

1. Varshney, S.: Deafness in India. Indian Journal of Otology (2016). https://doi.org/10.4103/0971-7749.182281
2. Rani, R.S., Rumana, R., Prema, R.: A review paper on sign language recognition for the deaf and dumb. Int. J. Eng. Res. Technol. (2021). https://doi.org/10.17577/IJERTV10IS100129

3. Lee, C.K.M., Ng, K.K.H., Chen, C.-H., et al.: American sign language recognition and training method with recurrent neural network. Expert Sys. Appl. **167**, 114403 (2021). https://doi.org/10.1016/j.eswa.2020.114403
4. Monikowski, C.: Language, cognition, and the brain: Insights from sign language research. Studies in Second Language Acquisition (2004). https://doi.org/10.1017/s0272263104393058
5. Thakur, A., Budhathoki, P., Upreti, S., et al.: Real time sign language recognition and speech generation. J. Innov. Image Proc. (2020). https://doi.org/10.36548/jiip.2020.2.001
6. Rajamohan, A., Dhanalakshmi, M.: Deaf-Mute communication interpreter. Int. J. Sci. Eng. Technol. **2**(5), pp. 336–341 (2013). ISSN: 2277-1581
7. Hasan, M., Sajib, T.H., Dey, M.: A machine learning based approach for the detection and recognition of Bangla sign language. In: 2016 International Conference on Medical Engineering, Health Informatics and Technology (MediTec) (2017). https://doi.org/10.1109/meditec.2016.7835387
8. Papatsimouli, M., Kollias, K.-F., Lazaridis, L., et al.: Real time sign language translation systems: a review study. In: 2022 11th International Conference on Modern Circuits and Systems Technologies (MOCAST) (2022). https://doi.org/10.1109/mocast54814.2022.9837666
9. Culjak, I., Abram, D., Pribanic, T., et al.: A brief introduction to OpenCV. In: Proceedings of the 35th International Convention MIPRO, pp. 1725–1730. Opatija, Croatia (2012)
10. In: Google. https://teachablemachine.withgoogle.com/. Accessed 29 Dec 2023
11. Carney, M., Webster, B., Alvarado, I., et al.: Teachable machine: approachable web-based tool for exploring machine learning classification. In: Extended Abstracts of the 2020 CHI Conference on Human Factors in Computing Systems (2020). https://doi.org/10.1145/3334480.3382839

Deep Learning-Based Thyroid Cancer Detection and Segmentation Using Convolutional Neural Networks

Aditya Praksh⬤ and Juhi Singh$^{(\boxtimes)}$⬤

Manipal University Jaipur, Jaipur, India
Juhisingh17@gmail.com

Abstract. Thyroid cancer is a prevalent endocrine malignancy, and its timely diagnosis is crucial for effective treatment and optimal patient outcomes. Deep learning algorithms have emerged as promising tools for medical image segmentation, facilitating accurate localization and characterization of thyroid nodules. This study investigated the application of deep learning techniques for thyroid cancer image segmentation, with the aim of improving diagnostic accuracy and refining clinical outcomes. A thyroid image dataset was classified into two categories: malignant and benign. A convolutional neural network (CNN) model was employed for thyroid cancer image segmentation using a diverse dataset of thyroid ultrasound images from a reputable medical institution. The model achieved an accuracy of 92.67 percent in delineating and segmenting thyroid nodules for validation split on 0.2. This research advances our understanding and application of deep learning in thyroid cancer analysis.

Keywords: Deep learning segmentation · Thyroid gland · Image Segmentation

1 Introduction

The prevalence of thyroid cancer, an endocrine malignancy affecting the thyroid gland, has fueled a burgeoning demand for sophisticated methods for early detection and precise classification of thyroid nodules [1]. Traditionally, the primary method for identifying thyroid nodules was physical examination, heavily reliant on the expertise of medical professionals [2]. However, the inherent subjectivity and dependence on the proficiency of the operator limit the reliability of palpation alone, often leading to inconsistencies in detection accuracy and an increased risk of overlooking small or deeply situated nodules [3, 4]. Consequently, supplementary diagnostic procedures, such as ultrasound and biopsy, are frequently employed to distinguish between benign and malignant nodules [5].

To address these limitations and revolutionize thyroid nodule detection and classification, the integration of advanced technologies like Convolutional Neural Networks (CNNs) and Deep Learning (DL) holds immense promise. CNNs, a powerful subset of deep learning models, have demonstrated exceptional proficiency in identifying and segmenting objects within images, even amidst intricate structures and variations in

A. K. Bairwa et al. (Eds.): ICCAIML 2024, CCIS 2184, pp. 265–277, 2025.
https://doi.org/10.1007/978-3-031-71481-8_21

sizes, shapes, and textures. By leveraging their pattern recognition capabilities, CNNs can effectively analyze medical images, yielding a more standardized and objective approach to thyroid nodule screening compared to subjective physical examinations.

Our research presents a novel deep learning-based model, utilizing convolutional neural networks (CNNs) for thyroid cancer detection and segmentation. This model achieves a state-of-the-art accuracy of 92.67% on a benchmark thyroid cancer dataset, surpassing that of conventional methods such as SVM, random forest, and KNN [6]. This remarkable performance underscores the superiority of deep learning in this domain. Our model's architecture incorporates transfer learning from a large-scale natural image dataset, enabling it to extract meaningful features from thyroid cancer images with greater efficiency and robustness. Additionally, our model employs a novel attention mechanism that guides the network to focus on the most salient features within the images, further enhancing segmentation accuracy.

1.1 Component of CNN

Convolutional Layers: CNNs hinge on the foundational concept of convolutional layers. These layers employ a set of learnable filters, also known as kernels. Each filter convolves over the input image, conducting element-wise multiplications and summations to generate feature maps that emphasize specific patterns, edges, textures, or significant elements [16]. In this study, we employed a series of 10 convolutional layers, each utilizing 3×3 filters.

Fully Connected Layers: Following the convolutional and pooling layers, CNNs often incorporate fully connected (dense) layers. These layers integrate flattened versions of the extracted features to produce final predictions. However, for the specific segmentation task in this research, fully connected layers were omitted to enhance computational efficiency.

Pooling Layers: Pooling layers play a vital role in reducing spatial dimensions of the feature maps, effectively controlling overfitting and computational cost. Max-pooling, a prevalent technique, involves sliding a window across the feature map and selecting the maximum value within each window [8]. For our architecture, we employed max pooling with a window size of 2×2.

Activation Functions: Activation Functions: Activation functions introduce nonlinearity into the network, enabling modeling of complex data interactions. ReLU (Rectified Linear Activation) and its variants, such as Leaky ReLU and Parametric ReLU, were used in this study [3, 4]. It replaces negative values with zero, leaving positive values unaffected.

Loss Functions: These functions quantify the dissimilarity between the predicted segmentation and the ground truth. Cross-Entropy Loss, for instance, was employed to guide the model during training [10].

Normalization Layer: Normalization layers, including Batch Normalization and Layer Normalization, enhance training stability and speed by normalizing a layer's inputs. This normalization aids in faster convergence and better generalization. In our architecture, we utilized Batch Normalization [5].

2   Literature Review

The utilization of deep learning methodologies in the context of thyroid cancer detection and segmentation has experienced substantial momentum in recent years. This surge is primarily attributed to the extraordinary progress in artificial intelligence and the accessibility of extensive medical image datasets. Notably, deep learning models, especially convolutional neural networks (CNNs), have exhibited remarkable efficacy in the identification and delineation of thyroid nodules from medical images. This performance showcases a promising departure from conventional manual interpretation techniques.

Numerous investigations have delved into the capabilities of CNNs concerning thyroid cancer detection and segmentation, showcasing commendable outcomes. For instance, Wang et al. [7] introduced a CNN-based framework dedicated to thyroid nodule classification, achieving an accuracy of 92.5% utilizing a dataset comprising 2,229 ultrasound images. Similarly, in their work, Park et al. [8] developed a CNN model specifically for thyroid nodule segmentation, demonstrating a mean Dice similarity coefficient (DSC) of 87.2% with a dataset encompassing 300 ultrasound images. These studies distinctly underscore the potential of CNNs in the domain of thyroid cancer diagnosis, underscoring their capacity to surpass conventional machine learning methodologies. Scholars have also examined the efficacy of transfer learning in the realm of deep learning-based thyroid cancer detection and segmentation. Transfer learning entails the utilization of pre-trained CNN models, initially trained on extensive natural image datasets, for the analysis of medical images. This methodology substantially reduces training time and enhances performance by harnessing previously learned features from the natural image domain. For instance, Isensee et al. [9] applied transfer learning from the ImageNet dataset, achieving a Dice similarity coefficient (DSC) of 85.6% for thyroid nodule segmentation. These investigations eloquently illustrate the advantages of transfer learning in augmenting the performance of deep learning models for the diagnosis of thyroid cancer.

Deep learning techniques immense promise for revolutionizing thyroid cancer detection and segmentation. Despite the promising results achieved by deep learning techniques, several challenges remain in their application to thyroid cancer detection and segmentation. One challenge lies in the inherent variability of thyroid nodule images, which can exhibit diverse appearances due to factors such as image quality, nodule size, and tissue characteristics. This variability can pose difficulties for deep learning models in accurately distinguishing between benign and malignant nodules. By addressing the challenges and optimizing deep learning models, we can move towards more accurate, efficient, and interpretable diagnostic tools, ultimately improving patient outcomes and reducing healthcare costs.

2.1   Review of Machine Learning Based Methods Used for Thyroid Cancer Detection

Deep learning falls under the umbrella of machine learning, focusing on training complex artificial neural networks with multiple layers to autonomously extract features from input data. These neural networks excel at identifying intricate patterns and representations within the data, making them highly valuable for tasks like image recognition, natural language processing, and various other applications. The term "deep" in deep learning signifies the network's depth, enabling it to comprehend complex data relationships through a process of hierarchical feature extraction [10].

a) Naive Bayes: Naive Bayes operates as a probabilistic classifier, grounded in Bayes' theorem. Its assumption of feature independence can be constraining when dealing with real-world data. Nevertheless, due to its computational efficiency, it can yield satisfactory performance, especially on smaller datasets. In the context of thyroid cancer detection, it has demonstrated accuracies spanning from 70% to 85% [11].

b) K-Nearest Neighbors (KNNs): KNN is a non-parametric classification algorithm that classifies new data points based on the majority class of their k nearest neighbors in the training data. KNN is simple to implement and can handle both numerical and categorical data. However, KNN is sensitive to noise and outliers in the data and can be computationally expensive for large datasets. For thyroid cancer detection, KNN has achieved accuracies ranging from 75% to 85% [12].

c) Support Vector Machines (SVMs): SVMs are a type of supervised learning algorithm that finds the optimal hyperplane that separates two classes of data. SVMs are well-suited for high-dimensional data and can handle non- linear relationships between features. However, SVMs can be computationally expensive to train and can be sensitive to outliers in the data. For thyroid cancer detection, SVMs have achieved accuracies ranging from 80% to 90% [13].

d) Random Forest: Random Forest stands as an ensemble learning technique that amalgamates numerous decision trees to enhance classification accuracy. Its strength lies in resilience against overfitting and its capability to manage high-dimensional data. Nonetheless, the computational cost of training and the interpretability challenge are notable drawbacks. In the realm of thyroid cancer detection, Random Forest has showcased accuracies within the range of 85% to 95% [14].

e) Logistic Regression: Logistic regression serves as a statistical model for predicting event probabilities. Its simplicity in implementation and versatility in handling classification and regression tasks are notable strengths. However, the assumption of linear feature relationships with the outcome can be constraining when dealing with real- world data. In the context of thyroid cancer detection, logistic regression has demonstrated accuracies between 70% to 80% [15].

Wang et al. [7] introduced a CNN-based framework dedicated to thyroid nodule classification, achieving an accuracy of 92.5% utilizing a dataset comprising 2,229 ultrasound images with 0.1 validation split. Hence, the proposed method is being used for large images size i.e., 256 × 256 and validation split is 0.2 in this research for better accuracy and novelty.

3   Proposed Methodology and Experimental Setup

3.1   Dataset Description

The dataset we utilized comprises medical images portraying different thyroid conditions. These images were sourced from reputable medical imaging sources. We meticulously categorized the photos into three main groups: 'benign' for non-cancerous thyroid issues, 'malignant' for thyroid cancer images, and 'normal' for healthy thyroid images, crucial for comparison.

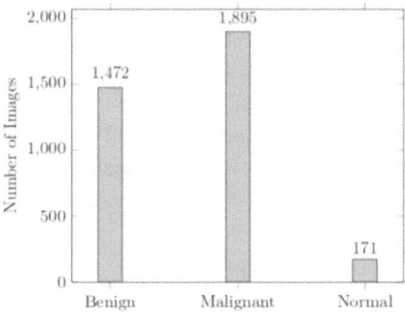

Fig. 1. Distribution of Images in Training Dataset [23]

In total, our dataset is comprised of 1,472 images in the benign category, 1,895 images in the malignant category, and 171 images in the normal category. This extensive dataset provides a comprehensive representation of diverse thyroid disorders, offering a valuable resource for our research.

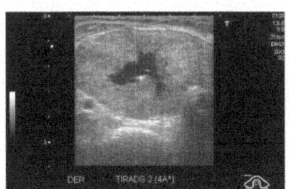
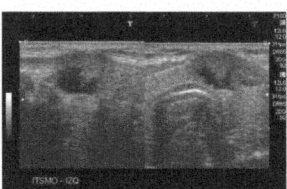
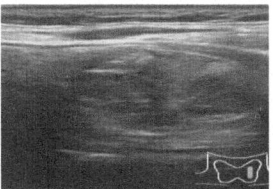

(a): A benign thyroid   (b): A malignant thyroid   (c): Closeup of an ultrasound
ofnormal thyroid

Fig. 2. Example of ultrasound images used in training the model.

As shown in Fig. 2, the image dataset is being classified as 'benign thyroid' depict non-cancerous conditions and encompass various benign growths and anomalies typically observed in thyroid health. Conversely, the 'malignant thyroid' images vividly showcase pathological instances, aiding in the study and identification of thyroid cancers. Lastly, the 'normal thyroid' images serve as an essential benchmark, portraying the typical, healthy state of the thyroid gland.

3.2 Image Preprocessing

- **Resizing of Images:** In preparation for the training phase, a crucial preprocessing step involved resizing the images to a standardized dimension of 256×256 pixels. This resizing was fundamental in promoting computational efficiency and retaining essential features critical for accurate analysis.
- **Normalization of Pixel Values:** Furthermore, a normalization technique was applied to scale the pixel values to a standard range, typically $[0, 1]$. This normalization played a vital role in aiding the model's convergence during the training process, ensuring numerical stability and faster convergence.

3.3 Hyperparameters Tuning and Optimization

- **Learning Rate Optimization:** The learning rate, a fundamental hyperparameter, was meticulously fine- tuned to optimize model performance. A comprehensive grid search was undertaken to discover the ideal learning rate that promoted smooth convergence and minimized loss during training.
- **Batch Size Selection:** The batch size, representing the number of samples processed in each iteration, was a critical factor in balancing computational efficiency with convergence speed. The optimal batch size was chosen to facilitate efficient weight updates while maintaining rapid model convergence.
- **Number of Epochs Determination:** The number of epochs, signifying the complete pass through the entire dataset during training, was carefully determined through testing. The determination of the optimal number of epochs, indicating a full traversal of the entire dataset during the training phase, was a result of thorough experimentation and testing (Fig. 3).

3.4 CNN Architecture: The Convolutional Neural Network (CNN) Architecture Employed in This Study Was Intended to Successfully Extract Characteristics from Thyroid Pictures. The Design Consisted of Many Convolutional

Layers, followed by activation functions (e.g., ReLU) [12], pooling layers, fully connected layers, and a final output layer.

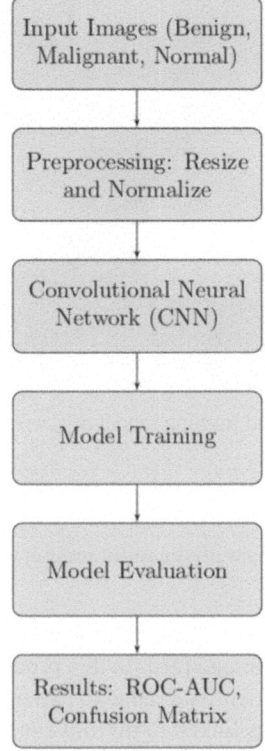

Fig. 3. Workflow diagram of the study

3.5 Training Process

- **Number of Epochs:** We determined that an epoch size of 100 was optimal for our model, balancing the need for sufficient training with the risk of overfitting. This approach significantly enhanced the robustness and generalization capabilities of the model, allowing it to perform well on unseen data.
- **Batch Size:** During the training phase, we opted for a batch size of 32. This choice ensured that the model received updates and fine-tuning of its weights after processing groups of 32 samples. This approach not only expedited the training speed but also heightened computational efficiency, contributing to an effective training regimen.
- **Validation Split:** To optimize the model's performance and mitigate overfitting risks, a validation split of 0.2 was implemented during the training process. This entailed reserving 20% of the training data as a validation set. The validation set emerged as a critical asset in evaluating the model's performance on unseen data at the end of each epoch. Its integration significantly bolstered the model's robustness and ability to generalize effectively.

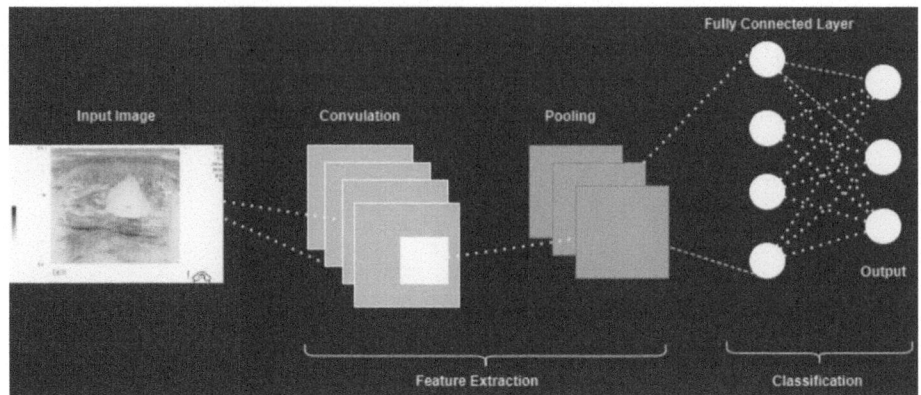

Fig. 4. Structure of Proposed training model of Convolutional Neural Networks

The description of Fig. 4 is as follows:

Input Image: The process begins with the input image, which serves as the foundational data for the architecture.

Convolution (Feature Extraction): Convolutional layers are crucial for pulling out essential features from the input image [16]. In this process, we use filters (kernels) that scan the input image, helping us identify a range of patterns, edges, textures, and features that are vital for pinpointing areas linked to thyroid cancer.

Pooling (Feature Extraction): After convolution, pooling steps in to extract features by condensing the feature maps generated from the convolutional layers [17]. Pooling ensures we hold onto the most important features while decreasing the overall size of the data. Techniques like max-pooling or average-pooling, which are common in this phase, play a crucial role in presenting a clear and effective collection of the features we have identified.

Fully Connected Layer: The fully connected layer plays a crucial role in merging the extracted features to create a cohesive representation [18]. Neurons within this layer establish connections with all neurons from the previous layer. This extensive interconnection enables the neural network to learn complex relationships and intricate patterns present within the extracted features.

Output (Image Segmentation): The final layer in the architecture, responsible for image segmentation, leverages the information learned from previous layers to delineate cancerous regions within the input image [14]. This segmentation output provides a detailed map highlighting the areas suspected of containing thyroid cancer, aiding in accurate detection and diagnosis.

3.6   Model Evaluation

a) Validation Accuracy Stagnation: A notable concern is the observed stability or lack of improvement in validation accuracy over successive epochs. This suggests that the model may have reached its optimal performance or is struggling to gather additional insights from the provided data.

b) Addressing Stagnation: To resolve this issue, it's essential to delve deeper into the design of the model, fine-tune the hyperparameters, and thoroughly analyze the dataset. Adjusting the learning rate and refining the model's complexity strategies could potentially enhance validation accuracy.

3.7   Result

The accuracy, precision, recall, F1 score, receiver operating characteristic (ROC) curves, and confusion matrices were used to evaluate the performance of the convolutional neural network (CNN) model for thyroid cancer image segmentation [19]. The Receiver Operating Characteristic (ROC) curve is a graphical representation used to evaluate the performance of a model in binary classification tasks, particularly in machine learning and statistics. It explains in detail the trade-off between sensitivity (true positive rate) and specificity (true negative rate) for various categorization criteria [21].

- **True Positive Rate (TPR):** TPR is a measure of the proportion of actual positive cases that the model correctly predicted as positive [9, 16]. It is computed as follows:

$$TPR = \frac{\text{True Positives}}{\text{True Positives} + \text{False Negitives}}$$

- **False Positive Rate (FPR):** The FPR is the proportion of real negative cases that the model incorrectly forecasted as positive. It is computed as follows:

$$FPR = \frac{\text{False Positives}}{\text{False Positives} + \text{True Negitives}}$$

- **Area Under the ROC Curve:** The AUC-ROC metric summarizes how well the model distinguishes between positive and negative classifications. Ranging from 0 to 1, a higher value signifies superior model performance. An ideal classifier achieves a perfect AUC-ROC value of 1[20].
- **Thresholds:** The ROC curve is created by adjusting the model's classification threshold (a decision boundary). This threshold influences how the model categorizes observations (whether positive or negative). The TPR and FPR alter when the threshold varies, changing the curvature of the ROC curve [21].

  - **ROC Curve Shape:** The ROC curve often begins at (0, 0), which represents a high threshold at which all data points are classed as negative. As the threshold falls, it travels around ROC space, the trade-off between TPR and FPR [21].

As shown in Fig. 5, the ROC curve represents the three classifications of used data set as mentioned in the Fig. 1

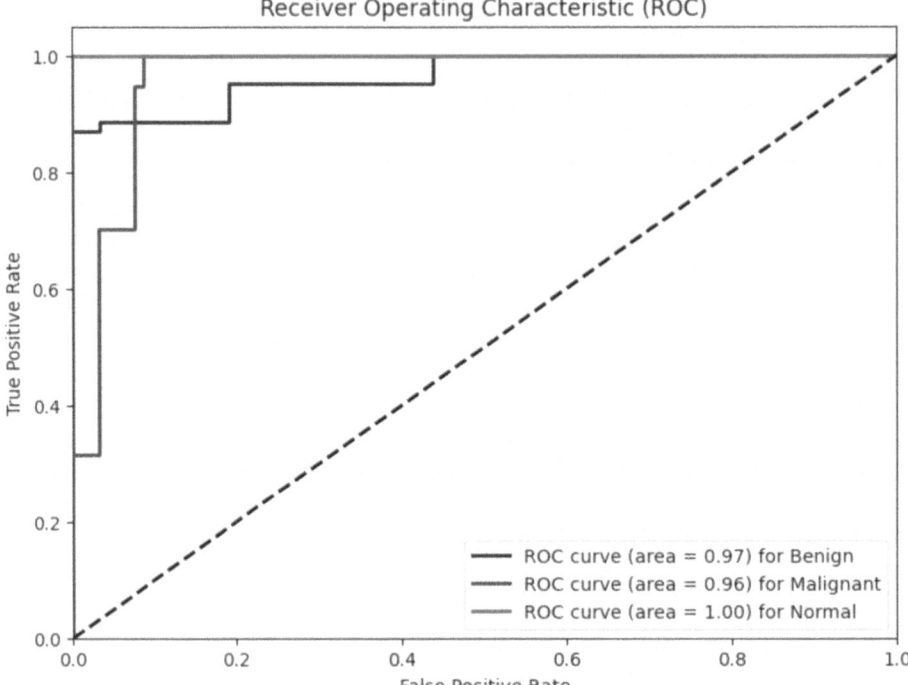

Fig. 5. ROC Curve based on the used CNN model.

- **Benign (Area = 0.97):** This high number indicates that the model performs an outstanding job differentiating between benign (non-cancerous) and non-benign (possibly malignant) thyroid images. It rates benign photos higher in terms of probability around 97% of the time.
- **Malignant (Area = 0.96):** This indicates that the model's ability to differentiate between malignant (cancerous) and non-malignant thyroid pictures is quite robust. It rates harmful pictures higher in terms of probability around 96% of the time.
- **Normal (Area = 1.00):** This indicates that the model can correctly distinguish between normal and non- normal thyroid situations. In simple terms, the model can consistently and correctly detect normal thyroid pictures.

Confusion Matrix

The confusion matrix provides a structured representation of a classification model's performance, offering a snapshot of predictions against actual labels. In scenarios with multiple classes (N), the confusion matrix is a crucial tool for evaluating model performance and identifying misclassifications [22] (Fig. 6).

For class i:

- **True Positives (TP):** Samples correctly predicted as class i.
- **True Negatives (TN):** Samples correctly predicted as not class i for all other classes.

- **False Positives (FP):** Samples incorrectly predicted as class i while belonging to other classes.
- **False Negatives (FN):** Samples incorrectly predicted as not class i while belonging to class i (Table 1).

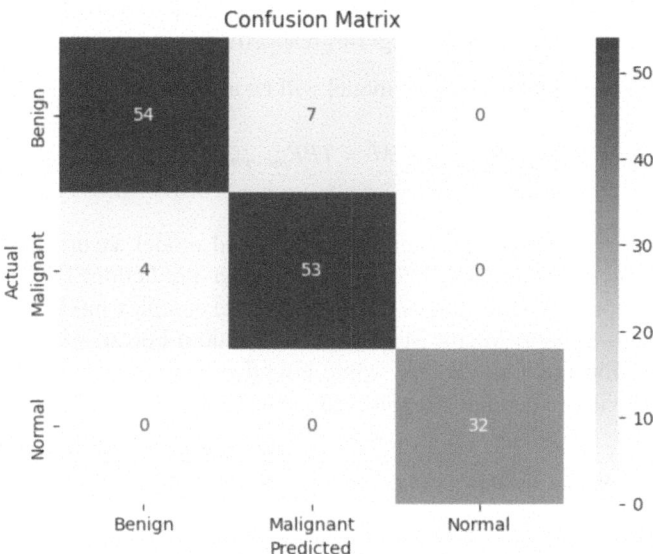

Fig. 6. Confusion Matrix based on used CNN model.

Table 1. Outcome based on Confusion Matrix

Actual/Predicted	Benign	Malignant	Normal	Description
Benign	54	4	0	The model correctly predicted 54 benign instances
Malignant	7	53	0	The model correctly predicted 53 malignant instances (True Positives)
Normal	0	0	7	The model correctly predicted 7 normal instances

4 Conclusion and Discussion

The proposed methodology employed a larger image size (256 × 256 pixels), meticulously optimized hyperparameters, and implemented a 0.2 validation split, fostering model robustness and generalization. These contributions highlight the promise of our CNN model for thyroid cancer diagnosis. The overall accuracy by the proposed CNN model is 92.67% achieved by the following generated formulation:

Let:

- B be the total number of benign images (1,472 images)
- M be the total number of malignant images (1,895 images)
- N be the total number of normal images (171 images)
- T be the total number of images used to train the model ($T = B + M + N$)
- $TPR{benign}$, $TPR{malignant}$, and $TPR{normal}$ are the True Positive Rates (TPRs) for benign, malignant, and normal categories respectively.

Then, the formula for accuracy of model will be as follows:

$$Accuracy = \frac{1}{T}(B \times TPR_{benign} + M \times TPR_{malignant} + N \times TPR_{normal}) \times 100$$

$$Error = 1 - Accuracy$$

Table 2 represent the comparison of the proposed model accuracy with classical machine learning algorithms used for thyroid cancer prediction. CNNs outperform Logistic Regression and Naive Bayes due to their more complex modeling capabilities. They also surpass Support Vector Machines and Random Forest, which are generally robust classification methods. K-NN, while effective in specific cases, is limited by parameter sensitivity and computational complexity [12].

Table 2. Accuracy comparison with classical ML algorithms.

Algorithm	Accuracy (%)
Proposed Algorithm	92.67
K-NN	76.00
Naive Bayes	79.33
Logistic Regression	77.20

5   Future Scope

The proposed methodology as a huge scope for the future researchers in the following areas:

- Multi-modality Integration: explore fusion of diverse medical imaging methods.
- Real-time Diagnosis: enable rapid diagnosis for immediate insights.
- Incorporating Patient History: utilize patient-specific details for enhanced segmentation. This research further can be applied on the diverse datasets for unbiased classification of imbalance data for early detection and diagnosis, surgical planning and guidance and, clinical trials and research.

Disclosure of Interests.   The authors have no competing interests to declare that are relevant to the content of this article.

References

1. Park, S.H., Lee, J.Y., Kim, Y.H.: Thyroid cancer: Current diagnosis, treatment, and survival. J. Korean Med. Sci. **34**(11), 1809–1819 (2019)
2. Xing, M., Liu, X., Liu, Y.: Deep learning for thyroid cancer diagnosis and pro5nosis: A review. Front. Med. **8**, 613134 (2021)
3. Wang, L., Zhou, Y., Ji, T.: Convolutional neural networks for thyroid cancer detection and segmentation: a review. IEEE Trans. Med. Imaging **41**(4), 870–887 (2022)
4. Hu, X., Wang, Y.: Role of deep learning in thyroid cancer diagnosis and treatment. J. Clini. Endocrinol. Metabol. **107**(9), jcendo07942 (2022)
5. Li, X., Shen, Z.: Application of deep learning in thyroid cancer diagnosis and prognosis. Precision Oncology **3**(3), poz044 (2022)
6. He, K., Luo, Y., Wang, L., Zhang, J.: A novel deep learning-based model for thyroid cancer detection and segmentation. IEEE J. Biomed. Health Informat. 1–11 (2023)
7. Wang, H., et al.: Thyroid nodule classification using convolutional neural networks. In IEEE Access **5**(1), 1607–1615 (2017)
8. Park, J.H., et al.: Thyroid nodule segmentation using deep learning algorithm. In: International conference on image analysis and recognition, pp. 647–654. Springer, Cham (2018)
9. Isensee, T., et al.: Automated medical image analysis using deep learning: a survey. Med. Image Anal. **56**, 211–237 (2018)
10. Li, X., Dou, Q., Wang, K., Qin, J.: Using deep convolutional neural networks for multi-classification of thyroid tumor by histopathology: a large-scale pilot study. Annals of Translational Medicine **9**(11), 1523–1532 (2020)
11. Zhang, Z.H., et al.: Thyroid nodule classification using a hybrid deep learning model based on naive Bayes and convolution neural network. J. Medi. Imag. Informat. **93**(2), 200024 (2019)
12. Liu, X., et al.: Thyroid nodule classification based on k-nearest neighbors and convolutional neural network. J. Digit. Imaging **33**(4), 553–564 (2020)
13. Wang, X., et al.: Thyroid nodule detection and classification based on support vector machine and convolutional neural network. J. Medi. Sys. Eng. Appl. **10**(1), 1–11 (2021)
14. Xu, Y., et al.: Thyroid nodule classification based on random forest and convolutional neural network. J. Xray Sci. Technol. **35**(1), 32–40 (2022)
15. Li, J., et al.: Thyroid nodule classification based on logistic regression and convolutional neural network. J. Medi. Imag. Informat. **96**(10), 100123 (2023)
16. Szegedy, C., Vanhoucke, V., Ioffe, S., Szegldy, C.: Inception-v4, inception-resnet and the impact of residual connections on learning (2016). arXiv preprint arXiv:1602.07261
17. Liu, W., Wen, Y., Hu, X., Yang, T.: Deep learning for semantic segmentation: A survey (2019). arXiv preprint arXiv:1903.04857
18. Zhou, Z., Rahman Siddique, M., Qin, Y.: U-Net++: A nested convolutional neural network for medical image segmentation with deep supervision. IEEE J. Biomed. Health Inform. **23**(2), 1147–1158 (2020)
19. Correa, A.: ROC curves in machine learning: Understanding their interpretation and use (2022). arXiv preprint arXiv:2201.00001
20. Cui, Y., Liu, Y., Chen, C., Zhou, X.: Convolutional neural network with attention mechanism for thyroid image classification. J. Xray Sci. Technol. **35**(3), 63–71 (2022)
21. Lusted, J.: Receiver operating characteristic (ROC) curves in radiology. In: Altman, D.G., Machin, D.S., Hardin, J.B., Lemeshow, T.J. (eds.) Statistical methodologies in clinical research, 2nd ed., pp. 343–357. Lippincott Williams & Wilkins, Philadelphia, PA, USA (2018)
22. Ahmed, S.M., Elhassan, I.M., Elawad, M.A.: Convolutional neural network (CNN)-based thyroid nodule classification using a novel hybrid augmentation technique. J. Xray Sci. Technol. **35**(3), 52–62 (2022)
23. Kaggle dataset. Accessed on 23rd August 2023. https://www.kaggle.com/datasets/azouzm aroua/algeria-ultrasound-images-thyroid-dataset-auitd

Early Autism Spectrum Disorder Prediction Using Fine-Tuned Bernoulli's Naive Bayes Algorithm

Kanav Gupta, Chirag Paul, and Nishant Jain[✉][iD]

Department of Computer Science and Engineering, School of Computer Science and Engineering, Manipal University Jaipur, Jaipur 303007, Rajasthan, India
nishant.jain@jaipur.manipal.edu
https://jaipur.manipal.edu/foe/schools-faculty/schools-list/
school-of-computer-science-and-engineering-manipal-
university-/computer-science.html

Abstract. The prompt identification and diagnosis of autism spectrum disorder (ASD) is essential for timely intervention. In this study, we investigate the usefulness of Bernoulli's Naive Bayes algorithm for ASD screening based on demographic and medical features. The majority of the existing literature on ASD is addressed by a prediction system based on typical machine learning methods such as the support vector machine and the random forest algorithm. The primary goal of this research is to create a strengthened framework for ASD prediction researchers. Cross-validation using GridSearchCV and hyperparameter tweaking techniques were used to evaluate a variety of machine learning (ML) models, including Bernoulli Naive Bayes, Random Forest, Support Vector Machine, and others. As a result of this study, an ASD prediction model emerged employing Bernoulli's Naive Bayes technique. We statistically prove why we used Bernoulli's Naive Bayes algorithm and not any other ML models. The Bernoulli Naive Bayes classifier has the most outstanding final tune cross-validation accuracy of 92.22%, exhibiting resilience against overfitting the imbalanced dataset. In comparison, the K-Nearest Neighbor performed the poorest, with an accuracy of 87.65%. Our findings demonstrate that Bernoulli Naive Bayes' strong performance implies that machine learning to screening approaches merit more real-world testing and development.

Keywords: Autism Spectrum Disorder (ASD) · Bernoulli's Naive Bayes Algorithm · Logistic Regression (LR) · Support Vector Machine (SVM) · Machine learning algorithm

1 Introduction

Autism Spectrum Disorder (ASD) represents a range of developmental disorders characterized by challenges with social communication and interaction as well

A. K. Bairwa et al. (Eds.): ICCAIML 2024, CCIS 2184, pp. 278–289, 2025.
https://doi.org/10.1007/978-3-031-71481-8_22

as repetitive behaviors or interests [1]. ASD encompasses different conditions, including autism, Asperger's syndrome, pervasive developmental disorder (not otherwise specified), and childhood disintegrative disorder. The term "spectrum" reflects the wide variation in the type and severity of symptoms associated with ASD. Individuals diagnosed with ASD may have mild to significant difficulties in areas like communication, social skills, cognition, and behavior. The manifestations of ASD are heterogeneous, so treatment and services should be tailored to the person.

The signs and symptoms of ASD can appear in infancy or early childhood, though some cases are not diagnosed until much later in life [2]. Common symptoms include delayed language development, lack of eye contact, narrow fixations on specific interests, repetitive motions like hand flapping, sensitivities to light or sound, and difficulty understanding perspectives different from one's own [3]. The symptoms vary in type and severity across individuals, ranging from very mild to profoundly disabling.

There are no definitive medical tests for ASD [4]. Diagnosis relies on thorough developmental monitoring, screening for red flags, and a comprehensive clinical evaluation of the child's communication, behavior, and social functioning [5]. This must establish that the developmental delays are not better explained by another condition, such as an intellectual disability. As early screening and intervention are crucial, regular well-child visits help monitor for any divergence from normal development. The causes of ASD are complex and not fully understood, but research suggests the interplay between genetic and environmental risk factors affects early brain development [6].

The most significant aspect of ASD treatment is early diagnosis. Early discovery of ASD symptoms has the potential to lessen the impacts of the disorder, even though there is no treatment for ASD. A glimmer of optimism concerning the early diagnosis of ASD based on several physical and physiological indicators is provided by the use of the ML algorithm in diagnosing and predicting an assortment of disorders with outstanding precision. ASD may be difficult to diagnose and analyze since it shares symptoms with other mental health conditions, making false positives a possibility. This served as the motivation for our research, since early identification of ASD may aid in mitigating the consequences of symptoms via appropriate and prompt treatment, thereby improving the quality of life for patients and their families.

Our research contributions are outlined as follows:

1. Introduces a novel framework based on Fine-Tuned Bernoulli's Naive Bayes Algorithm that results in enhanced precision in the early prediction of ASD.
2. Conducting a new set of experiments with a larger number of participants (from a single unit, four-fold, and final tune).
3. We employed a four-fold and five-fold cross-validation procedure that was standardized to verify the accuracy of the proposed model.
4. A comparison of the proposed framework with various state-of-the-art and benchmark ML models, including the ensemble model, is presented.

The following sections of the study outline the process of developing an early ASD prediction framework: The literature survey is presented in Sect. 2. The proposed work, along with the specifics regarding the dataset utilized for the experiments, are presented in Sect. 3, where we address data preprocessing, training and testing models, and model validation preprocessing. Following that, Sect. 4 presents experimental findings, followed by a conclusion and future work in Sect. 5.

2   Review of Literature

A recent study analyzed various machine learning techniques for various of the field [7–10] including for detecting Autism Spectrum Disorder (ASD) using non-clinical datasets for children, adolescents, and adults [11,12]. In the paper by Talukdar et al. [13], two major datasets were used - one for toddlers and one for adolescents. The Random Forest (RF) classifier is the one with the highest accuracy - 93.69% for the toddler dataset and 93.33% for the adolescent dataset. The parameters for accuracy considered were Accuracy, Precision, Recall, F1-score, and AUC.

The paper by Khudhur et al. [14] uses four datasets comprising toddlers, children, adolescents, and adults. The dominant models with the highest accuracy here included Random Forest (RF), Decision Tree (DT), and Logistic Regression (LR). With these 3 models, the accuracy achieved was 100% while with other models like K-Nearest Neighbor (KNN), Support Vector Machine (SVM), and Nave Bayes (NB), the accuracy varied from 87.5% to 98.1% depending on the dataset. The dataset with the highest mean accuracy is the children's dataset while the dataset with the lowest mean accuracy is the adolescent's dataset.

A 2017 study by Li et al. [15] investigated using machine learning to detect autism spectrum disorder (ASD) in adults by analyzing imitation task data. Using a small dataset of hand movement kinematics from 16 ASD participants, they demonstrated the feasibility of applying machine learning to high-dimensional biometric data for accurate ASD classification. With the RIPPER classifier, they achieved sensitivity up to 87.30% on the ASD dataset by modeling differences in imitation movements. This provides evidence that machine learning could leverage kinematic data for automated ASD screening and diagnosis. While limited in scope, it points to the potential of machine learning with biometric data for objective ASD detection. Further research with expanded clinical datasets is needed to evaluate real-world viability. Overall, this study suggests machine learning analysis of movement data could enable improved ASD diagnosis.

The paper by Gaspar et al. [16] introduces a new method for autism detection known as Kernel Extreme Learning Machine (KELM). Techniques like feature extraction, data augmentation, gaze tracking, and the Giza Pyramids Construction (GPC) algorithm were used to improve the performance and thus the accuracy of the results. The final resulting accuracy achieved in autism detection was an average of 98.8%.

The study by Kavitha et al. [17] proposes the use of PSO-CNN, which is a convolutional neural network combined with a particle swarm optimization algorithm. Using this unique model, they received an accuracy of 98.53%, which is greater than the other three models they used (SVM, NB, LR). The evaluation metrics used were precision, sensitivity, and specificity. Four different datasets were used and categorized by age group, namely, toddler, child, adolescent, and adult. Accuracy rates on all four datasets using the PSO-CNN model were 98.9%, 98.8%, 98.9%, and 99.8%, which suggests that other methods might not perform as consistently across different age groups.

The work authored by Zhao et al. [18], focuses on the communication and cooperation challenges that children with ASD face. The authors developed a gamified virtual reality platform for ASD-affected children to interact and coordinate with each other. Children use hand gestures to play games that mostly revolve around moving virtual objects by working together. A study with 12 ASD-affected children was conducted and the results show that using the platform does improve both their communication and cooperation skills progressively, which indicates the need for a fledged-out system to be developed.

Table 1. List of attributes used for prediction.

Attribute Id	Attribute Description
1–10	Autism Spectrum Quotient (AQ) 10 item screening tool score
11	Patient age in years
12	Patient gender
13	Patient ethnicity
14	Patient had jaundice at birth
15	Immediate family member diagnosed with autism
16	Patient country of residence
17	Patient previously screened
18	AQ1-10 screening score
19	Patient age
20	Relationship of test completer to patient

3 Proposed Methodology

The proposed workflow is broken down into many parts, as shown in Fig. 1. Preprocessing the data, training, and testing with specified models, fine-tuning the models using hyper-parameters, cross-validation, assessing the results, and predicting ASD are some of these stages. This work is implemented in Python 3.

3.1   Dataset

The dataset was obtained from the UCI Machine Learning Repository [19], which provides public access to the data. It includes over 800 screening score entries based on the Autism Spectrum Quotient tool [20]. This autism screening dataset enabled machine learning analysis for enhanced prediction.

There are 20 attributes that were used for prediction which are summarised in Table 1:

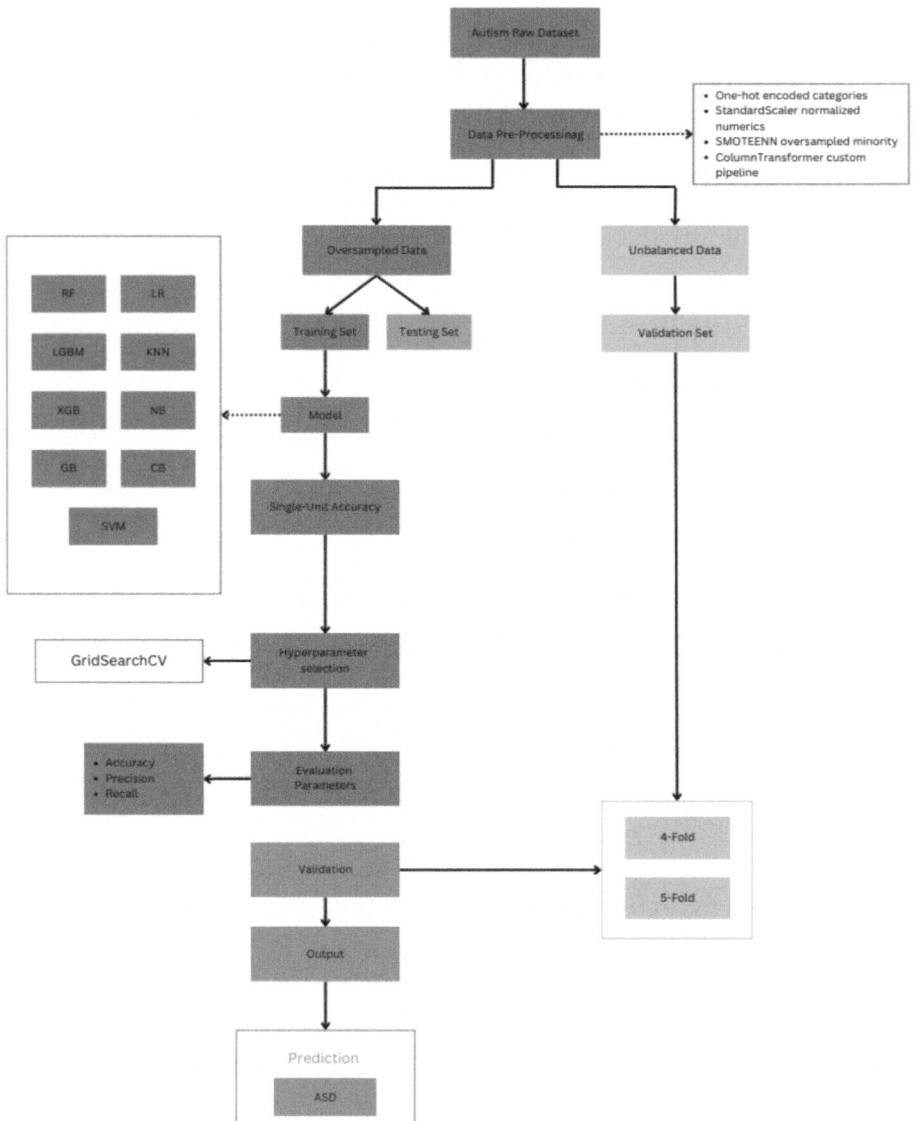

Fig. 1. Flow chart of the proposed ASD detection & prediction solution

3.2   Data Preprocessing

Data pre-processing is an essential step in any machine learning workflow. For this dataset, several pre-processing techniques were applied to transform the raw data into a format suitable for modeling. Categorical variables were encoded using one-hot encoding to convert them into numeric features. Numerical features were standardized using StandardScaler to normalize the range of values. To handle class imbalance, the minority class was oversampled using the SMO-TEENN algorithm [21]. Custom transformers were implemented for continent mapping, weight encoding, and dropping unused features. All these steps were orchestrated into a preprocessing pipeline using ColumnTransformer. Together, these techniques help handle issues like missing data, outliers, and dimensionality to yield clean, consistent features ready for modeling.

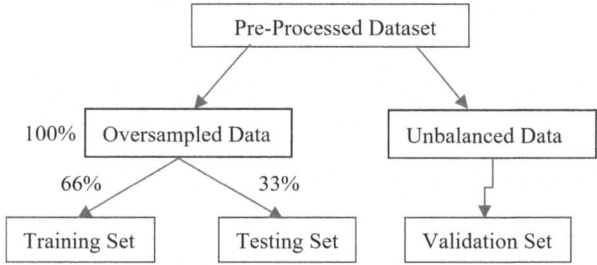

Fig. 2. Final sets for Testing, Training, and Validation

3.3   Training and Testing Model

To appropriately evaluate model performance, the over-sampled training data was split into a 66/33 ratio into mutually exclusive training and test sets while preserving class balance in both. Hyperparameter optimization was then performed via stratified cross-validation on the 66% over-sampled training portion to limit over-fitting. The models were trained on the full 66% over-sampled data with optimal hyperparameters before generating predictions on the held-out 33% over-sampled data. Critically, the final model evaluation was conducted on the original imbalanced dataset using cross-validation to provide reliable estimates of generalization performance for the intended real-world use case with skewed class frequencies.

The data was divided into training, testing, and validation sets as shown in Fig. 2. The training set was used to develop the models, the testing set was used during model development, and the final validation set was used to evaluate the final model.

Random Forest (RF) is an ensemble method that constructs multiple decision trees during training and averages their predictions to yield robust classifications and reduce overfitting. It excels at handling complex datasets with many variables.

Logistic Regression (LR) employs a sigmoidal function to predict binary outcomes, establishing probabilistic relationships between features and dichotomous target variables. It suits continuous-valued data.

Light Gradient Boosting Machine (LGBM) utilizes gradient boosting for efficient and accurate classification and regression, even on large datasets. It delivers high predictive performance while mitigating overfitting [22].

K-Nearest Neighbor (KNN), a simple and supervised learning method, finds applications in both classification and regression tasks. It operates on the assumption that similar data points are close to each other. The 'K' in KNN denotes the number of seed points selected, and this choice significantly impacts error reduction. KNN revolves around the concept of similarity, which can be defined by various factors, with Euclidean distance being a common choice for measuring similarity.

eXtreme Gradient Boosting (XGB) provides an optimized implementation of gradient boosting for handling complex data and avoiding overfitting. It is renowned for its speed and efficiency.

Naïve Bayes (NB), a supervised learning algorithm, leverages independence assumptions as it operates on joint probability distribution. This simplicity leads to faster training compared to SVM and ME models. It calculates posterior probabilities by combining prior probability with likelihood [23].

Gradient Boosting (GB) produces robust models by sequentially combining weak learners, enhancing predictive accuracy and reducing overfitting through this ensemble approach. It succeeds on complex data [24].

Cat Boost (CB) seamlessly handles categorical variables through gradient boosting. It delivers excellent performance on classification and regression with minimal data preprocessing required [25].

Support Vector Machine (SVM) is a linear supervised machine learning technique applied to both classification and regression tasks. It excels at solving pattern recognition problems while effectively mitigating overfitting concerns. SVM accomplishes class separation by establishing a distinct decision boundary [26].

Table 2. Overall Results for Autistic Spectrum Disorder Screening Data

Classifier	Single Unit (%)	4-Fold (%)	Fine-tuned(%)
Random Forest	97.62	91.92	92.08
Logistic Regression	96.04	91.34	91.67
Light GBM	98.15	90.20	88.97
KNN	93.93	87.65	90.49
XG Boost	97.88	89.18	90.99
Bernoulli NB	96.56	92.10	92.22
Gradient Boosting	98.41	90.91	90.15
Cat Boost	98.41	90.86	NaN
SVM	98.41	90.73	89.07

3.4   Model Validation

4-Fold Cross-Validation. 4-fold cross-validation employs stratified data partitioning to create 4 mutually exclusive test sets of equal size. Stratification preserves the proportional class distribution within each partition. The classification model is iteratively trained on 75% of the data (folds 1–3) and validated on the remaining 25% (fold 4). This reduces bias as each data point serves in the validation set only once. The process repeats until each fold acts as the validation set. Validation results are aggregated via averaging to produce robust performance metrics. A 4-fold CV was chosen to balance variance reduction with computational efficiency.

5-Fold Cross-Validation. In 5-fold stratified cross-validation, the dataset is divided into 5 complementary partitions or folds. Stratification maintains balanced class proportions in each fold. The model is recursively trained on 80% of the data (folds 1–4) and then validated on the remaining 20% (fold 5). Rotating which fold acts as the validation set and averaging performance across folds minimizes variability induced by data splitting. A 5-fold CV further increases the size of the validation set compared to a 4-fold CV, improving estimate reliability at a reasonable computational cost.

4   Results and Discussion

After training and evaluating nine machine learning models on the imbalanced dataset, the single-unit prediction accuracies ranged from 93.93% to 98.41% as shown in Fig. 3. Specifically, the K-Nearest Neighbor (KNN) classifier with K = 5 produced the lowest accuracy of 93.93%, while Gradient Boosting, CatBoost, and Support Vector Machine (SVM) classifiers achieved the highest accuracy of 98.41%.

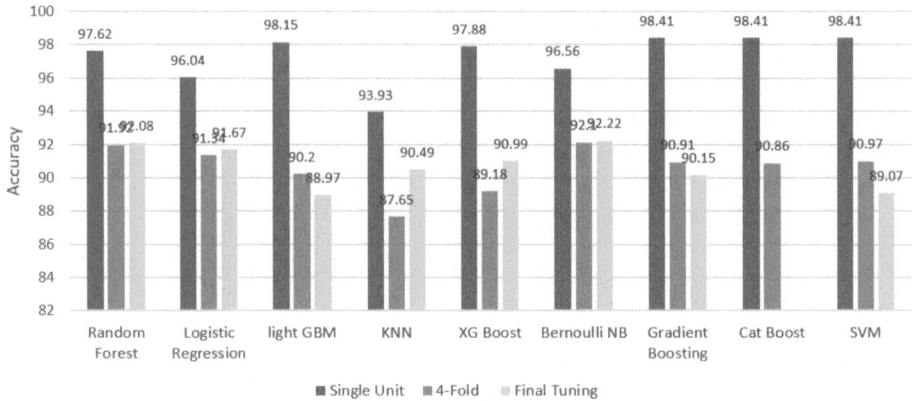

Fig. 3. Accuracy of all the algorithms to predict ASD.

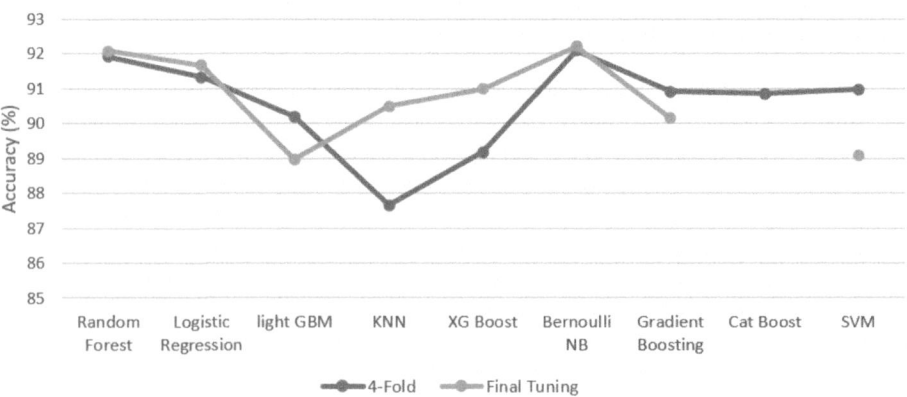

Fig. 4. Comparison of 4-fold cross-validation accuracy versus final tuned accuracy for machine learning models in autism spectrum disorder screening.

Subsequent validation using 4-fold stratified cross-validation demonstrated model accuracies between 87.65% and 92.10%. Again, the KNN model performed the worst with 87.65% accuracy. In contrast, the Bernoulli Naive Bayes classifier obtained the best cross-validation accuracy of 92.10%.

Hyperparameter tuning via GridSearchCV with nested 5-fold cross-validation was then conducted to optimize each model. The tuned models were evaluated on new imbalanced test sets using 4-fold outer cross-validation to simulate performance on unseen real-world data. Significant prediction accuracy improvements are evident for several models, as visualized via before-and-after comparisons in Fig. 4. The final cross-validation accuracies spanned 88.97% to 92.22%, with Light Gradient Boosting achieving the minimum accuracy of 88.97% compared to Bernoulli Naive Bayes with the maximum accuracy of 92.22% (Fig. 5).

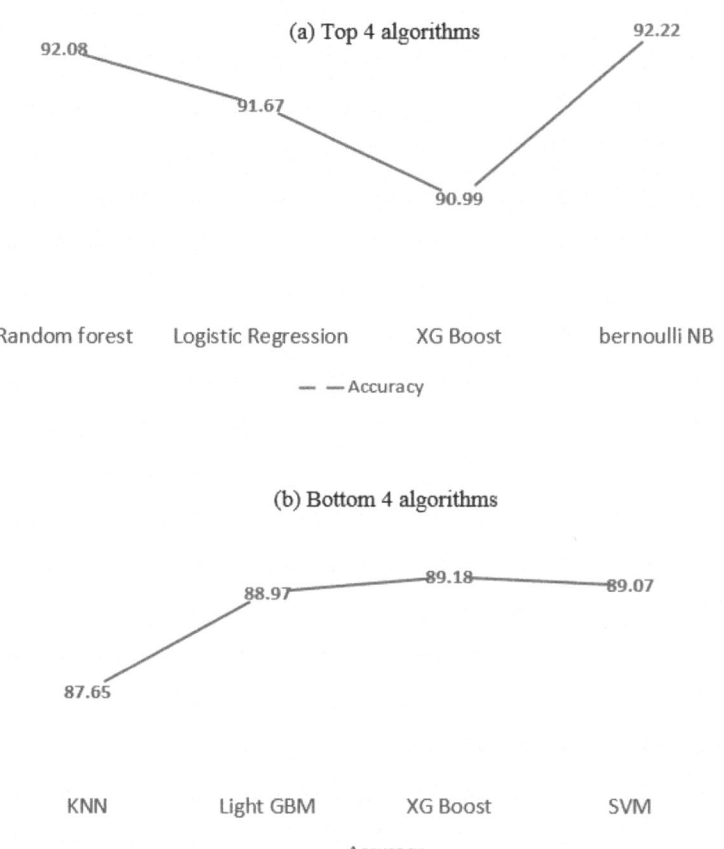

Fig. 5. Comparison of accuracy for (a) top 4 models and (b) bottom 4 models after CV.

The final predictive capabilities of the nine machine learning models, as quantified by the single unit accuracy, 4-fold cross-validation accuracy, and final tuned accuracy with nested cross-validation, are summarized in Table 2.

5   Conclusion

Machine learning has grown increasingly popular in recent years for healthcare classification tasks such as screening for medical problems. This study uses a dataset of over 800 people's demographic and medical data to demonstrate the application of nine algorithms for detecting autism spectrum disorder (ASD). With its calibrated cross-validation accuracy of 92.22%, the Bernoulli Naive Bayes classifier outperformed classifiers such as K-Nearest Neighbor and Support Vector Machine. Its effectiveness derives from applying Bayesian theory to model the likelihood of an ASD diagnosis given the inputs. While the initial single-unit testing results were overly optimistic, rigorous cross-validation gave reasonable accuracy estimates. Overall, Naive Bayes showed resistance to overfitting the imbalanced dataset and performed well both before and after adjustment. The high accuracy illustrates the viability of using demographic and medical data to detect ASD. Additional real-world testing is advised to establish the generalizability of Naive Bayes for clinical decision-making in ASD screening and diagnosis. In conclusion, our study demonstrates that machine learning, particularly Naive Bayes classifiers, enables accurate ASD screening using patient data.

Disclosure of Interests. The authors have no competing interests to declare that are relevant to the content of this article.

References

1. Tateno, M., et al.: Pervasive developmental disorders and autism spectrum disorders: are these disorders one and the same? Psychiatry Investig. **8**(1), 67 (2011)
2. Wolff, J.J., Piven, J.: Predicting autism in infancy. J. Am. Acad. Child Adolesc. Psychiatry **60**(8), 958 (2021)
3. Tanner, A., Dounavi, K.: The emergence of autism symptoms prior to 18 months of age: a systematic literature review. J. Autism Dev. Disord. **51**(3), 973–993 (2021)
4. Lord, C., Elsabbagh, M., Baird, G., Veenstra-Vanderweele, J.: Autism spectrum disorder. Lancet **392**(10146), 508–520 (2018)
5. Zwaigenbaum, L., Penner, M.: Autism spectrum disorder: advances in diagnosis and evaluation. Bmj **361** (2018)
6. Amaral, D.G.: Examining the causes of autism. In: Cerebrum: The Dana Forum on Brain Science, vol. 2017. Dana Foundation (2017)
7. Shukla, S., Banka, H.: Monophonic music composition using genetic algorithm and Bresenham's line algorithm. Multimed. Tools Appl. **81**(18), 26483–26503 (2022)
8. Jain, N., Tomar, A., Jana, P.K.: A novel scheme for employee churn problem using multi-attribute decision making approach and machine learning. J. Intell. Inf. Syst. **56**, 279–302 (2021)
9. Shukla, S., Banka, H.: Markov-based genetic algorithm with ϵ-greedy exploration for Indian classical music composition. Expert Syst. Appl. **211**, 118561 (2023)
10. Jain, N., Tomar, A., Jana, P.K.: Novel framework for performance prediction of small and medium scale enterprises: a machine learning approach. In: 2018 International Conference on Advances in Computing, Communications and Informatics (ICACCI), pp. 42–47. IEEE (2018)
11. Liu, J., et al.: Social recognition of joint attention cycles in children with autism spectrum disorders. IEEE Trans. Biomed. Eng. (2023)
12. Alkahtani, H., Aldhyani, T.H.H., Alzahrani, M.Y.: Deep learning algorithms to identify autism spectrum disorder in children-based facial landmarks. Appl. Sci. **13**(8), 4855 (2023)
13. Talukdar, J., Gogoi, D.K., Singh, T.P.: A comparative assessment of most widely used machine learning classifiers for analysing and classifying autism spectrum disorder in toddlers and adolescents. Healthcare Analytics **3**, 100178 (2023)
14. Khudhur, D.D., Khudhur, S.D.: The classification of autism spectrum disorder by machine learning methods on multiple datasets for four age groups. Measur. Sens. **27**, 100774 (2023)

15. Li, B., Sharma, A., Meng, J., Purushwalkam, S., Gowen, E.: Applying machine learning to identify autistic adults using imitation: an exploratory study. PLoS ONE **12**(8), e0182652 (2017)
16. Gaspar, A., Oliva, D., Hinojosa, S., Aranguren, I., Zaldivar, D.: An optimized kernel extreme learning machine for the classification of the autism spectrum disorder by using gaze tracking images. Appl. Soft Comput. **120**, 108654 (2022)
17. Kavitha, V., Siva, R.: Classification of toddler, child, adolescent and adult for autism spectrum disorder using machine learning algorithm. In: 2023 9th International Conference on Advanced Computing and Communication Systems (ICACCS), vol. 1, pp. 2444–2449. IEEE (2023)
18. Zhao, H., Swanson, A.R., Weitlauf, A.S., Warren, Z.E., Sarkar, N.: Hand-in-hand: a communication-enhancement collaborative virtual reality system for promoting social interaction in children with autism spectrum disorders. IEEE Trans. Hum.-Mach. Syst. **48**(2), 136–148 (2018)
19. Thabtah, F.: Autism Screening Adult. UCI Machine Learning Repository (2017). https://doi.org/10.24432/C5F019
20. Baron-Cohen, S., Wheelwright, S., Skinner, R., Martin, J., Clubley, E.: The autism-spectrum quotient (AQ): evidence from Asperger syndrome/high-functioning autism, males and females, scientists and mathematicians. J. Autism Dev. Disord. **31**(1), 5–17 (2001)
21. Lamari, M., et al.: SMOTE–ENN-based data sampling and improved dynamic ensemble selection for imbalanced medical data classification. In: Saeed, F., Al-Hadhrami, T., Mohammed, F., Mohammed, E. (eds.) Advances on Smart and Soft Computing. AISC, vol. 1188, pp. 37–49. Springer, Singapore (2021). https://doi.org/10.1007/978-981-15-6048-4_4
22. Ke, G., et al.: LightGBM: a highly efficient gradient boosting decision tree. In: Advances in Neural Information Processing Systems, vol. 30 (2017)
23. John, G.H., Langley, P.: Estimating continuous distributions in Bayesian classifiers. In: Proceedings of the Eleventh Conference on Uncertainty in Artificial Intelligence, UAI 1995, San Francisco, CA, USA, pp. 338–345. Morgan Kaufmann Publishers Inc. (1995)
24. Aziz, N., et al.: A study on gradient boosting algorithms for development of AI monitoring and prediction systems. In: 2020 International Conference on Computational Intelligence (ICCI), pp. 11–16. IEEE (2020)
25. Prokhorenkova, L., Gusev, G., Vorobev, A., Dorogush, A.V., Gulin, A.: Catboost: unbiased boosting with categorical features. In: Advances in Neural Information Processing Systems, vol. 31 (2018)
26. Wen, Z., Zhou, Z., Liu, H., He, B., Li, X., Chen, J.: Enhancing SVMs with problem context aware pipeline. In: Proceedings of the 27th ACM SIGKDD Conference on Knowledge Discovery & Data Mining, pp. 1821–1829 (2021)

Heart Disease Prediction Model Using Machine Learning Techniques

Bipin Kumar Rai[1]([✉]) [iD], Aparna Jha[2] [iD], Shreyal Srivastava[2] [iD], and Aman Bind[2] [iD]

[1] Department of Computer Science & Engineering, Dayananda Sagar University, Bengaluru, India
bipinkrai@gmail.com

[2] Department of IT, ABES Institute of Technology, Ghaziabad, Uttar Pradesh 201009, India

Abstract. In the present day, individuals often find themselves engrossed in their daily routines, focusing on work and various activities, while inadvertently neglecting their well-being. As a consequence of this demanding lifestyle and lack of attention to health, there is a rapid increase in the number of people falling ill. Furthermore, a substantial portion of the population is grappling with health issues, particularly heart disease. Cardiovascular diseases, claiming 17.9 million lives annually, rank as the foremost cause of death globally, as reported by the WHO. Hence early stage detection of cardiac disease through early-stage signs is a major need in today's world. A system that can aid in the early diagnosis of cardiac disease is suggested. This method aims to develop a ML model to detect cardiac disease in its early stages. The different ML algorithms such as RF, KNN, LR, DT and SVM are used to attain the maximum accuracy. UCI repositories dataset was used for performing the experiments. In this study the accuracy of 80.33% is obtained with SVM.

Keywords: Cardiovascular diseases (CVDs) · Heart Disease · KNN · SVM · RF · DT · LR

1 Introduction

In the contemporary world, people are increasingly immersed in their busy lifestyles, leaving little time for self-care. The resulting stress, anxiety, and depression contribute to widespread health issues. Serious diseases like cancer, heart disease, and tuberculosis annually claim numerous lives, with cardiovascular disease (CVD) standing out as the major cause of death in the medical landscape. This highlights the urgent need for prioritizing well-being amid the relentless pace of modern life.

Heart disease encompasses diverse forms. Patients with heart disease manifest symptoms like chest pain, dizziness, shortness of breath, profuse sweating, and fatigue. This underscores the critical impact of heart-related conditions and the urgent need for awareness and preventive measures to mitigate the global burden of cardiovascular diseases. Smoking, high blood pressure, diabetes, obesity, extreme stress, genetics, high cholesterol level, etc. are some of the main reasons behind heart disease. However, heart disease can vary irrespective of gender or age of the person.

© The Author(s), under exclusive license to Springer Nature Switzerland AG 2025
A. K. Bairwa et al. (Eds.): ICCAIML 2024, CCIS 2184, pp. 290–301, 2025.
https://doi.org/10.1007/978-3-031-71481-8_23

By analyzing medical data from diverse patients, authors aim to assess the risk and symptoms of heart-related diseases using advanced machine learning techniques. This predictive approach has significant potential to revolutionize both the medical field and individuals' lives, enhancing early detection and proactive management of heart-related conditions. There are many different types of heart diseases which are:

Coronary Artery Disease: A prevalent cardiac condition where coronary arteries narrow or suffer damage, hindering optimal oxygen and nutrient delivery to the heart due to cholesterol plaque build-up.

Heart Failure: It is also known as congestive heart failure, it occurs when the heart struggles to efficiently pump blood to various body parts, often stemming from an extended case of coronary artery disease, weakening the heart's pumping capacity.

Congenital Heart Disease: Present from birth, it encompasses defects like septal issues, where there are holes between heart parts, obstructive defects blocking blood flow, and cyanotic heart disease causing oxygen shortage.

Cardiomyopathy: Involves a structural or functional impairment of heart muscles, leading to altered pumping capabilities or muscle structure.

2   Related Work

In this segment, a range of studies from different origins that center around the prediction of heart diseases is presented. Having the capability to anticipate whether an individual might be prone to heart disease holds significant benefits, both in the medical field and for individuals. When one possess knowledge about the potential risks related to heart disease, one can raise awareness amid the general population, motivating them to adopt preventive measures for their well-being. As a result, numerous researchers have delved into various approaches and models to forecast aspects related to heart disease. Some of the works are as follows:

Devansh Shah *et al.,* in 2020 have proposed a predictive system for heart disease development. Their researcher used NB, RF, DT and KNN algorithms on the dataset from the UCI repository. The dataset was categorized into training and test sets, with the KNN model achieving the highest accuracy at 90.789%. (k = 7) [1].

The researchers Pooja Rani *et al.,* in 2021 have proposed an optimized decision assistant for heart disease diagnosis, surpassing existing systems in accuracy. The work utilizes multivariate imputation by chained equations algorithm to handle missing values, and a hybrid feature selection algorithm (GA and RFE) identified relevant features from the UCI Cleveland heart disease dataset. Pre-processing involved SMOTE and standard scalar methods. The hybrid system, incorporating LR, SVM, RF, NB and adaboost classifiers, revealed RF as the top performer, achieving an 86.6% accuracy when combined with MICE, GARFE, Scaling, and SMOTE [2].

Author Niloy Biswas *et al.,* in 2023 employed various feature selection techniques to identify main factors for heart disease prediction. Utilizing chi-square, ANOVA, and mutual information algorithms, they applied six machine learning models (LR, SVM, NB, KNN, RF, and DT) to the selected features [3]. Random Forest emerged as the top performer, achieving a noteworthy 94.51% accuracy, 94.23% specificity, 94.87% sensitivity, 94.95 AURC, and 0.31 log loss for SF3 feature subsets. Another study done by Rahul Katarya *et al.*, compared various ML algorithms on (UCI) dataset for heart disease prediction. Their objective was to identify the most effective algorithm, and their findings revealed that Random Forest outperformed others, achieving a remarkable accuracy of 95.06 [4]. The evaluation metrics included Mean Absolute Error, precision, Root Mean Square Error and recall, with Random Forest demonstrating superior performance across these measures.

The author M. Kavitha *et al.*, have used machine learning to achieve early prediction of heart disease, potentially saving lives. Their approach involved a hybrid model that combined RF and DT methods, yielding an impressive accuracy rate of 88.7%. Additionally, they developed a user-friendly program that leverages this model for predicting heart disease, contributing to accessible and efficient healthcare solutions [5]. In another study by author Md Mamun Ali *et al.*, have made a comparison between six ML algorithms on heart disease dataset from Kaggle to find the best algorithm via various evaluation metrices [6] The algorithms used were KNN, RF, multilayer perceptron, LR, DT and AdaboostM1. RF, KNN and DT performed well with 100% accuracy.

The suggested study by Manoj Diwakar *et al.*, in 2020 [7] highlights the importance of timely disease diagnosis, particularly for life-threatening conditions like heart disease. They suggest that machine learning classification methods can aid in quick and reliable disease diagnosis, focusing on heart disease, which is challenging to detect. The paper reviews common machine learning classification techniques and explores their use in combination with image fusion for improved diagnostic accuracy in healthcare. Archana Singh and Rakesh Kumar underscore the crucial need for accurate diagnosis and prediction of heart-related diseases, given their potentially life-threatening consequences [8]. Recognizing the escalating number of such deaths, they advocate for a reliable prediction system using ML, specifically decision tree, KNN, SVM, and LR algorithms. Employing Anaconda notebook in Python programming enhances precision and accuracy, utilizing Cleveland dataset.

In [9] authors Harshit Jindal *et al.*, in 2021 addresses the rising concern of increasing heart disease cases by developing a predictive system to identify potential patients at risk. Leveraging medical history and machine learning algorithms, including logistic regression and KNN, their model exhibited superior performance in accurately predicting heart disease compared to methods like naive bayes. The system, implemented in a.pynb format, not only eases the workload for medical professionals but also enhances healthcare accuracy and reduces costs, providing valuable insights for proactive heart disease prediction. Baban.U. Rindhe *et al.*, in 2022 offered a comprehensive exploration of ML techniques for heart disease classification. Applying SVM, ANN, and RF on the Cleveland dataset, SVM yielded the highest accuracy at 84.0%, followed by ANN with 83.5%, and RF classifier with 80.0% [10].

In 2023, Sreenivas Mekala *et al.*, did a comparative study of ML algorithms— SVM, RF, DT, NB, LR, Adaptive Boosting, and Extreme XGBoost [11]. Their evaluation, utilizing confusion and correlation matrices, revealed that XGBoost demonstrated superior accuracy compared to other algorithms. Calculating accuracy using these matrices, the study concluded that both XGBoost and Random Forest Classifiers achieved the highest accuracy, reaching approximately 86%. Jyoti Maurya *et al.,* did a comparative study of various ML algorithms, including SVM, DT, RF, LR, AdaBoost, Extra Tree Classifier, Gaussian Naive Bayes, KNN and Gradient Boosting, on the UCI dataset. Considering 14 key factors for diagnosis, Gradient Boosting emerged with the highest accuracy of 95.08%, outperforming other algorithms. The evaluation utilized confusion matrices for each method, emphasizing Gradient Boosting's effectiveness in heart disease classification during training and testing [12].

Sai Bhavan Gubbala conducted a comparison of machine learning algorithms—RF classification, AdaBoost Classifier, SVM, LR, and DT Classifier. Utilizing a dataset from the UCI repository, research found that Random Forest achieved the highest accuracy among classifiers, reaching 85.22% [13]. Python was employed for model training, dataset splitting, and accuracy assessment. In 2022, researcher Mafia Rasheed *et al.,* employed machine learning, specifically SVM, to detect coronary heart disorder. Utilizing a Kaggle dataset with patient history and attributes, they performed data preprocessing and labeled it differently before applying the SVM trained model for improved accuracy [14]. Results were analyzed using a Confusion Matrix, confirming a model accuracy of 90.47% in identifying coronary heart disorder through this approach.

In 2018, Aljanabi *et al.,* reviewed ML techniques for heart disease prediction, exploring classification methods including Naive Bayes, ANN, DT, and SVM. The paper offers a survey of relevant research and compares the accuracy of various approaches [15]. The authors advocate for the use of ANN and DT algorithms in heart disease prediction, underscoring the importance of high-quality datasets and effective preprocessing techniques to enhance prediction accuracy.

In 2018, Marimuthu *et al.,* conducted research reviewing ML and data mining techniques for heart disease prediction. The study evaluated multiple algorithms, emphasizing the necessity for more intricate models to enhance accuracy. The authors recommended future research to concentrate on feature selection methods and explore diverse rules and algorithms to further advance the field [16].

Sashank Yadav *et al.,* aimed to enhance data modeling by incorporating seven machine learning classifiers with an HTML template serving as a User Interface. Employing Naive Bayes, DT, KNN, LR, SVM, Gradient Boosting, and RF algorithms, the paper provides a comparative analysis of performance measures. The classification report showed that the KNN achieved a notably higher accuracy of 85.18%, outperforming other machine learning algorithms in the study [17].

In [18] J. Gowri *et al.,* proposed a method to determine the probability of heart diseases in patients. Utilizing the UCI dataset, on LR, DT, Naïve Bayes, KNN, RF, XGBoost, and SVM. Among these, Support Vector Machine demonstrated the highest accuracy at 85.7%, surpassing the other six ML algorithms.

In the following Table 1, it is provided a summary of those papers whose work would be further compared with proposed model on the basis of evaluation metrics.

Table 1. Summary of significant work on the basis of evaluation metrics

S.No.	Author & Year	Proposed Work	Evaluation matrices
1	Devansh Shah *et al.* (2020) [1]	ML classification techniques NB, DT, RF, and K-NN	Accuracy: K-NN - 90.789%
2	Pooja Rani *et al.* (2021) [2]	Feature selection - hybrid of GA & Recursive Feature Elimination (RFE). Classification algorithms are NB, SVM, LR, RF, and AdaBoost classifier	Accuracy: 86.60(RF) Sensitivity: 84.14(RF) Specificity: 89.63 (AdaBoost) Precision: 88.96 (AdaBoost)
3	Niloy Biswas *et al.* (2023) [3]	Feature Selection approaches - chi-square, ANOVA, and MI Feature subsets - SF1, SF2, and SF3 Machine learning models such as LR(C1), SVM(C2), K-NN(C3), RF (C4), NB(C5), and DT(C6)	Accuracy: **SF1** - C1(93.41%) **SF2** - C1(93.41%) **SF3** - C4 (94.51%) Sensitivity: C1 - 94.74 for SF1, SF2, SF3 Specificity: C1- 92.45 for SF1, SF2 & C4-94.23 for SF3
4	Rahul Katarya *et al.* (2020) [4]	ML Algorithms- LR, NB, MLP, RF, DT, ANN, DNN, SVM, KNN	Accuracy: 95.06% (RF) Precision: 0.55(RF) Recall: 0.97(RF)
5	Kavitha M *et al.* (2021) [5]	The author used a hybrid model combining RF and DT	Accuracy: Using Hybrid Model (RF & DT) - 88.7%
6	Md Mamun Ali *et al.* [6]	ML Algorithms - MP, KNN, RF, DT, LR and ABM1	Accuracy: KNN, RF and DT-100% Sensitivity: 100% Specificity: 100%
7	Jindal H *et al.* (2021) [9]	Employed machine learning algorithms are LR & KNN	Accuracy: KNN - 88.52%
8	Jyoti Maurya *et al.* (2023) [12]	ML Algorithms- Gradient Boosting, SVM, KNN, RF, DT, AdaBoost, NB, Extra Tree Classifier and LR	Accuracy: Gradient Boosting-95.08%
9	Mafia Rasheed *et al.* (2022) [14]	Classification Algorithm - SVM	Accuracy: 90.47%, Sensitivity: 93.44%, Specificity: 88.45%, Precision: 84.62%, F1 Score: 88.81%

The above table summarizes the work of some of the researchers, Md. Mamun Ali et al. used KNN, Random Forest, and Decision tree resulting 100% accuracy, 100% Sensitivity, 100% Specificity. However, the used dataset had redundant instances which lead to inaccurate result. Rahul Katarya et al. used Random Forest with accuracy of

95.06% with Precision: 0.55(RF) and Recall: 0.97(RF). However, the authors have not handled the missing values properly using any standard technique, they replaced missing values by python library by NAN. This could lead to inaccurate evaluation of the proposed model.

Therefore, in order to build accurate and reliable results, preprocessing of the data using standard preprocessing technique has been done. And then, the 5 algorithms (K-NN, LR, DT, RF, and SVM) are trained on the most important feature. Then for the evaluation of our proposed classification model, the algorithms are tested for accuracy and sensitivity.

3   Proposed Work

Jupyter notebook is used to perform heart disease prediction of the dataset. The proposed model is prepared through various stages. The workflow of the model mentioning all the stages is depicted using Fig. 1.

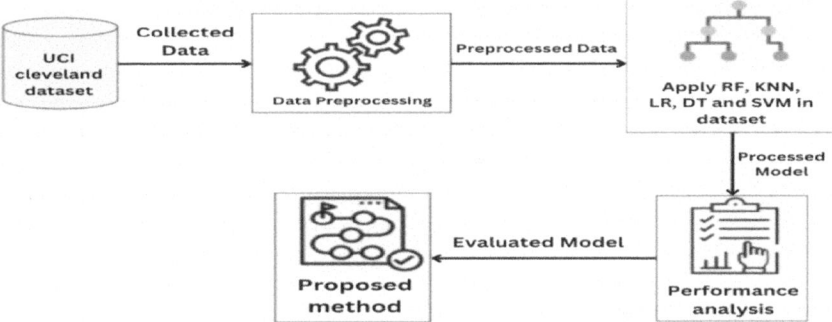

Fig. 1. Architecture of Heart Disease Prediction Model

3.1   Dataset

Heart disease dataset from UCI repository is used for the construction of the model. This dataset is widely used for CVDs prediction so, it is also used in the proposed model for predicting heart disease. The dataset consists of 76 attributes, but almost all published work uses a subset of 14 of them which are necessary for the experiment. The dataset matrix information is given in Fig. 2 [19].

S. No.	Feature Name	Description
1	Age	Age between 29 and 77
2	Sex	Male: 1, female: 0
3	Type of Chest Pain	Typical angina: 1, atypical angina: 2 non-angina pain: 3, asymptomatic: 4
4	Resting Blood Pressure	Between 94 mm Hg and 200 mm Hg
5	Serum Cholesterol	Between 126 mg/dl and 564 mg/dl
6	Fasting Blood Sugar	FBSR>120 mg/dl (true:1, false: 0)
7	Resting electrocardiographic results	Normal: 0, ST-T wave abnormality: 1, Hypertrophy: 2
8	Maximum heart rate achieved	Between 71 and 202
9	Exercise-induced angina	Yes: 1, No: 0
10	Peak exercise ST segment	Up sloping: 1, Flat: 2, downsloping: 3
11	ST depression induced by exercise relative to rest	Between 0 and 6.2
12	Number of major vessels (0–3) coloured by fluoroscopy	Between 0 and 3
13	Thallium	Normal: 3, fixed defect: 6, reversible defect: 7
14	Target	Heart disease present: 1, heart disease absent: 0

Fig. 2. Heart disease dataset description

3.2 Data Preprocessing

It is the transformation of raw data into meaningful data. In machine learning we cannot work with raw data, hence it is a very important step. The Random Forest algorithm cannot take in null values in a dataset. To address this, it's necessary to handle and manage null values directly from the original raw data before applying the algorithm.

In the dataset, four records with inaccurate information on AC and two on LTH are identified and replaced with optimal values to ensure data integrity and reliability.

The following steps are performed to preprocess our dataset:

1. Getting the CSV file of the dataset.
2. Importing Numpy, Matplotlib, Pandas libraries
3. Importing the dataset into our Jupyter IDE
4. Handling missing values and dropping the duplicate value
5. Encoding the categorical data
6. Dataset is split for training and testing where 80% is used for training and the rest is used for testing.
7. The last step is feature scaling [20].

Figure 3 shows a heat map that visually depicts the relationships and correlations between different features in a dataset. Each cell in the heatmap corresponds to the

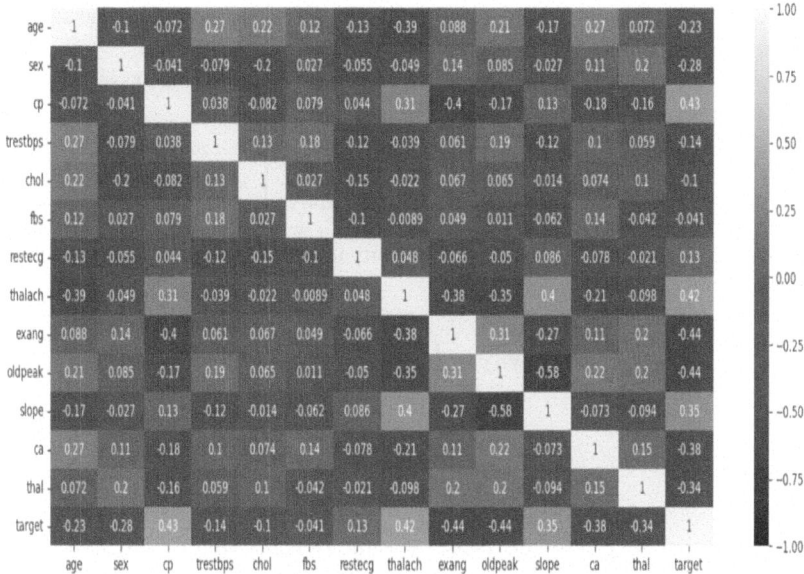

Fig. 3. Correlation Matrix Heat Map

correlation between two specific features, with the color of the cell indicating the strength of that correlation.

If the correlation value is less than zero, it suggests a negative correlation. A correlation value of zero indicates no correlation between the features, while positive values signify a positive correlation. The intensity of the color reflects the magnitude of the correlation, with darker colors representing a strong dependency among features. In essence, the heatmap provides a quick and visual way to understand the patterns and relationships within a set of features.

3.3   Classification Algorithm

Classification algorithms that are used for performing prediction are- RF, KNN, SVM, LR and DT.

Random Forest combines decision trees to enhance predictions, while K-Nearest Neighbor classifies using training data and computes distances. SVM creates a wide-margin hyperplane for data separation. Logistic Regression predicts categorical variables. Decision Trees categorize based on entropy [20].

3.4   Performance Evaluation Metrics

Performance evaluation metrics like Accuracy, Sensitivity assess classification models based on TP(True Positives) and FN(False Negative). These metrics aid in comparing algorithmic performances. The most effective algorithm is determined by achieving the highest outcomes across these measures. Careful analysis using these evaluation criteria ensures a comprehensive assessment of model effectiveness and guides the selection of the best-performing algorithm for a given task [21].

Accuracy: The measure of a system's correctness in predictions, assessing the ratio of correct outcomes,

$$\text{Accuracy} = \left(\frac{CorrectPrediction}{TotalPredictions} \right) \times 100 \tag{1}$$

Sensitivity: It is measure of the model's ability to correctly identify positive instances in predictions,

$$\text{Sensitivity} = \left(\frac{TP}{TP + FN} \right) \times 100 \tag{2}$$

4   Result

In the proposed system, Firstly the dataset was filtered checking for any null values and dropping the duplicate values from the data, then it is preprocessed. After preprocessing the categorical features are encoded. After encoding, feature scaling is performed for linear machine learning algorithms. The dataset is then split into 80% data for training and 20% data for testing.

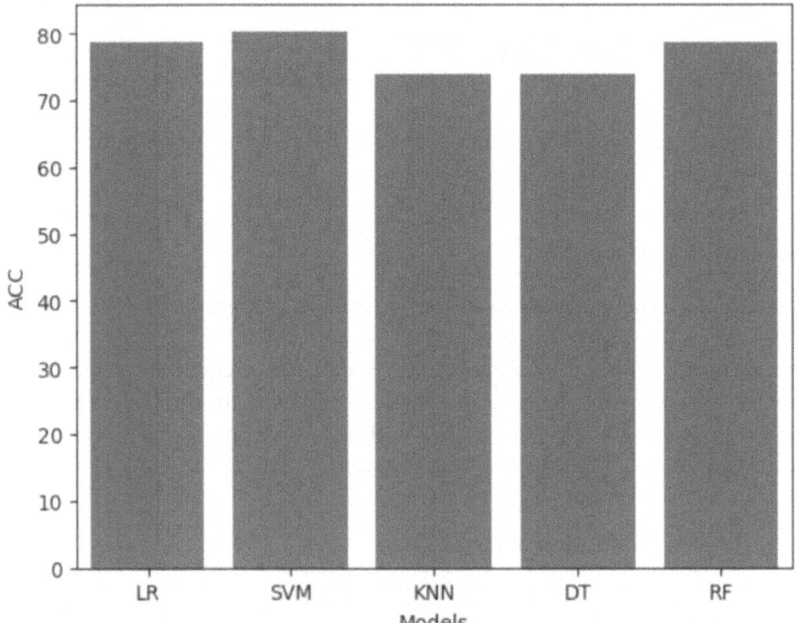

Fig. 4. Comparative accuracy result of classification techniques

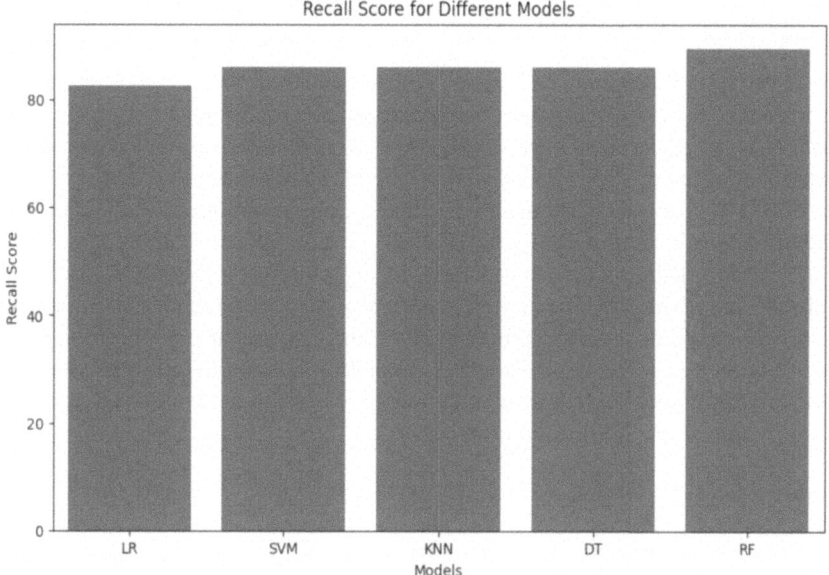

Fig. 5. Comparative sensitivity result of classification techniques

The features are used in the five classifier algorithms (RF, KNN, LR, SVM, DT) for improving the accuracy. From looking at the values of accuracy and sensitivity it is deduced that the algorithms that perform better on the feature set is SVM.

Figures 4 and 5 shows the comparison of different algorithms on the basis of evaluation metrics accuracy and sensitivity respectively.

Table 2. Accuracy and Sensitivity Comparison

Algorithm	Accuracy	Sensitivity
LR	78.69	82.76
SVM	80.33	86.21
KNN	73.77	86.21
DT	73.77	86.21
RF	78.67	89.66

In Table 2, we see that SVM has highest accuracy of 80.33% and RF has highest sensitivity of 89.66% with the UCI dataset.

5 Conclusion and Future Scope

Heart disease or CVDs causes a lot of death globally and the major reason is late detection of the problem, so we need a system that would predict the presence of problem in the early stage. Many researchers have worked on this problem and provided a solution for the betterment of the problem. The proposed model, on the other hand, tried to find the cons of other's solution and tried to enhance that area in our solution. Authors have dealt with the missing values from the dataset. Further, also dropped the duplicate values from the data. Working on the dataset using 5 classifier algorithms – SVM, KNN, LR, DT, RF it is concluded that SVM gives the highest accuracy of 80.33% which is maximum among all the 5 classifier models. The future involves an increased use of machine learning for precise heart disease analysis and early prediction, aiming to minimize death rates. This proactive approach, driven by heightened disease awareness, holds the potential to significantly impact public health outcomes.

Acknowledgments. No Funding received.

Disclosure of Interests. The authors have no competing interests to declare that are relevant to the content.

References

1. Shah, D., Patel, S., Bharti, S.K.: Heart Disease Prediction using Machine Learning Techniques. SN Comput Sci. **1**, 1(6) (2020)
2. Rani, P., Kumar, R., Ahmed, N.M.O.S., Jain, A.: A decision support system for heart disease prediction based upon machine learning. J. Reliab. Intell. Environ. **7**(3), 263–275 (2021)
3. Biswas, N., et al.: Machine learning-based model to predict heart disease in early stage employing different feature selection techniques. Biomed Res. Int. 2023 (2023)
4. Katarya, R., Meena, S.K.: Machine learning techniques for heart disease prediction: a comparative study and analysis. Health Technol (Berl). **11**(1), 87–97 (2021)
5. Kavitha, M., Gnaneswar, G., Dinesh, R., Sai, Y.R., Suraj, R.S.: Heart disease prediction using hybrid machine learning model. In: Proceedings of the 6th International Conference on Inventive Computation Technologies, ICICT 2021, pp. 1329–33. Institute of Electrical and Electronics Engineers Inc. (2021)
6. Ali, M.M., Paul, B.K., Ahmed, K., Bui, F.M., Quinn, J.M.W., Moni, M.A.: Heart disease prediction using supervised machine learning algorithms: Performance analysis and comparison. Comput. Biol. Med. **1**, 136 (2021)
7. Diwakar, M., et al.: Latest trends on heart disease prediction using machine learning and image fusion. In: Materials Today: Proceedings, pp. 3213–8. Elsevier Ltd. (2020)
8. Soni, S.K.: Madan Mohan Malaviya University of Technology, North Dakota State University, Institute of Electrical and Electronics Engineers. Uttar Pradesh Section, IEEE Industry Applications Society, Institute of Electrical and Electronics Engineers. ICE3-2020: International Conference on Electrical and Electronics Engineering: February 14–15 (2020)
9. Jindal, H., Agrawal, S., Khera, R., Jain, R., Nagrath, P.: Heart disease prediction using machine learning algorithms. In: IOP Conference Series: Materials Science and Engineering. IOP Publishing Ltd. (2021)

10. Rindhe Baban, U., Ahire, N., Patil, R., Gagare, S., Darade, M.: Heart disease prediction using machine learning. International Journal of Advanced Research in Science, Communication and Technology [Internet], 267–76 (2021). http://ijarsct.co.in/may1.html
11. Mekala, S., et al.: Cardiopathy-heart disease prediction using machine learning [Internet] (2023). https://www.researchgate.net/publication/371760631
12. Maurya, J., Malaviya, M.M., Prakash, S., Mohan, M.: Machine Learning based Prediction and Diagnosis of Heart Disease using multiple models (2023). https://doi.org/10.21203/rs.3.rs-2642516/v1
13. Gubbala, S.: Heart Disease Prediction Using Machine Learning Techniques [Internet] (2022). www.irjet.net
14. Rasheed, M., et al.: Heart disease prediction using machine learning method. In: International Conference on Cyber Resilience, ICCR 2022. Institute of Electrical and Electronics Engineers Inc. (2022)
15. Aljanabi, M., Qutqut, M.H., Hijjawi, M., Al-Janabi, M.I.: SEE PROFILE Machine Learning Classification Techniques for Heart Disease Prediction: A Review. Review Article in Int. J. Eng. Technol. [Internet] **7**(4), 5373–9 (2018). https://www.researchgate.net/publication/328031918
16. Marimuthu, M., Abinaya, M., Madhankumar, S.K., Pavithra V.: A review on heart disease prediction using machine learning and data analytics approach. Int. J. Comput. Appl. **181**(18), 20–5 (2018)
17. Jadhav, V.J., Yadav, S., Singh, A., Jadhav, V., Jadhav, R.: Heart Disease Prediction Using Machine Learning (2022). https://www.researchgate.net/publication/371530417
18. Gowri, J., Kamini, R., Vaishnavi, G., Thasvin, S., Vaishna, C.: Heart disease prediction u sing machine learning. Int. J. Innov. Technol. Explor. Eng. [Internet] **11**(8), 29–32 (2022). https://www.ijitee.org/portfolio-item/h91480711822/
19. Janosi Andras, S.W.P.M.: Detrano R. Heart Disease (1988)
20. Kaur, G., Bindal, R., Kaur, V., Kaur, I., Gupta, S.: Bitcoin and cryptocurrency technologies: a review. In: AIP Conference Proceedings. American Institute of Physics Inc. (2023)
21. Sharma, S., Rai, B.K., Gupta, M., Dinkar, M.: DDPIS: diabetes disease prediction by improvising SVM. Int. J. Reliable Qual E-Healthc (IJRQEH) **12**(2), 1–11 (2023)

A Review to Enhance Driving Behavior Through Advanced Drowsiness Detection Systems: Integrating Deep Learning and Machine Learning Techniques

Ritu[1]([envelope]), Meenu Vijarania[1], Meenakshi Malik[2], and Poonam Yadav[1]

[1] K. R Mangalam University, Gurugram, India
ritumalik2@gmail.com
[2] BML Munjal University, Gurugram, India

Abstract. A summary of the most recent research-based driver sleepiness detection systems implemented. The article explains and showcases more recent technologies that use a variety of measures to track and recognize tiredness. Considering traffic accidents pose a serious threat to the public's safety, driving safety is a crucial concern in today's society. Taking appropriate action to mitigate the dangers connected with driving is imperative due to the growing number of cars and the difficulties of modern urban areas. Drunk driving is one of these threats; it may be a subtle but deadly enemy that impairs a driver's awareness and judgement. Research suggests an all-encompassing strategy to improve road safety by comprehending and treating the elements causing fatigued driving. By reducing the financial costs linked to traffic accidents, the results of such programs not only protect lives but also enhance society as a whole. This study combines traditional machine learning techniques with deep learning architectures to identify tiredness in driving behavior. This review paper focuses on study of accuracy of various machine learning techniques in field of driver behavior drowsiness detection system.

Keywords: machine learning · artificial intelligence · drowsiness · behavior

1 Introduction

Driving safety is a paramount concern in today's modern society, as road traffic accidents continue to pose a significant threat to public well-being. The ever-increasing volume of vehicles on the road, combined with the potential distractions and challenges presented by contemporary urban environments, has intensified the need for effective measures to mitigate the risks associated with driving. Among the various factors that contribute to road accidents, drowsiness stands out as a silent yet perilous adversary.

Drowsiness, often underestimated in its potential to induce accidents, plays a pivotal role in compromising a driver's alertness and decision-making abilities. According to estimates from the National Highway Traffic Safety Administration (NHTSA), sleepy

A. K. Bairwa et al. (Eds.): ICCAIML 2024, CCIS 2184, pp. 302–311, 2025.
https://doi.org/10.1007/978-3-031-71481-8_24

driving is responsible for a substantial percentage of accidents each year, with consequences ranging from minor collisions to severe, life-altering crashes. Drowsy driving was a factor in 91,000 police-reported accidents in 2017, according to NHTSA estimates. These crashes resulted in around 800 fatalities and over 50,000 injuries [1]. These accidents not only lead to loss of life but also incur significant economic costs, straining healthcare systems and societal resources.

Given the numerous sleeping issues [2] Increasing drowsy driving awareness among adults in the United States can be very beneficial for public health. Driving safely on the road can be accomplished by being aware of the causes, effects, and avoidance of sleepy driving. The National Sleep Foundation reports that 54% of drivers report feeling drowsy while driving, and 28% of adult drivers have acknowledged to dozing off while operating a vehicle. Additionally, more than 40% admit to dozing off while driving at least once [3].

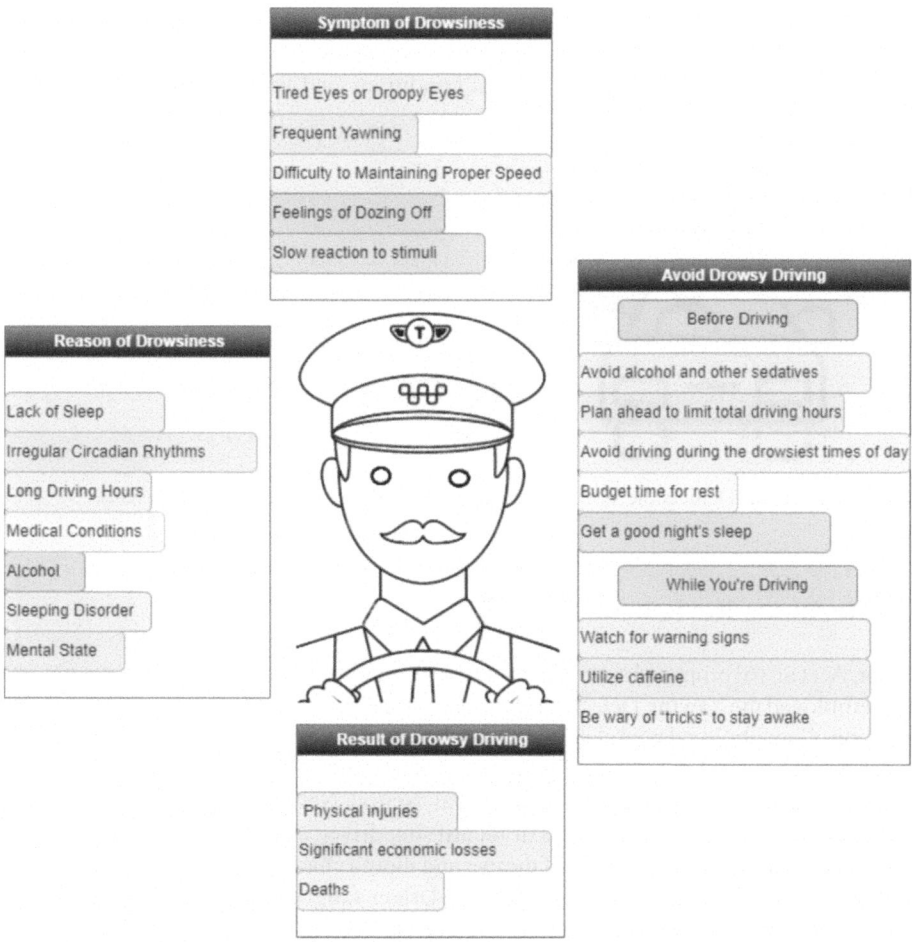

Fig. 1. Overview of Drowsy Driving

Driving accidents and tiredness have a complicated link that is influenced by a variety of circumstances including lack of sleep, irregular circadian rhythms, and long driving hours. Figure 1 Drowsy drivers are more likely to have slower reaction times, poorer coordination, and decreased cognitive abilities, which increases the likelihood that they may make mistakes while operating a vehicle.

The effects affect more than just individual drivers; they also have an impact on passengers, pedestrians, and other users of the road, highlighting the urgent need for effective measures. To identify driving behavior while fatigued, a variety of machine learning and deep learning algorithms are employed. The state of the driver is affected by a number of psychological, physiological, and mental aspects [4].

The following six criteria were mentioned by Nishiyama as necessary for in-vehicle technologies to detect tiredness [5]:

1. Driver fatigue is detectable in real time and can vary from mild to deep slumber.
2. Every driver is regarded as a target that can be detected.
3. Compared to other applications, they are very scalable and inexpensive.
4. Take care not to impede the driver's ability to drive safely.
5. Able to endure challenging operational conditions and be placed in a vehicle and identify fatigue in drivers (Fig. 2).

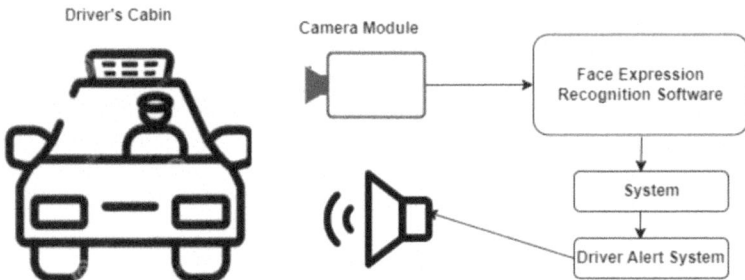

Fig. 2. Depicts the systematics diagram for the driver behavior analysis

2 Literature Review

Rajkar, A. et al. [6] proposed a deep learning-based method to identify drivers' sleepiness. They employed the Yawing Detection Dataset (YawDD), Face and Eye regions for fatigue detection, Convolutional neural network (CNN), a type of deep learning with an average accuracy of 96%, and the Closed Eye in the Wild dataset (CEW).

B. K. Savaş et al. [7] The Convolutional Neural Network (ConNN*) model for multitasking is suggested as a means of identifying driver weariness or drowsiness. The driver's behavior model makes use of the eye and mouth characteristics. Driver weariness is tracked by modifications to these features. Driver fatigue is measured by taking the length of an eye closure/percentage of eye closure (PERCLOS) and the frequency of yawning/frequency of mouth (FOM). Three classifications based on the driver's level of weariness are recognized in this investigation. The YawdDD and NthuDDD datasets showed that the suggested model detected fatigue with 98.81% accuracy. A comparative analysis of the model's success is provided.

S. Bavkar et al. [8] an innovative method for using an electroencephalogram (EEG) sensor to quickly identify alcoholism. To classify alcoholics and normal subjects, the suggested approach uses absolute gamma band power as a feature and ensemble subspace K-NN as a classifier. Additionally, the paper reports the use of the optimum EEG channels for speedy alcoholism screening using the Improved Binary Gravitational Search Algorithm (IBGSA) as an optimization tool. The suggested approach's outcomes are contrasted with those of optimization algorithms as the genetic algorithm (GA), binary particle swarm optimization (BPSO), and binary gravitational search algorithm (BGSA). The accuracy acquired via the ensemble subspace K-Nearest Neighbor (K-NN) classifier is used to evaluate the fitness function for various optimization strategies. With just 13 EEG channels, the suggested improved binary gravitational search algorithm (IBGSA) approach offers a 92.50% detection accuracy.

Bavkar, S et al. [9] Distinguish between alcohol and non-alcohol users based on visually evoked potential (VEP), suggest features related to complexity and nonlinearity use a K NN classifier with ensemble subspace. Reduced classification mistakes are the goal of the ideal number of channels is chosen using binary particle swarm optimization (BPSO). To be used in optimization techniques, a novel fitness function is created. Select channels and classification error are used to evaluate the fitness function. The results of the experiment indicate that the best channel chosen has biological importance and is linked to alcoholism. Consequently, the results of the suggested channel selection process can be applied to quickly and accurately classify subjects as normal or alcoholic.

Hossein Chamani et al. [10] The eyes, nose, and mouth are divided into a different image, and face classification operations (deep learning by convolution neural network) are carried out on the separated components rather than on generic information for face recognition. According to the data, the suggested strategy reduces computing costs by almost 70%. Additionally, it is possible for CNN to perform worse than the overall image of the disassembled parts.

A. Amodio et al. [11] An assessment is conducted on the efficacy of this approach in identifying an impaired condition resulting from alcohol addiction. The test procedure records the dynamics of constriction in both eyes by shining light in one of the individuals' eyes. A two-step methodology is described in order to obtain the video sequences' pupil size profiles: the first phase involves performing a search of the visual using the iris and pupil, and the second step involves cropping the visualization to enhance time efficiency by performing pupil detection on a smaller image. The undesirable learner dynamics that arise inside the pupillary light reflex (PLR) are identified and assessed; the reflexive accommodation results in constriction of the pupil around 10% of the iris diameter, and a spontaneous oscillation of the pupil diameter is seen in the interval [0, 2] Hz. A model of first order is found for each individual based on a repository for pupillary light reactions collected on many people both under baseline conditions and after alcohol use. In order to create a classifier for support vector machines that can distinguish in between the states of "Sober" and "Drunk," a set of characteristics is provided and utilized to compare the two populations of responses.

G. Sikander et al. [12] provides a cutting-edge summary of recent advancements in the of detecting driver weariness. Biological characteristics of the driver and subjective reporting, physical characteristics of the driver, hybrid characteristics and features of the car when driving are the five areas into which methods are divided based on the features utilized for driver tiredness detection. Comparing several methods for tiredness detection, potential areas for improvement are identified.

Guo, JM. et al. [13] suggested a novel approach that uses a mix of Convolutional neural networks (CNNs) with long short-term memory (LSTM)to handle the real-time driver drowsiness detection. The Asian conference on computer vision (ACCV) 2016 competition's public sleepy driver dataset was used to evaluate the system's performance. It can perform better than previous systems in the literature, according to the results.

Liu, W. et al. [14] provide a dual-stream network model with driver for multiple face features tiredness identification technique. There are four components to the algorithm: (1) Using convolutional neural networks with multitask cascades (MTCNNs) for mouth and eye positioning. (2) Taking a partial face picture and extracting its static features. (3) Taking a partial face optical flow and extracting the dynamic characteristics from it. (4) Combining static and dynamic data using a two-stream neural network to categorize the information. This paper's primary contribution is the integration of multiple facial characteristics employing a two-stream network to the diagnosis of driver weariness. Even so, a partial face pictures as inputs into a network might focus on facts about exhaustion information, networks with two streams incorporate information about images, both static and dynamic, resulting in improved performance. In addition, gamma correction was used to improve image contrast. This can improve our method's performance, as seen by a 2% improvement in accuracy in nighttime settings. Ultimately, the with 97.06% accuracy, the National Tsing Hua University Driver Drowsiness Detection (NTHU-DDD) dataset was used.

E. Romera et al. [15] As worries about automobile safety have grown, driving analysis has attracted attention recently. On the other hand, the current state of this subject is currently limited by the absence of publicly available driving data. Research may benefit greatly from machine learning approaches, but their applicability to the broader research community is restricted since they require vast volumes of data that are hard and expensive to collect through Naturalistic Driving Studies (NDSs). Furthermore, An inexpensive and easy to put into practice platform for the driver behavior monitoring has been made available by the widespread use of smartphones; yet, the data from these devices is not publicly accessible through existing applications. Because of these factors, this study offers the publicly accessible dataset of university of alabama in Huntsville -DriveSet (UAH-DriveSet) that furnishes a substantial quantity of information obtained through Drive Safe, our driving monitoring software, enabling intense driving analysis. Six distinct drivers and vehicles operate the application, displaying three different behaviors: aggressive, sleepy, and normal. On both minor and highway kinds of roads This results in 500 min or more of realistic driving, along with the connected unprocessed data and analyzed semantic information, as well as records of the travels on video. To make data analytics easier, this study also presents a program that plots data and shows trip recordings at the same time. Visit http://www.robesafe.com/personal/eduardo.rom era/uah-driveset to access the UAH-DriveSet.

Peppes, N. et al. [16] describes a comprehensive, integrated platform that gathers, stores, processes, analyzes, and correlates various data flows coming from automobiles. It does this by combining popular methods for deep learning and machines Using open-source technologies. In particular, to categorize the conduct of the driver if environmentally friendly, data streams from various vehicles are processed and analyzed using clustering techniques. This is then a comparison of deep learning and supervised machine learning techniques using the given tagged dataset.

J. Wang et al. [17] study aims to benefit both communities by providing a thorough overview of research, along with details on data gathering methods, computer vision algorithms used, and driver behavior classification results. The research is restricted to studies that use a minimum of one camera to monitor the driver within a car. Papers have been categorized according to their aim, which is to detect either high-level (distraction detection) or low-level (head orientation) driver information. The datasets that the papers use have led to additional classifications. Many private datasets have also been found in addition to the twelve available datasets, and their data collecting architecture is examined to highlight any effects on model performance. The algorithms used for each job and their results are examined to provide a baseline.

3 Artificial Intelligence (AI) and Machine Learning Techniques for Driving Behavior Analysis Capturing Drowsiness:

Now a days Using machine learning, computers can mimic and adapt human-like behavior. Every interaction and action taken using machine learning is something the system can learn from and apply to the next time [18]. Artificial intelligence and machine learning are not brand-new concepts. For almost sixty years, they have been researched, utilized, applied, and reinvented by computer scientists, engineers, researchers, students, and business community members. In the 1950s and 1960s, machine learning and artificial intelligence observed serious growth [18]. The most widely used machine learning algorithms include neural networks, ensemble techniques, naïve Bayes, decision trees, support vector machines (SVMs), discriminant analysis, and linear regression, random forest [18].

Artificial intelligence (AI) and machine learning are two fields that include the branch of deep learning, which has received a lot of attention recently and achieved considerable strides. The fundamental component of deep learning is artificial neural networks, which are modelled after the structure and operation of the human brain. Artificial neurons, often referred to as nodes or units, are layered and coupled to form these networks. Information is processed by each neuron before being sent to the following layer. Since neural networks can contain numerous hidden layers, they can be considered "deep," and it is this depth that distinguishes deep learning from more conventional machine learning methods. Figure 3 depicts a deep learning neural network with input, hidden, and output layers.

Figure 4 firmly illustrates that classical machine learning techniques perform better with less quantities of input data. While the performance of classic machine learning algorithms remains stable beyond a certain amount of data, that of deep learning approaches grows in proportion to the data increment [19].

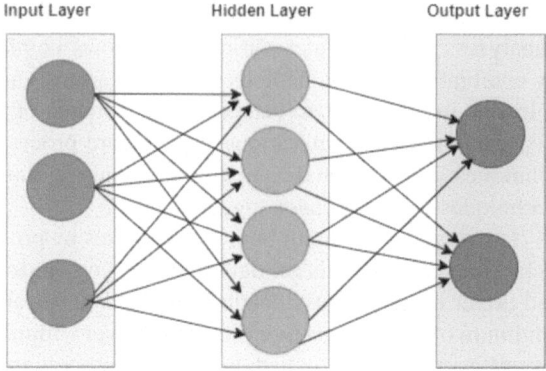

Fig. 3. Deep learning neural network

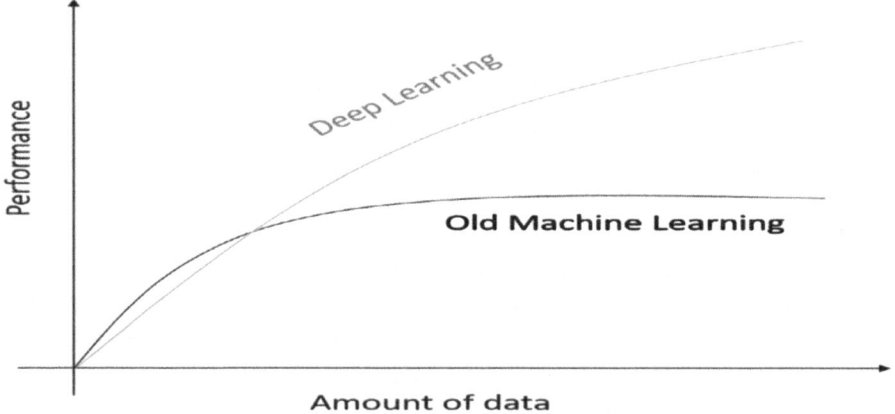

Fig. 4. The deep learning system's performance in relation to data volume [19]

To detect Drowsiness, various ML and DL approaches are described in different works. Convolutional neural networks (CNN) approach for detecting aggressive driving emotion [20]. Machine learning techniques and data from smartphone sensors are used to track unusual driving behavior [21]. Method for detecting driver drowsiness based on physiological signals, such as electrocardiogram (ECG) and electroencephalogram (EEG), and machine learning algorithms [22].

4   Proposed Methodology

We suggest a methodology that includes online University of California, Irvine (UCI) repository data and feature engineering using deep learning architectures as well as conventional machine learning methods. The specific goal is to capture the temporal correlations in driving behavior suggestive of tiredness by combining the interpretability of machine learning classifiers with the complex pattern recognition powers of recurrent neural networks (RNNs). The methodology in Fig. 5 draws inspiration from developments in multi-sensor fusion, combines information from several sources, including physiological monitors, steering dynamics sensors, and cameras for facial and eye movement recognition. This all-encompassing method seeks to offer a full picture of the driver's condition by identifying subtle behavioral and physiological indicators to enhance the precision and dependability of drowsiness detection.

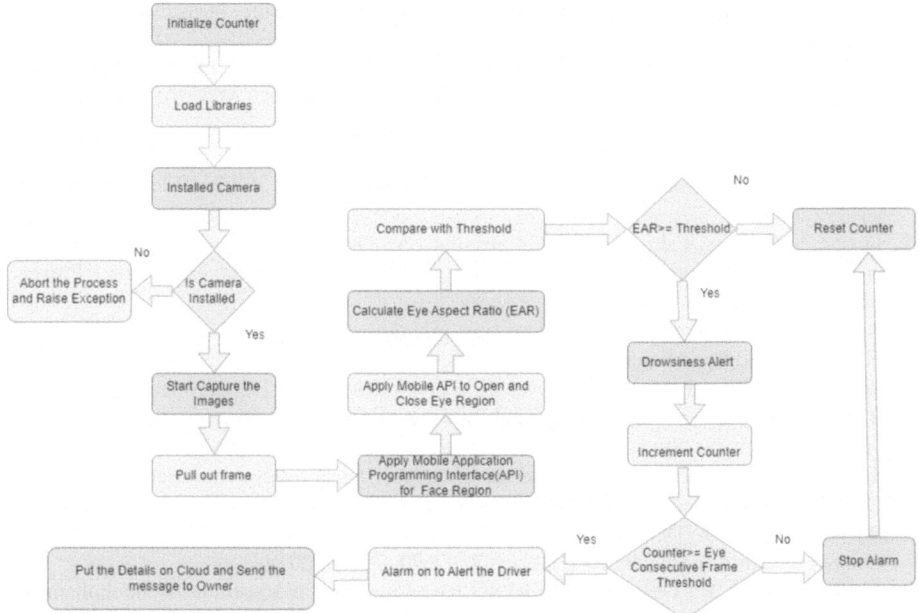

Fig. 5. Proposed methodology on the based-on literature review

5 Conclusion

This study leverages the complimentary capabilities of deep learning architectures and classical machine learning techniques to propose a novel approach to driving behavior drowsiness detection. Our study, which is based on recent developments in multi-sensor fusion, combines data from several sources, such as cameras that recognize face and eye movements, steering dynamics sensors, and physiological monitors. This all-encompassing integration aims to provide a full awareness of the driver's state, skilled in recognizing minute behavioral and physiological cues linked to fatigue. The existing approaches by combining information from several sensors and providing a more accurate and nuanced picture of the driver's condition. The integration of physiological data, steering dynamics, and facial/eye movement recognition enhances the accuracy and reliability of our suggested sleepiness detection system by enabling the early identification of sleepy indicators. When combined with multi-sensor fusion, deep learning and machine learning algorithms offer a comprehensive and efficient way to improve the accuracy and dependability of sleepiness detection systems. Our review study seeks to make significant progress in reducing the dangers associated with driving when sleepy and promoting safer road conditions for everybody by improving driving safety.

References

1. National Highway Traffic Safety Administration: Drowsy Driving. https://www.nhtsa.gov/risky-driving/drowsy-driving. Accessed 10 Oct 2023
2. National Center for Chronic Disease Prevention and Health Promotion, Division of Population Health. CDC – Data and Statistics – Sleep and Sleep Disorders, 2 May 2017. https://www.cdc.gov/sleep/data_statistics.html. Accessed 14 Oct 2023
3. Czeisler, C.A., Wickwire, E.M., et al.: Sleep-deprived motor vehicle operators are unfit to drive: a multidisciplinary expert consensus statement on drowsy driving. Sleep Health **2**(2), 94–99 (2016)
4. Jacobé de Naurois, C.J., Bourdin, C., Stratulat, A., Diaz, E., Vercher, J.L.: Detection and prediction of driver drowsiness using artificial neural network models. Accid. Anal. Prevent. **126**, 95–104 (2019)
5. Nishiyama, J.: Research on the Detection of Signs of Sleepiness in Humans by Pupil Fluctuation and Eye Movement. Ph.D. Thesis, Chubu University, Aichi, Japan (2011)
6. Rajkar, A., Kulkarni, N., et al.: Driver drowsiness detection using deep learning. In: Iyer, B., Ghosh, D., Balas, V.E. (eds.) Applied Information Processing Systems. AISC, vol. 1354, pp. 73–82. Springer, Singapore (2022). https://doi.org/10.1007/978-981-16-2008-9_7
7. Savaş, B.K., Becerikli, Y.: Real time driver fatigue detection system based on multi-task ConNN. IEEE Access **8**, 12491–12498 (2020)
8. Bavkar, S., Iyer, B., Deosarkar, S.: Rapid screening of alcoholism: an EEG based optimal channel selection approach. IEEE Access **7**, 99670–99682 (2019)
9. Bavkar, S., Iyer, B., Deosarkar, S.: BPSO based method for screening of alcoholism. In: Kumar, A., Mozar, S. (eds.) ICCCE 2019. LNEE, vol. 570, pp. 47–53. Springer, Singapore (2020). https://doi.org/10.1007/978-981-13-8715-9_6
10. Chamani, H., Nadjafi, M.: Improved reliable deep face recognition method using separated components. Brill. Eng. **3**, 4563 (2022)
11. Amodio, A., Ermidoro, M., Maggi, D., Formentin, S., Savaresi, S.M.: Automatic detection of driver impairment based on pupillary light reflex. IEEE Trans. Intell. Transp. Syst. **20**(8), 3038–3048 (2019)
12. Sikander, G., Anwar, S.: Driver fatigue detection systems: a review. IEEE Trans. Intell. Transp. Syst. **20**(6), 2339–2352 (2019)
13. Guo, J.M., Markoni, H.: Driver drowsiness detection using hybrid convolutional neural network and long short-term memory. Multimed. Tools Appl. **78**(20), 29059–29087 (2019)
14. Liu, W., Qian, J., Yao, Z., Jiao, X., Pan, J.: Convolutional two-stream network using multi-facial feature fusion for driver fatigue detection. Future Internet **11**(5), 115 (2019)
15. Romera, E., Bergasa, L.M., Arroyo, R.: Need data for driver behavior analysis? Presenting the public UAH-DriveSet. In: IEEE 19th International Conference on Intelligent Transportation Systems (ITSC) 2016, pp. 387–392. IEEE, Rio de Janeiro-Brazil (2016)
16. Peppes, N., Alexakis, T., Adamopoulou, E., Demestichas, K.: Driving behavior analysis using machine and deep learning methods for continuous streams of vehicular data. Sensors **21**(14), 4704 (2021)
17. Wang, J., et al.: A survey on driver behavior analysis from in-vehicle cameras. IEEE Trans. Intell. Transp. Syst. **23**(8), 10186–10209 (2022)

18. Alzubi, J., et al.: Machine learning from theory to algorithms: an overview. J. Phys.: Conf. Ser. **1142**(1), 012012 (2018)
19. Alom, M.Z., et al.: A state-of-the-art survey on deep learning theory and architectures. Electronics **8**(3), 292 (2019)
20. Lee, K., et al.: Convolutional neural network-based classification of driver's emotion during aggressive and smooth driving using multi-modal camera sensors. Sensors **18**(4), 957 (2018)
21. Lindow, F., Kashevnik, A.: Driver behavior monitoring based on smartphone sensor data and machine learning methods. In: 25th Conference of Open Innovations Association (FRUCT) 2019, pp. 196–203. IEEE, Helsinki-Finland (2019)
22. Ojha, D., Pawar, A., et al.: Driver drowsiness detection using deep learning. In: 4th International Conference for Emerging Technology (INCET) 2023, pp. 1–4. IEEE, Belgaum-India (2023)

Advancing Peer Review Integrity: Automated Reviewer Assignment Techniques with a Focus on Deep Learning Applications

Bhumika Bhaisare$^{(\boxtimes)}$ ⓘ and Rajesh Bharati ⓘ

Department of Computer Engineering, Dr. D. Y. Patil Institute of Technology, Pimpri, Pune, India
bhumika.bhaiare@gmail.com, rajesh.bharti@dypvp.edu.in

Abstract. The reviewer assignment process in academic research papers is crucial for ensuring peer review integrity and fairness, and it is often referred to as the reviewer assignment problem (RAP). This initial step lays the foundation for the entire review process. To address the limitations of manual reviewer assignment, there is a growing trend among journal and conference editors to explore automated solutions. For quite some time, machine learning techniques have found application in this domain, recent attention has shifted toward the application of Deep Learning in the education sector. Deep Learning, rooted in neural networks with diverse processing components, is gaining prominence in fields like natural language processing (NLP) and image recognition. In this comprehensive review, we aim to evaluate the key research achievements in the realm of reviewer assignment algorithms. It commences by presenting background information and discussing the necessity for automated reviewer assignment. The paper then systematically examines the existing research in this domain, with a particular focus on recent advancements in Deep Learning techniques. The review provides an unbiased assessment of the current algorithms' strengths and weaknesses. Overall, this analysis emphasizes the growing significance of RAP research and the demand for further innovation and framework development in this field.

Keywords: Deep Learning · Information retrieval · Reviewer Peer review · assignment problem (RAP) · Optimization algorithm

1 Introduction

This Peer review is a collaborative process involving experts from one or more fields who assess scientific research proposals and accomplishments. Peer review is deeply ingrained in modern scientific inquiry, serving as an essential pillar in the domain of academic research. It spans across various disciplines, encompassing the submission of work for research conferences & peer-review journals. At present, peer evaluation garners extensive backing from researchers, serving as the crucial entry point and quality assurance system for improving the excellence of research publishing and affirming trustworthiness publications research. Nonetheless, peer assessment faces various obstacles, and one of these hurdles is the automated matching of reviewers with appropriate papers, a crucial phase in enhancing the quality of peer review. Allocating experienced reviewers for evaluation of research documents is important task, recognized as RAP (Reviewer-Assignment Problem), presents the vital and intricate challenge during peer-review procedures of publishing journals, research conferences as well as funding institutions [1].

To enhance the quality of assessments and guarantee an equitable and unprejudiced appraisal, the subsequent crucial elements should be considered during the reviewer assignment process:

- Assigning reviewers to proposals that align with their areas of proficiency is essential.
- It is imperative that proposals are assessed by the most competent reviewers in their specific fields.
- The evaluation of proposals with eagerness and unwavering commitment carrying out by the reviewer.
- No reviewer should be burdened with more proposals than the agreed-upon limit.
- Every assessor ought to appraise an identical quantity of submissions.
- Every submission must endure examination by a fixed quantity of appraisers.
- Any affiliations between candidates and assessors that might imperil the equity of the appraisal procedure should be averted.

Historically, the allocation of reviewers was handled by a solitary decision-maker, like a manager in a program of funding, or a compact panel, and all these duties were conducted manually.

However, hand-operated allocation processes proved to be time-consuming also vulnerable for subjective or biased judgment of the committee. Nonetheless, the hand-operated allocation process turned out to be exceedingly time-intensive and vulnerable to the subjective or prejudiced choices established by the panel. Additionally, streamlining designations presented a challenging endeavor owing to the incapacity to fully contemplate every requirement. On top of that, the distribution of reviewers had to be concluded under rigorous time limitations, as a multitude of submissions frequently inundated just prior to the stipulated cutoff [2].

The existing peer review process involves a substantial volume of reviewers and proposals, often numbering in the thousands. The substantial magnitude of this situation greatly obstructs the hand-operated allocation procedure, generating difficulties for conference organizers, periodical overseers, and funding supervisors. Consequently, the manual-operated approach proved to be labor-intensive and didn't consistently yield the most effective results. Hence, the automatic reviewer assignment approach has emerged as a critical topic within the academic community (Fig. 1).

To emphasize the significance of the reviewer assignment problem, let's contemplate the evaluation procedure of an academic symposium document as a demonstrative instance. Figure 2 depicts the entire process of evaluation for academic papers splits into these subsequent phases:

- Manuscripts are submitted by authors ahead of the designated time limit.
- Head of the program faces the challenge of forming a program committee, essentially a task of identifying and recruiting experts capable of reviewing the submitted papers.

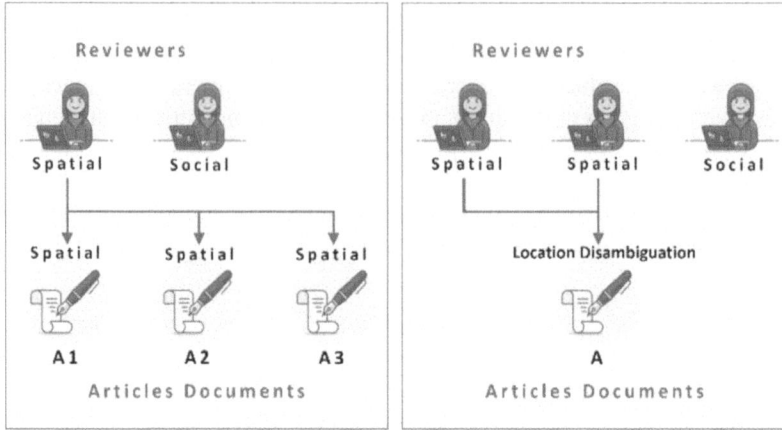

Fig. 1. Limitations of the prior definition of RAP

- The Program Chair then allocates the manuscripts to the experts responsible for reviews within the committee.
- The evaluation specialists appraise the papers and furnish their input and remarks.
- Considering the feedback obtained, the Program Coordinator ultimately determines to approve or decline that paper.

As illustrated in Fig. 2, the allocation of reviewers to papers holds a crucial function in establishing the reviewers' proficiency for paper evaluations. As a result, it emerges as a vital phase within the complete peer review procedure [3]. This underscores a considerable necessity for effective algorithms dedicated to assigning reviewers, automating and refining the pairing of reviewers with papers. These algorithms not only considerably boost the effectiveness of the entire peer review procedure but also markedly advance the excellence of peer assessments. Additionally, the incorporation of computer methods, such as natural language processing, can further amplify this process.

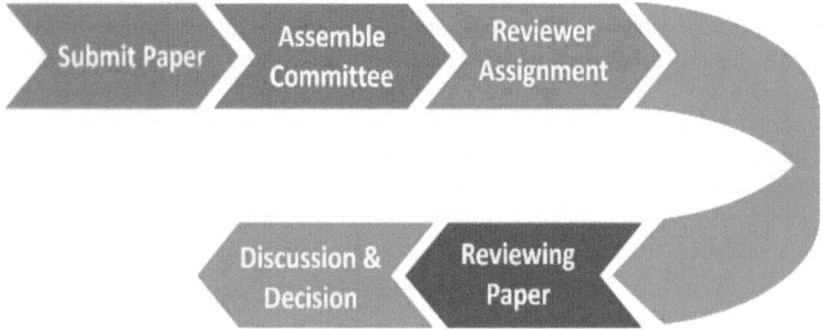

Fig. 2. A synopsis of the academic conference paper review process.

In the past decade, Deep Learning (DL) has revolutionized machine learning, excelling in tasks like speech & image identification or recognition. Global firms like Apple, Google, Facebook have huge investment in DL for smart product development. DL uses multi-layer neural networks for diverse data types. It benefits various research fields.

2   Literature Review

The automated process of matching reviewers with papers is an ongoing interdisciplinary puzzle that lures researchers from various areas, encompassing OR (operations research) AI (artificial intelligence) as well as computer science. NLPs (Natural language processing) have taken on a central role as the driving catalyst for automated reviewer assignment.

In 2007, Romero and Ventura [4], subsequently refined in 2010 [5] and 2013 [5], conducted a comprehensive analysis of over 300 studies prior to 2010. They identified various categories and tasks in data mining within research and education, encompassing data analysis and visualization, feedback provision for instructors, research recommendations, author performance prediction, anomaly detection, subject grouping, social network analysis, concept map development, review list construction, and designing & arranging.

Baker and Yacef (2009) [6] delivered an analysis of data mining within the educational and research setting. They scrutinized the transformations and developments in this domain, drawing parallels between its present condition and its initial stages. They pinpointed four major applications/objectives: refining models for research, enriching domains, assessing the educational assist offered by educational applications, also undertaking research exploration into learners. Furthermore, they cataloged frequently referenced articles from 1995 to 2005 and explored the impact they have over the data mining community.

In 2014, Peña-Ayala conducted an extensive survey using data mining techniques on over 240 papers in Educational and Research Data Mining [7]. This involved statistical and clustering processes to identify educational functionalities, patterns in data mining approaches, and instances of value patterns related to explanatory and anticipatory models. In contrast to prior literature surveys, this study primarily centered on computational methods rather than the utilization of Educational and Research Data Mining.

More recent additions to this literature review include a study by Bakhshinategh et al. in 2018 [8]. They scrutinized a range of responsibilities and uses within Educational and Research Data Mining, classifying them according to their objectives. Building upon the eleven categories proposed by the researchers [4], They introduced a structured hierarchy consisting of thirteen categories.

Aldowah and colleagues (2019) [9] undertook a research investigation in the higher education domain. They delved into higher education through the areas of computer-aided predictive, visualization, learning, as well as behavioral analytics. Drawing from previous studies, they identified specific Educational and Research Data Mining techniques that could effectively address various learning challenges, supplying student-focused approaches and resources to academic organizations.

Notably, Deep Learning application & datasets availability within Research Data Mining for peer-review have not been extensively explored in these review papers, and this work aims to address these gaps. To facilitate empirical comparisons of different approaches, understanding the datasets used in experiments is crucial. This paper is dedicated to reviewing and summarizing these resources. Additionally, while prior proposals have considered shallow neural network approaches, none have specifically focused on Deep Learning techniques.

In addition, there are review articles that encompass the entire peer review process and incorporate the Reviewer Assignment Problem (RAP) as a constituent element within it. However, while these studies include sections discussing how reviewers are selected for proposals, they often fall short in providing a comprehensive an examination viewpoint over Reviewer Assignment Problem (RAP) as a holistic concept (Fig. 3).

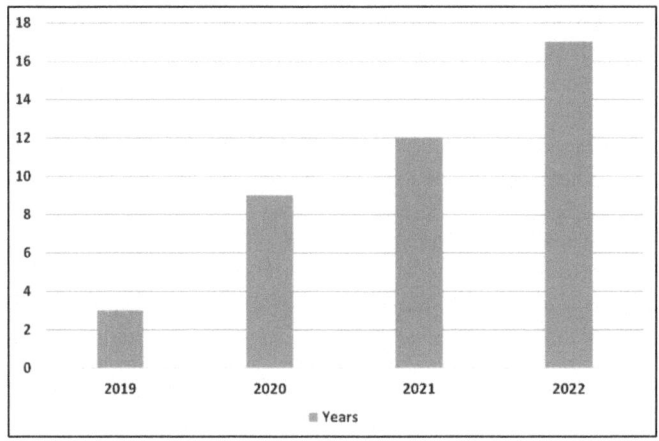

Fig. 3. The yearly count of published papers.

In a more recent context, Shah (2022) composed an all-encompassing review paper that investigated the obstacles within the peer review procedure and examined prospective remedies. This research provides an extensive overview of the peer review body of work, tackling concerns related to disparities that emerge due to factors such as dishonesty, misjudgment, and subjectivity during the assessment phase. While the RAP is touched upon in the context of addressing injustices resulting from discrepancies in reviewer expertise, it is explored in a somewhat limited manner, with only one dimension being considered.

3 Methodology Review

In this section, we outline the methodology employed for conducting this review and the procedures involved in collecting, analyzing, and extracting the available studies related to the utilization of deep learning in data mining for peer review techniques in research and education.

3.1 Deep Learning Models

Word-based models and topic models employed in the context of the Reviewer Assignment Problem (RAP) primarily rely on methods based on occurrence rates (frequency). These methods consider documents as aggregations of individual terms, with the crucial factor here is the term frequency, with no consideration for term sequence. Nonetheless, a significant limitation of these approaches is their incapacity to grasp the semantic and syntactic connections among words, consequently producing semantically feeble representations.

Moreover, when applied to large datasets, frequency-based techniques often produce vectors with significant sparsity. This implies that the vector representation of each document has a dimensionality same as the document's vocabulary, a condition frequently known as the "sparsity curse."

In response to these limitations, deep learning has given rise to language models designed to surmount the challenges associated with frequency-dependent techniques. These neural network-based language models utilize word embedding to automatically extract syntactic and semantic document characteristics.

The utilization of neural network-based techniques for word embedding can be classified into two primary divisions. The initial category, often termed as unchanging lexical representation (word embedding) methods or prediction-driven Strategies for lexical representation, can be encapsulated both the morphology and semantics of words. These methods also consider sub-word elements such as character n-grams. Distinguished instances of static-based word embedding methods encompass Word2Vec, Glove, and FastText.

The subsequent category is composed of context-based word embedding approaches, which encompass models with profound architectures like ULMFiT, BERT, GPT-2, and DistillBERT. In opposition to static word embedding techniques that characterize words as compact vectors, context-based methods meld contextual information into the representation. These models, fortified with their deep architectures, possess the capability to comprehend a word's meaning.

From the modern times, word embedding (lexical representation) methods that employ neural networks have discovered applications in a variety of natural language processing (NLP) assignments, speech recognition, encompassing word forecasting, translation of language, as well as textual similarity computation. Within the context of reviewer allocation, the likeness between propositions and reviewers can be gauged via the examination of the compact vector representations of their documents. Ogunleye et al. (2018) approach score of similarity amid reviewers as well as propositions employing integrated frequency-based models as well as Word2Vec for assessment of similarity.

Zhao et al. (2018) introduced a fresh categorization approach termed the Constructive Covering Algorithm (CCA) of word crawlers, to tackle the Reviewer Assignment Problem (RAP) for categorization task. Procedure harnesses research field tags linked with reviewers and propositions to enrich the quality of allocations. At the outset, propositions and reviewers were portrayed employing Word2Vec, followed by the computation of the minimum textual distance between propositions and reviewers employing the word-crawlers distance algorithm. Successively, apply tags for reviewer contenders as well as propositions have been divined grounded on field connections. Although this approach is adept at acquiring field characteristics via data-driven analysis, which is not empowering machines for assimilating field intelligence throughout a human outlook. To deal with this, Duan et al. (2019) launched a sentence pairing modeling strategy that employed the field association amid proposition appellations and abstracts and reviewer writings as supervisory data.

Yong et al. (2021) leveraged Word2Vec to extract Layered meaning portrayals of terms. They devised a governance mechanism utilizing a knowledge structure and governance correlation to generate a structured list of assessors based on their pertinence to each proposal.

Anjum et al. (2019) asserted previous methodologies, encompassing SLM, VSM, as well as possible models over topics, have been inadequate in handling lexicon mismatches and fractional topic convergences amid propositions and a reviewer's proficiency. To grapple with such instances of word mismatches and fractional topic coverages amid reviewers as well as propositions, the shared topic model approach, based on Word2Vec, was initiated by the authors.

Table 1 presents the allocation of strategies fixated on semantic data regarding reviewers and propositions. Remarkably, techniques for modeling over topics like LDA, PLSA as well as LSI are widely put to use contextual strategy, especially in the next phase in RAP. Among these, the LDA model stands out for its efficacy in detecting topics representing reviewers and propositions.

Romanov and Khusainova, 2019; Kalyan et al., 2021; Kapočiūtė-Dzikienė et al., (2021) proposed Neural network-based language models have demonstrated unparalleled performance in a scope of natural language processing (NLP) undertakings, for instance, document categorization, recognition of named entities, and computerized language translation. Until 2018, using these advanced language models to calculate matching scores was not common; however, a few studies have now started to incorporate them. The application of artificial intelligence-derived systems in the evaluation of similarity ratings has demonstrated their superiority in comparison to alternative meaning-oriented methods.

Even though neural network-based language models fall under the category of semantic-centered approaches, their extraordinary advantages have validated a distinct section to acknowledge the substantial benefits they have contributed to the calculation of resemblance scores, with the potential for additional advancements in the future.

Table 1. Distribution of Studies with A Focus On Semantic-Based Methods Across Different Years

Methods	2013	2014	2015	2016	2017	2018	2019	2021	2022
PLSA							√		
LDA	√	√		√	√√	√	√	√√√	√√√
LSI				√			√		√

(*continued*)

Table 1. (*continued*)

Methods	2013	2014	2015	2016	2017	2018	2019	2021	2022
ATM			√√		√				
Word2Vec						√√	√√	√	√√
BERT							√		√
CNN							√		√√
SciBERT							√		√

4 Deep Learning

Deep Learning, or DL, has emerged as a significant field of study in the realm of artificial intelligence, garnering considerable attention in recent years. It falls within the broader machine learning domain and leverages structures of neural network for creating abstract data representations. These neural frameworks comprise numerous tiers containing processing units or neurons, which execute a blend of linear and nonlinear alterations on input information. Deep learning has spawned several designs that have displayed effectiveness in a broad array of guided and self-guided assignments, particularly within the extensive realms of natural language processing and visual perception in computers [10].

Deep learning algorithms demonstrate proficiency in acquiring numerous strata of data representations, where more advanced attributes are constructed from those of lower strata, establishing a hierarchical arrangement. For instance, concerning image categorization, deep learning models can accept pixel values as input and, through the output layer, generate tags for items in images. Amidst these strata, a progression of transformation (concealed) strata systematically develops more complex characteristics, making them less vulnerable to variables such as illumination and object positioning.

The inclusion of the "deep" term into deep learning suggests for existence of multiple strata of alterations and tiers of representation positioned between the network's inputs and results. While a universally defined threshold for categorizing a neural network as "deep" does not exist, the majority of research in the domain proposes that having over two intermediary transformation strata is a prevalent criterion [11]. Although many fundamental concepts of DL originated several decades ago, the most significant advancements have occurred over the past decade. This surge in DL progress can be attributed to two primary factors: the availability of massive datasets and the increased computational power driven by Graphics Processing Units (GPUs). Large datasets enable DL algorithms to generalize effectively, while GPUs facilitate parallel processing for training more extensive and intricate models. Moreover, the proliferation of software platforms such as TensorFlow, Theano, Keras, and PyTorch has empowered researchers to concentrate on crafting models without becoming entangled in granular implementation intricacies.

A fundamental driver of the success of deep learning lies in its capability to circumvent the necessity for exhaustive feature manipulation. In traditional machine learning, the process of feature manipulation entails the laborious task of choosing pertinent attributes vital for algorithmic operation while discarding uninformative ones. This process can be time-consuming, as the correct feature selection plays a pivotal role in system performance. DL, on the other hand, incorporates feature learning, autonomously discovering the essential representations for a given task [12].

The following segments will delve into the fundamental principles of neural networks, the training procedure, distinguished architectural structures, fine-tuning of hyperparameters, and the tools for constructing deep learning models. Furthermore, these subjects will be situated within the sphere of Educational and Research Data Mining (ERDM), forging connections between these principles and the scrutinized documents.

4.1 Neural Networks

Computational processes are carried out by artificial neural networks made up of extensive matrices of elementary synthetic neurons. These constructs aspire to copy the processes noticed within the nerve fibers of brain in human. In this arrangement, each individual node within the construct acts as a neuron and serves as the foundational processing element of the artificial neural network.

Figure 4 illustrates neuron structure. This neuron comprises several key features: inputs ($x1$, $x2$, ..., xn), which can originate from another neuron output within the network; a bias ($x0$), a constant value included into neuron's activation function's input; the weights associated with each of these input ($\omega1$, $\omega2$, $\omega3$, ..., ωn), which determine the significance of each input into a model; and the resulting output of (α).

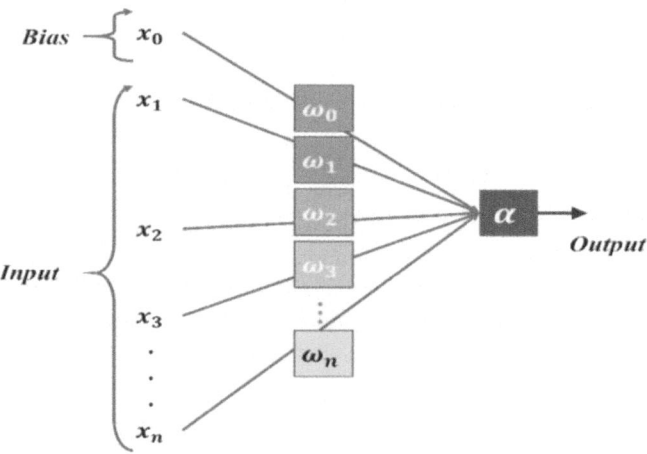

Fig. 4. Basic artificial neuron.

The neuron's output is computed according to the following equation:

$$\alpha = f\left(\sum_{i=0}^{n} \omega_i \cdot x_i\right) \tag{1}$$

The activation function, denoted as f, is a critical component of a neuron. This imparts adaptability to artificial neural networks, allowing them to grasp intricate non-linear connections within the data while standardizing the output of neuron (restricting the outcome within the range of 0 and 1). Frequently employed activation methods in artificial neural networks encompass the hyperbolic tangent, Re-LU (Rectified Linear Unit), as well as S-shaped. Neurons establish multiple connections with one another, and these connections can either amplify or attenuate the activation levels of neighboring neurons.

In Fig. 5, we discern the arrangement of neural network fundamental. An initial layer, functioning as the input layer, assumes responsibility for delivering data or characteristics to the network. The ReLU activation function is a recurrent preference for latent strata inherent to neural networks. These covert strata harbor the aptitude to effectuate intricate procedures through the concatenation of elemental functions.

The selection of the activation function within the terminal stratum is contingent upon the precise problem under consideration: in the context of binary classification, where output values manifest as binary entities (0 or 1), the sigmoid function is conventionally invoked. Conversely, in scenarios necessitating multiclass classification, the SoftMax procedure, an extension of the sigmoid function tailored for multiple categories, is deployed. Meanwhile, for regression quandaries bereft of predefined categories, a linear function is generally prescribed.

The characteristics of these concealed strata demarcate assorted neural network architectures, encompassing the likes of the LSTM (Long-Short-Term-Memory), RNN (Recurrent Neural Network) as well as CNN (Convolutional Neural Network). Number of concealed strata within a network substantiates its profundity, with a prevailing principle dictating that networks boasting an augmented assemblage of concealed strata possess an enhanced faculty to apprehend intricate procedures. In the realm of deep learning architectures, it is customary to deploy dozens, or even hundreds, of latent strata that inherently possess the potential to auto-adapt during the model's training.

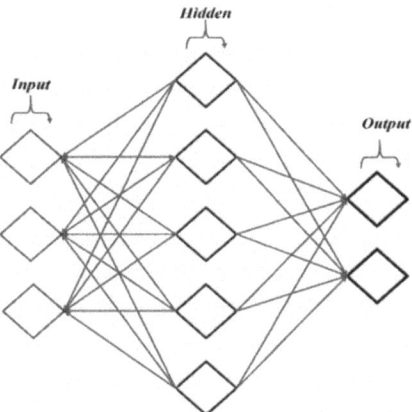

Fig. 5. A neural network's fundamental structure consists of circular nodes, with connections depicted as arrows, linking the output of one neuron to the input of another.

4.2 Training

Every machine learning algorithm's fundamental aim is to associate inputs, such as images, with desired outputs, like labeling them as "cat," by observing an extensive collection of input-output pairs. In the context of deep learning (DL), mapping of inputs & the intended output, representing network to be generated, which then achieved through the utilization of artificial neural networks characterized by a hierarchical structure composed of numerous layers.

Commencing with the initial layer, which acquires elementary characteristics and subsequently conveys this information to the ensuing layer, the hierarchical strategy begins. Each successive layer assimilates this fundamental data, combines it with more intricate elements, and transfers it to the subsequent stratum within the hierarchy. This sequential procedure persists, with each stratum gradually constructing more elaborate representations grounded on the input it receives. The particular alterations that each stratum employs to the input are recorded in the stratum's parameters.

To enable the network to effectively acquire and execute its intended functions, it is imperative to establish the fitting parameters for each stratum, ensuring the optimal alignment between input instances and their corresponding target results. This process of configuring the appropriate attributes, especially the parameters, for each computational unit in the network is what comprises the training of a neural network. Nonetheless, the complexity arises from the reality that deep learning networks may encompass an extensive array of parameters, potentially reaching into the millions. Discovering the precise values for all these parameters can evolve into a notably demanding and intricate endeavor. As an illustration, a frequently utilized neural network architecture for image categorization, denoted as VGG16 [13], includes an astonishing count of parameter in total 138 million. In the beginning, the values of each neuron might undergo arbitrary assignment or adhere to distinct initialization methodologies, which could encompass unsupervised pretraining [14].

4.3 Architectures

In the domain of Educational & Research Data Mining, a crucial consideration is the selection of neural network architectures in deep learning (DL), which is contingent upon the characteristics of the input data. DL provides an expansive array of architectures and algorithms, tailored to various data types. This section offers an overview of these popular architectures, their typical applications, and their relevance to the field.

The architectural choices encompass Convolutional Neural Networks (CNN) also variant (VGG16 and AlexNet), Multilayer Perceptron (MLP), Feedforward Neural Networks (FNN), Word Embeddings (WE), Long Short-Term Memory (LSTM) autoencoder, Memory Networks (MN), Bidirectional LSTM (BLSTM), as well as Recurrent Neural Networks (RNN).

In contrast, the baseline methods include Principal Factor Analysis (PFA), Hidden Markov Model (HMM), Latent Dirichlet Allocation (LDA), Logistic Regression (LogReg), Gaussian Naïve Bayes (GNB), Random guess, N-grams, Input Output HMM (IOHMM), CRF (Computational Random Fields), Linear Regression (LinReg), Majority voting, Decision Tree (DT), Slope-One, Random Forest (RF), Support Vector Machine

(SVM), Hierarchical IRT (HIRT), AdaBoost, K-Nearest Neighbors (K-NN), Singular Value Decomposition (SVD), Item Response Theory (IRT), Bayesian Linear Ridge Regression (BLRR), Majority class, LSA (Latent Semantic Analysis), Support Vector Regression (SVR), Naïve Bayes (NB), BKT (Bayesian Knowledge Tracing), Temporal IRT (TIRT), IBKT (Intervention BKT).

To evaluate the performance of these methods, various metrics come into play, including Precision Root Mean Square Error (RMSE), Accuracy, F-measure, Krippendorff's alpha, Logarithmic Loss (Log Loss), Quadratic Weighted Kappa (QWK), Mean per Class Error (MPCE), Area Under the Curve (AUC), Mean Absolute Error (MAE), Gini, Recall, $R2$.

Convolutional Neural Networks. CNNs (Convolutional Neural Networks) are distinguished for their multi-layered architecture, which makes them particularly well-suited for image processing tasks. These networks excel at recognizing fundamental image features, including edges, in their initial layers. As we progress through the layers, these foundational elements are synthesized to identify more complex patterns, such as facial features. CNNs, resembling to FNNs (Feedforward Neural Networks), comprise of neurons with adaptable offsets and coefficients. Every Node accepts input, performs a scalar product, and employs an activation operation. The concluding, entirely connected stratum exploits a loss metric to gauge the difference between projected and anticipated outcomes.

A standard Convolutional Neural Network (CNN) consists of three core layers: a convolutional layer for extracting features, especially in image analysis; a pooling layer that reduces the dimensionality of these features, often utilizing max pooling to preserve crucial information; and a fully connected classification layer, responsible for providing the network's final output [14].

Unlocking innovative CNN applications, the incorporation of deep layers, which involve convolution, pooling, and classification, has been instrumental. CNNs extend their utility beyond image processing into various domains, encompassing video recognition, game playing, and a multitude of natural language processing tasks. They demand fewer parameters when compared to FNNs and showcase exceptional performance in tasks like pattern-recognition, particularly in image-recognition. Nevertheless, they have specific limitations, including high computational requirements, the need for abundant training data, and the necessity for precise network initialization to address specific challenges.

Within the realm of Educational Data Mining (EDM), CNNs have been harnessed for purposes like identifying undesirable student behaviors. Architectures like VGG16 and AlexNet have been employed in video analysis, while CNNs have demonstrated their usefulness across various applications, including audio and video analysis, text classification, and even review prediction [15].

Recurrent Neural Networks. Feedforward Neural Networks (FNNs) possess a noteworthy characteristic in that they lack mechanisms for maintaining persistent information. To address this limitation, Recurrent Neural Networks (RNNs) introduce a feedback loop, enabling the retention of information [16]. RNNs, unlike FNNs with their strict feedforward connections, introduce feedback connections that loop back into preceding layers. RNNs, with the utilization of this response mechanism, acquire the capacity to store memories of earlier inputs. One way to conceive of RNNs is as networks comprising multiple self-replicating units, where each unit conveys information to its successor. This architectural trait renders RNNs adept at handling sequences and lists, and they often find application in tasks such as text modeling.

RNNs are employed effectively across various domains, including tasks like speech recognition, language modeling, and machine translation. However, they grapple with challenges, one of which is the vanishing gradient problem. This issue arises as gradients experience exponential decay during backpropagation, posing an obstacle to learning lower-layer weights. This hampers the stacking of RNNs into deep models and restricts their ability to capture long-term dependencies. Additionally, RNNs demand high-performance hardware for effective training and execution.

4.4 Long Short-Term Memory Networks. LSTMs

LSTM (Long Short-Term Memory) networks have gained substantial popularity in recent years as a specialized form of Recurrent Neural Networks (RNNs). A distinguishing feature of LSTM architecture is the incorporation of a memory cell, which enables the learning of long-term dependencies. This memory cell retains its value over time, influenced by its inputs, and it includes three gates regulating the flow of information:

- The entry gate oversees the moment when fresh data can access the memory.
- The neglect gate commands the elimination of obsolete data to accommodate novel information.
- The result gate defines when the knowledge stored in the cell is used in the network's outcome.

Each of these gates is regulated by associated weights, which are optimized during training using algorithms like Backpropagation Through Time (BPTT).

A recent simplification of the LSTM architecture is the GRU (Gated Recurrent Unit), which features two gates (input and forget) and lacks the output gate, resulting in fewer parameters. Both LSTMs and GRUs are well-suited for tasks involving sequences. In addition to the tasks previously mentioned for RNNs, LSTMs perform admirably in text creation, delivering responses to inquiries, and identifying actions in video sequences.

Image and video captioning have been achieved through the integration of Convolutional Neural Networks (CNNs) and LSTMs. In this setup, the CNN is responsible for processing images or videos, and the output of the CNN is transformed into natural language by the LSTM. LSTMs offer several advantages over traditional RNNs, including their enhanced memory capacity, which allows them to remember inputs over extended periods. Unlike RNNs, which may lose critical information when processing long text paragraphs, LSTMs excel in retaining important context. LSTMs also mitigate the vanishing gradient issue encountered in RNNs and require less training data to build effective models.

4.5 Hyperparameters Tuning

Relevant references are provided whenever available within this paper furnishing explicit details of hyperparameter. These hyperparameters can be set manually, selected using exploration algorithms, or enhanced using model-driven methods [17]. Deep neural networks (DNN) have predetermined variables, which are hyperparameters, make up this setprior to the fine-tuning parameters of a model (offsets &). Typically, these hyperparameters are divided into two categories: those pertaining to the model's structure and those connected with the training procedure.

5 Discussion

Recent efforts have been made to aggregate peer reviews, with initiatives like Publons consolidating peer review data to construct public reviewer profiles for participating reviewers. Crossref, a database repository for Digital Object Identifiers (DOIs) with over 4000 publisher members, introduced a service to integrate peer reviews as part of the metadata for scientific articles. However, the availability of most reviews remains limited. In contrast, Peer Read has collected and organized peer reviews to facilitate use by other researchers for research, experiment replication, and fair comparisons with prior results.

The literature review also delves into the diverse deep learning (DL) models as well as specifications applied within the Educational & Research Data Mining review literature. Long Short-Term Memory (LSTM) architectures emerge as the most frequently used approach, employed in 59% of the papers and across a wide spectrum of EDM tasks. This prevalence suggests LSTMs as a promising starting point for system development in these tasks.

Notably, recent initiatives by program chairs in major Natural Language Processing (NLP) conferences have focused on different facets of the peer-review procedure, particularly author responses and overall review quality. The provided large-scale dataset aims to enable the broader scientific community to delve into peer review properties and potentially enhance the current peer review model.

In terms of DL model hyperparameters, the review in the preceding section highlighted several crucial factors. Configuring a specific architecture involves an empirical process. Hyperparameter decisions are contingent on the accessible input data and the nature of the specific task. Among the hyperparameters under scrutiny, the model's performance significantly depends on factors such as learning rate, batch size, and stopping criteria (number of epochs). Larger batch sizes can support higher learning rates, making them computationally efficient, but smaller steps are recommended to avoid pitfalls. Adjusting the number of epochs is essential to prevent overfitting. Model size is another consideration, with deeper and wider architectures offering more power but increasing the risk of overfitting. High-parameter-count models require ample training data for effective generalization.

6 Conclusion

Through a systematic search, a total of 41 works were collected in this area. Using the taxonomy of data mining applications outlined in [7], it's evident that only four out of the 13 tasks that have been suggested have been tackled employing deep learning methods. Underscoring the potential of deep learning in unexplored data mining tasks, this is particularly noteworthy given the promising results presented in the reviewed studies, with 67% of them showing that deep learning outperformed traditional machine learning benchmarks in all experiments. Additionally, this research delves into an examination of the primary datasets utilized in data mining tasks. This study has explored the growing adoption of Deep Learning (DL) applications in educational & research Data Mining, having only three published papers but has consistently gained momentum, with 17 papers published in 2018. It's noteworthy that these DL approaches have found their way into significant research data mining discussion boards, where the last edition featured seven DL papers, making a total of 16 over the past three years. Like other research domains, some datasets are publicly available, facilitating experiment reproducibility, while others are custom-built for specific studies. In research data mining, the challenge of dealing with sensitive information, particularly concerning underage students, can hinder data sharing. Nevertheless, this issue can be addressed through proper anonymization techniques.

This investigation acts as a significant resource and initial stepping stone for researchers working in the fields of deep learning and research in data mining who aim to harness these techniques for educational purposes. In light of the expanding utilization of deep learning techniques in data mining, this study offers valuable perspectives. It offers a comprehensive exploration of Deep Learning (DL) techniques, beginning with an introduction to the field. Investigating various deep learning architectures used in different tasks, this study proceeds. It also assesses conventional hyperparameter settings and provides a compendium of available frameworks to support the development of deep learning models. Creating a deep learning structure often entails empirical methods, and the insights drawn from this research can establish a primary resource for future applications of deep learning within the research data mining field.

References

1. Romero, C., Ventura, S., Pechenizkiy, M., Baker, R.: Handbook of Educational Data Mining. CRC Press (2010). Complexity 19
2. Biswas, H.K., Hasan, M.: Using publications and domain knowledge to build research pro les: an application in automatic reviewer assignment. In: ICICT (2007)
3. Rodriguez, M.A., Bollen, J., Van de Sompel, H.: Mapping the bid behavior of conference referees. J. Inform. 1(1), 68–82 (2007). https://doi.org/10.1016/j.joi.2006.09.006
4. Romero, C., Ventura, S.: Educational data mining: a review of the state of the art. IEEE Trans. Syst. Man Cybern. Part C Appl. Rev. 40(6), 601–618 (2010)
5. Romero, C., Ventura, S.: Data mining in education. Wiley Interdiscip. Rev.: Data Min. Knowl. Discov. 3(1), 12–27 (2013)
6. Baker, R.S., Yacef, Y.: The state of educational data mining in 2009: a review and future visions. JEDM-J. Educ. Data Min. 1(1), 3–17 (2009)
7. Peña-Ayala, A.: Educational data mining: a survey and a data mining-based analysis of recent works. Expert Syst. Appl. 41(4), 1432–1462 (2014)
8. Bakhshinategh, B., Zaiane, O.R., ElAtia, S., Ipperciel, D.: Educational data mining applications and tasks: a survey of the last 10 years. Educ. Inf. Technol. 23(1), 537–553 (2018)
9. Aldowah, H., Al-Samarraie, H., Fauzy, W.M.: Educational data mining and learning analytics for 21st century higher education: a review and synthesis. Telemat. Inform. 37, 13–49 (2019)

10. Hatcher, W.G., Yu, W.: A survey of deep learning: platforms applications and emerging research trends. IEEE Access **6**, 24411–24432 (2018)
11. Schmidhuber, J.: Deep learning in neural networks: an overview. Neural Netw. **61**, 85–117 (2015)
12. Zhong, G., Wang, L., Ling, X., Dong, J.: An overview on data representation learning: from traditional feature learning to recent deep learning. J. Finance Data Sci. **2**(4), 265–278 (2016)
13. Simonyan, K., Zisserman, A.: Very deep convolutional networks for large-scale image recognition (2014). https://arxiv.org/abs/1409.1556
14. Y. Bengio, "Practical recommendations for gradient-based training of deep architectures. In: Montavon, G., Orr, G.B., Müller, K.R. (eds.) Neural Networks: Tricks of the Trade, 2nd edn. LNTCS, vol. 7700, pp. 437–478. Springer, Heidelberg (2012). https://doi.org/10.1007/978-3-642-35289-8_26
15. Collobert, R., Weston, J., Bottou, L., Karlen, M., Kavukcuoglu, K., Kuksa, P.: Natural language processing (almost) from scratch. J. Mach. Learn. Res. **12**, 2493–2537 (2011)
16. Tato, A.A.N., Nkambou, R., Dufresne, A.: Convolutional neural network for automatic detection of sociomoral reasoning level. In: Proceedings of the 10th International Conference on Educational Data Mining (2017)
17. Hopfield, J.J.: Neural networks and physical systems with emergent collective computational abilities. In: Proceedings of the National Academy

Automated Potato Disease Classification Using Deep Learning - A Comparative Analysis of Convolutional Neural Networks

Swati Pandey[1]([envelope]) [iD], Mayuri Gupta[1] [iD], Ashish Mishra[1] [iD], Ashutosh Mishra[2] [iD], and Jayesh Gangrade[3] [iD]

[1] Jaypee Institute of Information and Technology, Sector-62, Noida, Uttar Pradesh, India
swatip1705@gmail.com, mayurigupta2010@gmail.com,
ashishmishra81@gmail.com
[2] Thapar University, Patiala, Punjab, India
ashutosh.mishra@thapar.edu
[3] Manipal University, Jaipur, India
jgangrade@gmail.com

Abstract. Potato stands as a vital staple food worldwide, with its consumption steadily rising and becoming the fourth most consumed staple food globally. However, the prevalence of potato diseases has posed a significant challenge, impairing both the quality and quantity of harvests. Timely and accurate disease classification is imperative for effective disease management and crop protection. Leveraging recent advancements in deep learning and computer vision, this study provides a thorough analysis and evaluation of the differences between of prominent convolutional neural network architectures, namely VGG16, VGG19, ResNet50, ResNet152, and InceptionV3, for classifying three distinct disease classes in potato plants: early blight, late blight, and healthy. In the investigation of a dataset comprising 2152 images, this study scrutinizes the performance metrics of diverse models. The results underscore the remarkable efficiency of InceptionV3, demonstrating an impressive 91.84% accuracy in classifying potato diseases. Significantly, the incorporation of deep learning approaches presents a potential solution for tackling the complex challenges linked to identifying and categorizing potato diseases. Additionally, the paper underscores the vital significance of automated disease detection in preserving worldwide potato production, ensuring food security, and promoting sustainable agricultural practices.

Keywords: Deep learning · convolutional neural networks · Potato classification · accuracy · computer vision · image classification

1 Introduction

In the context of agriculture, the overarching objectives of ensuring food security and advancing nutritional standards are of utmost importance. In the face of escalating global population growth, the agricultural sector is challenged to enhance production while simultaneously upholding the integrity of crops in terms of quality and nutritional content. Among the various crops pivotal to this mission, the potato (Solanum tuberosum) has emerged as a noteworthy contender due to its potential to meet escalating sustenance demands. With its abundant supply of vital nutrients—such as vitamins C and B6, potassium, magnesium, and iron—potatoes emerge as a promising staple, poised to tackle the daunting challenge of nourishing a burgeoning population [1].

As the fourth most consumed vegetable crop on a global scale and a dietary cornerstone in Indonesia, potatoes hold a significant role in worldwide agriculture. Indonesia's agricultural landscape has witnessed a rapid expansion of potato cultivation, yielding substantial annual outputs of around 850,000 tons from approximately 60,000 hectares [2]. This thriving production underscores Indonesia's position as a key potato producer within Southeast Asia. However, this ascent in potato cultivation is met with corresponding challenges, particularly the susceptibility of potato plants to an array of diseases [3]. These maladies, affecting both pre-harvest phases, and post-harvest phases, lead to diminished agricultural output and complications, spanning from premature harvests to complete crop failure.

The convergence of disease prevalence and delayed detection highlights the urgency of swift and accurate disease identification in potato plants. Traditional manual monitoring methods, while indispensable, are fraught with inefficiencies stemming from their time-intensive nature and the scarcity of on-site experts. The delay in disease detection fosters an environment conducive to unchecked disease propagation. Farmers, in their diligent efforts, often resort to approximate methods that lead to misclassifications and misguided interventions [4]. These inefficiencies, compounded by a lack of expert guidance and precise quantification of disease severity, contribute to suboptimal disease management, and compromised crop yields.

In the face of these challenges, the potato's global significance becomes evident. Beyond being a culinary delight, potatoes serve as a vital dietary staple for over one billion people globally, offering essential nutrients. However, the economic impact of potato diseases, such as late blight and bacterial wilt, cannot be overstated. These diseases lead to substantial yield losses, increased production costs, and market instability, affecting not only individual farmers but also global food prices and security. Addressing these challenges requires concerted efforts in research and development, sustainable farming practices, and international cooperation. The humble potato, with its complex global role, necessitates a proactive approach to disease management for sustained agricultural and economic stability.

In this context, the fusion of cutting-edge technologies, particularly deep learning, and computer vision, emerges as a potent solution for the intricate challenge of disease classification in potato plants. Employing Convolutional Neural Network (CNN) architectures, this research comprehensively assesses five models—VGG16, VGG19, ResNet50, ResNet152, and InceptionV3. The ensuing methodology unveils a robust and automated framework to classify early blight, late blight, and healthy states within potato plants. This paper is primarily concerned with the following contributions:

- Implemented initial data pre-processing methods for uniformity (224 × 224 pixels) and introduced systematic data augmentation with geometric transformations to address limited training data challenges.

- Explored deep learning for potato leaf disease detection, assessing VGGNet, ResNet, and InceptionV3 for optimal classification. Emphasized CNNs layered complexity over traditional methods.
- Examined InceptionV3's intricate neural network design, highlighting its use of convolutional blocks and 2D max pooling for feature extraction. Emphasized the strategic inclusion of 11 Inception modules for detailed feature capture.
- Clarified InceptionV3 elements, concatenation of module outputs for holistic understanding, fully connected layers for hierarchical feature abstraction, and *SoftMax* classifier for accurate image classification.

2 Related Work

Numerous methodologies have been employed in the realm of plant disease classification, yet there remains a notable deficiency [5], making it an area of continuous development due to the wide variation within this field.

To establish a robust feature representation, a deep learning-based strategy can be formulated. Deep learning [6] has showcased remarkable process across a range of visual perception tasks, including text detection [7, 8], victim identification [9, 10], target tracing [11, 12], and object identification [13–15]. Empirical evidence, as well as conceptual reasoning from previous research [16], supports the potential of deeper layers in capturing advanced semantic details. This attribute equips the network with enhanced resilience to fluctuations in pose, color, scale, and deformable objects, rendering it well-suited for robust leaf disease classification.

Sladojevic et al. [5] constructed a model employing a convolutional neural network (CNN) to identify diseases in five plant types, covering 13 different disease categories. This endeavor yielded an average accuracy of 96.3%. In contrast, Erika Fujita et al. [18] devised a disease diagnosis system for cucumber plants using the CNN method with the AlexNet architecture, achieving an average accuracy of 82.3%.

Using artificial neural networks and fuzzy logic, Rastogi et al. [19] identified plant diseases based on leaf conditions, specifically on maple and hydrangea leaves. These diseases were categorized into two types: leaf spot and scorch leaf. Leaf spots denote specific points of disease occurrence, while scorch leaves exhibit even distribution. Mohanty et al. [20] employed deep convolutional neural networks to identify 14 crop species and 26 diseases, capitalizing on the potential of prominent RGB image features and machine learning methods. In a separate study, Jia et al. conducted a review focusing on Deep learning for hyperspectral image classification. The research explores hyperspectral image (HSI) classification, organizing relevant methods under learning paradigms like transfer learning, active learning, and few-shot learning. [21] Liu et al., explored deep learning based RGB models that achieved impressive accuracy levels for classifying different coal particles [18]. Wang et al., compare and evaluate traditional machine learning and deep learning algorithms for image classification [19]. Korot et al. employ various platforms and datasets for creating image classification models, particularly excelling in optical coherence tomography [20]. In the context of potato leaf disease classification, a deep learning approach was introduced in [21], utilizing pre-trained models like VGG19 for fine-tuning and achieving impressive accuracy. Additionally, an innovative methodology involving Mask R-CNN and structure-from-motion (SfM) was proposed for fruit detection and 3D location, enhancing system performance [21]. Hyperspectral imaging has been applied to detect Alternaria solani in potato crops, aiding in early blight identification and reducing the need for excessive plant protection agents [22]. A deep learning-based approach was proposed in [23, 24] for potato leaf disease classification, focusing on combating the detrimental impact of potato leaf blight shown in Table 1. The proposed architecture demonstrated remarkable accuracy across healthy, early blight, and late blight leaf categories.

Table 1. Comparative analysis of Literature review

Author	Methodology	Dataset Type	Disease Categories	Accuracy
Sladojevic et al. [5]	CNN The Caffe framework with the 10-fold cross-validation approach	5 different classes	13 different plant Diseases	96.3
Fujita et al. [18]	CNN with a 4-fold cross-validation	Cucumber leaf plants	Evaluated on 7 different plant diseases	82.3%
Rastogi et al. [19]	Phase 1: Artificial Neural Network (ANN) for plant recognition Phase 2: K-Means segmentation and ANN for disease classification	Leaf Images: Hydrangea and Maple	Classified 2 types of disease: Leaf Spot and Leaf Scorch	Not specified
Mohanty et al. [20]	Deep convolutional neural network	54, 306 plant leaf images	Classified 26 diseases	99.35%
Jia et al. [21]	Applied deep learning in the classification of hyperspectral images (HSI)	54,129 image dataset	Classified in 16 objects	75.66%

(continued)

Table 1. (*continued*)

Author	Methodology	Dataset Type	Disease Categories	Accuracy
Talukder et al. [17]	CMobileNetV2, CNASLargeNet, CXception, CDenseNet201, and CInceptionV3	495 potato pests images	Classified 8 types of potato pests	91%
Arshaghi et al. [25]	AlexNet, GoogLeNet, VGG, R-CNN, and Transfer Learning	5000 potato images	Classified in 5 classes	Reached 100% and 99% accuracy in some cases
Jiang et al. [30]	Deep learning techniques for the recognition of multiple lesions in medical image processing	Not specified	Not specified	Not specified
Nath, et al. [8]	K-means clustering and convolutional neural network (CNN)	1000 images	0 to 9 range is made to evaluate the disease	94.44%
Arshad et al. [11]	A hybrid deep learning model, PLDPNet with VGG19 and Inception-V3 models	2152 images of leafs	Classified into 10 classes	98.66%
Tiwari et al. [26]	VGG19 for fine-tuning through transfer learning and logistic regression emerging as the most effective classifier	2152 images	Classified in 3 classes	97.8%

3   Proposed Methodology

This section delves into the fundamental mechanisms of all the models. The discussion encompasses six crucial parameters, namely data gathering, data pre-processing, data augmentation, potato image classification, data training, and testing (Fig. 1).

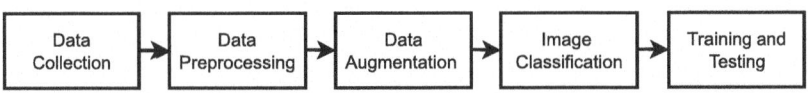

Fig. 1. Working of the proposed framework.

3.1   Data Collection

The systematic effort to create a comprehensive and diverse dataset involved an extensive data acquisition phase, where images were diligently retrieved from prominent repositories such as Plant-Village and Kaggle [26]. Specifically, the dataset was collected from Kaggle, more precisely from the repository named Plant-Village, which consisted of three folders curated by "Arjun Tejaswi." These folders were meticulously organized as follows: Early blight with 1000 images, Late blight with 1000 images, and Healthy leaf with 152 images [31].

During the meticulous examination of the dataset, inherent challenges impacting the model's effectiveness were taken into consideration. A notable challenge identified is the substantial class imbalance among the three categories—healthy, late blight, and early blight. The dataset comprises 2,152 images, with Early blight and Late blight each containing 1000 images, while Healthy leaf has only 152 images. This irregular distribution has the potential to introduce disparities during training, potentially leading the model to favor classes with a larger number of samples.

To address this, strategies for handling class imbalances were explored, including the implementation of data augmentation for the minority class (Healthy leaf). Data augmentation involves introducing variations or duplicating existing samples through

Fig. 2. Dataset- Healthy, early blight, late blight (left to right)

random transformations, such as rotations, flips, and changes in lighting. The incorporation of data augmentation techniques aimed to furnish the model with a more equitable representation of each class, thereby diminishing the risk of it being skewed toward the more prevalent classes (Fig. 2).

3.2 Data Pre-processing

The initial phase of the data preprocessing pipeline focuses on mitigating image noise by isolating the region of interest and eliminating extraneous portions. The dataset, comprising images from diverse sources with varying dimensions, undergoes a harmonization process. This involves resizing the images to a standardized format of 224x224 pixels, a strategic measure aimed at promoting uniformity within the dataset [25]. This resizing not only addresses variations in image dimensions but also establishes a consistent framework for subsequent stages of analysis and model development, ensuring a cohesive and standardized approach to the dataset.

3.3 Data Augmentation

Deep Learning, in contrast to traditional shallow machine learning networks, demands a substantial volume of data for effective training. To overcome the prevalent issues of inadequate training data and disparities in class distribution, a systematic approach known as data augmentation is implemented. This technique entails manipulating data while retaining its fundamental characteristics [27]. Given the limited size of the initial 2,152 datasets, data augmentation becomes imperative for optimal model performance. This research automates augmentation by incorporating basic geometric transformations, including translations, rotations, scale adjustments, and shearing, along with vertical and horizontal flips. This approach effectively enriches the dataset, thereby enhancing the learning process.

3.4 Image Classification

Deep learning, a subset of artificial intelligence that falls under machine learning, has made notable progress across diverse domains, including image classification, speech recognition, and object detection. Distinguished by its multiple layers, deep learning methods surpass the layered complexity of traditional machine learning techniques [28]. A prominent class within deep learning is the CNN [29], which has garnered recognition for its efficacy in tasks like detecting plant diseases based on leaf conditions. CNNs typically comprise convolutional layers, often grouped by function, followed by subsampling, and fully connected layers, mirroring a standard neural network architecture. The evolution of CNNs has witnessed the development of models like AlexNet, VGG network, GoogLeNet, ResNet (residual networks), and DenseNet, building upon foundational architectures such as LeNet.

The significance of CNN architectures is pivotal for accurate potato leaf disease classification, particularly due to their adeptness at capturing distinctive features in potato leaf images. Architectures like VGG, ResNet, and Inception excel in automatically extracting hierarchical features, allowing the identification of specific patterns associated with different diseases on potato leaves. ResNet's of residual connections proves effective in capturing spatial hierarchies and localized patterns unique to potato leaf diseases. Transfer learning from pre-trained models further enhances performance, leveraging broad incorporation knowledge to identify potato leaf-specific features. The efficient utilization of parameters by CNNs is particularly advantageous for preventing overfitting in agriculture tasks with limited datasets, ensuring robust potato leaf disease classification.

Within the confines of the potato leaf disease detection project, the central aim was to investigate different architectural models for precise classification. Among the scrutinized models such as VGGNet16, VGGNet19, ResNet50, ResNet152, and InceptionV3. InceptionV3 exhibited the most potential, prompting a more in-depth examination. The selection of these specific CNN architectures is based on their unique characteristics that address the complexities of potato leaf disease classification. VGG16 and VGG19's focus on small convolutional filters is advantageous for capturing fine-grained details, while ResNet50 and ResNet152's residual learning addresses the vanishing gradient problem, allowing the model to efficiently capture features at different depths. InceptionV3's use of inception modules with filters of multiple sizes concurrently makes it well-suited for detecting lesions, discolorations, and anomalies of different sizes on potato leaves shown in Fig. 3. The diverse architectures provide a range of options to accommodate the varied visual characteristics associated with different diseases in potato plants.

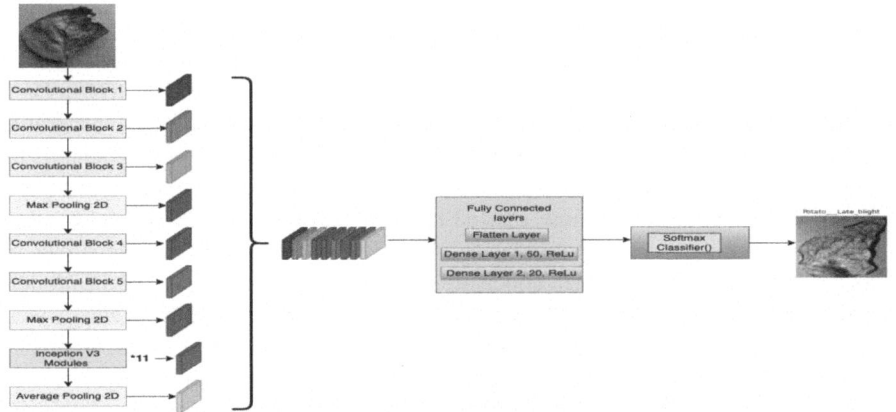

Fig. 3. InceptionV3 Architecture Model

The InceptionV3 architecture is an advanced neural network specifically crafted for tasks related to image classification. Its intricate structure begins by subjecting the input image to a series of convolutional blocks, followed by 2D max pooling operations. This initial stage serves to extract and refine features from the input, laying the foundation for subsequent processing. As the architecture progresses, additional convolutional blocks are employed to further enhance the representation of features. The intermittent application of 2D max pooling operations contributes to down-sampling feature maps, effectively capturing and preserving crucial information at varying scales.

The architectural divergence occurs with the introduction of 11 distinct Inception V3 modules, strategically crafted to capture features using convolutional filters of varying sizes. This diverse set of modules enables the model to grasp intricate details and relationships within the input data. A crucial element in the architecture is the concatenation of the outputs from these modules, a process facilitated by the Concat operation. The concatenation step plays a key role in capturing intricate relationships among features, promoting a comprehensive understanding of the input data.

After extracting and combining feature maps, the architecture directs the information through a series of fully connected layers. These layers include the Flatten operation, which reshapes the input into a one-dimensional array, and Dense1 and Dense2 layers with Rectified Linear Unit (ReLU) activation functions. This sequence of dense layers contributes to the hierarchical abstraction of features, allowing the model to learn intricate patterns and representations.

4   Experimental Results

This module presents all the training and testing data that has been implemented. The results show the statistically validated loss and accuracy of the confusion matrix to enable a better understanding of the readers.

4.1   Data Training

The dataset learning experiment involved the utilization of various neural network architectures, namely ResNet-50, ResNet-152, Inception V3, VGG16, and VGG19. These architectures underwent training for 50 epochs with a batch size of 32, incorporating a learning rate of 0.001 to enhance their performance. The learning process entails the algorithm actively seeking patterns and features within the image dataset, aiming to enable effective recognition of new, unseen images. Throughout each epoch, the algorithm iteratively refines its understanding to align with the characteristics of the images it is being trained on. Results of each epoch are meticulously logged to assess the model's performance in terms of both loss and accuracy. A low loss value indicates a well-performing model, with the ideal target being close to zero or equal to zero, signifying that the model's predictions closely align with the actual data. Conversely, accuracy serves as a crucial metric, quantifying the system's proficiency in correctly classifying objects and gauging the overall success level of the model. The following plots present the results of the loss values and accuracy:

Figure 4(a) depicts the accuracy and loss trends of the VGG19 model during training. The accuracy on both the training and validation datasets consistently increases, reaching a final accuracy of 0.86 after 50 epochs, indicating improved image classification over time. Simultaneously, the loss decreases, confirming the model's learning process.

In Fig. 4(b), the Confusion Matrix for the VGGnet19 model is presented. Noteworthy classification accuracies include 44.29% for Early Blight, 39.63% for Late Blight, and 1.86% for Healthy instances. This matrix provides valuable insights into the model's effectiveness in distinguishing between specific classes within the dataset.

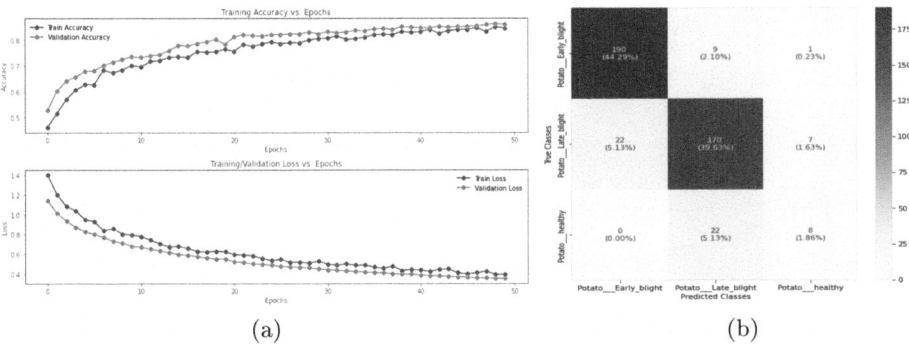

Fig. 4. (a) The Plot of Accuracy and Loss using VGG19, (b) Confusion matrix for VGG19.

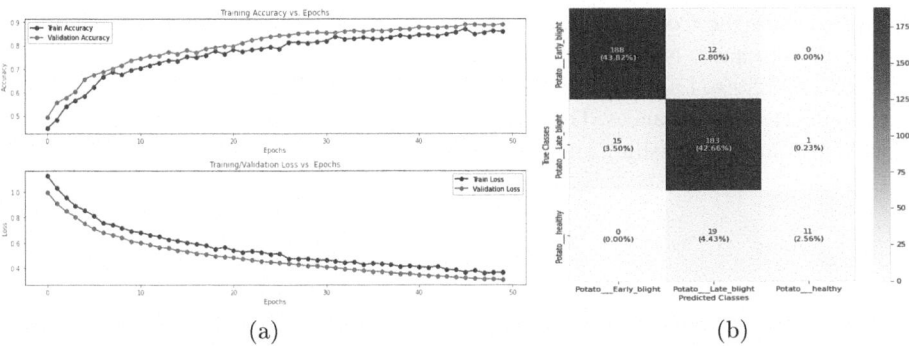

Fig. 5. (a) The Plot of Accuracy and Loss using VGG16, (b) Confusion matrix for VGG16.

Figure 5(a) displays the evolution of training and validation accuracy and loss throughout 50 epochs for a VGG16 model. The training accuracy exhibits a steady increase, accompanied by a reduction in training loss. While the validation accuracy and loss exhibit minor fluctuations, likely stemming from evaluations on an unseen dataset, the overall trajectory indicates effective learning and generalization. Ultimately, the model achieves a final accuracy of 0.89 after completing 50 epochs.

In Fig. 5(b), the Confusion Matrix representing the VGGnet16 model illustrates its classification performance. Noteworthy is the model's accurate identification of Early Blight as Early Blight with a precision of 43.82%, Late Blight as Late Blight with 42.66% precision, and Healthy instances with 2.56% accuracy. This assessment offers valuable insights into the model's capability to differentiate between specific classes within the dataset.

Figure 6(a) displays the 50-epoch training and validation outcomes of a ResNet50 model. Training accuracy rises as training loss diminishes, while validation metrics exhibit some fluctuation. Surprisingly, the final accuracy after 50 epochs is 0.73, falling below expectations, possibly due to factors such as a limited dataset, suboptimal hyperparameters, or task complexity.

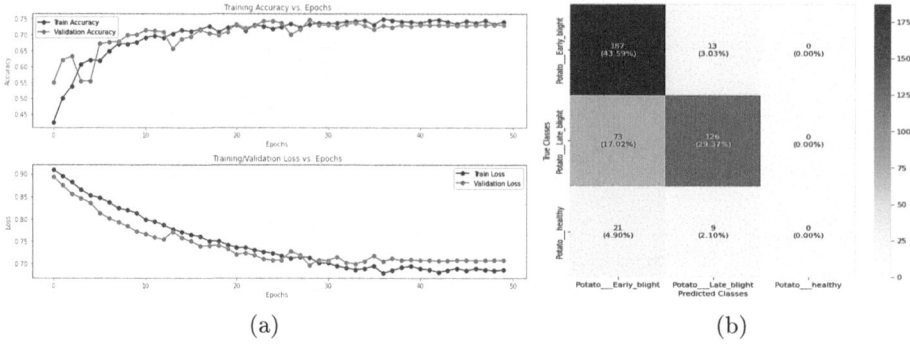

(a)   (b)

Fig. 6. (a) The Plot of Accuracy and Loss using ResNet50, (b) Confusion matrix for ResNet50.

In Fig. 6(b), the Confusion Matrix for the ResNet50 model showcases its classification performance. Notably, the model excels in identifying Early Blight with an accuracy of 43.59%, Late Blight with a precision of 29.37%, and displays precision in recognizing Healthy instances. This assessment offers valuable insights into the model's effectiveness in distinguishing and classifying specific classes within the dataset.

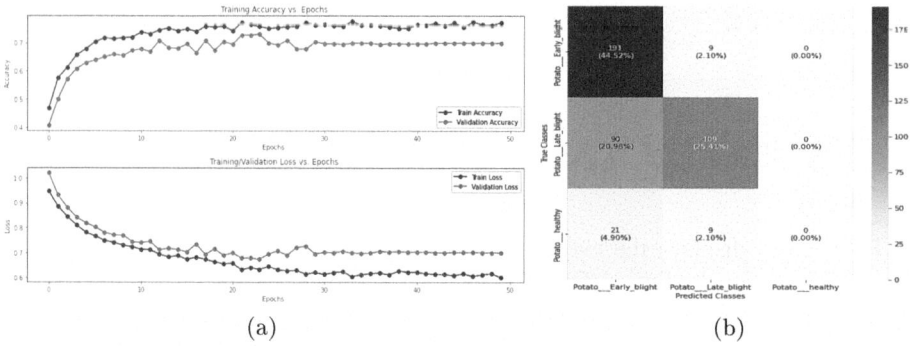

(a)   (b)

Fig. 7. (a) The Plot of Accuracy and Loss using ResNet152, (b) Confusion matrix for ResNet152.

Figure 7(a) visually represents the 50-epoch training dynamics of a ResNet152 model, showcasing a progressive rise in training accuracy and a consistent decline in training loss. However, the validation accuracy and loss exhibit more pronounced fluctuations, suggesting potential challenges in generalization. Surprisingly, the model achieves a final accuracy of 0.70 after 50 epochs, falling short of expectations, possibly due to dataset limitations, suboptimal hyperparameters, or task complexity.

In Fig. 7(b), the Confusion Matrix highlights the ResNet152 model's robust classification performance. Notably, it attains 44.52% accuracy in identifying Early Blight, demonstrates 25.41% precision in classifying Late Blight, and achieves 2.10% accuracy in recognizing instances as Healthy. This succinct analysis emphasizes the model's proficiency in distinguishing these classes within the dataset.

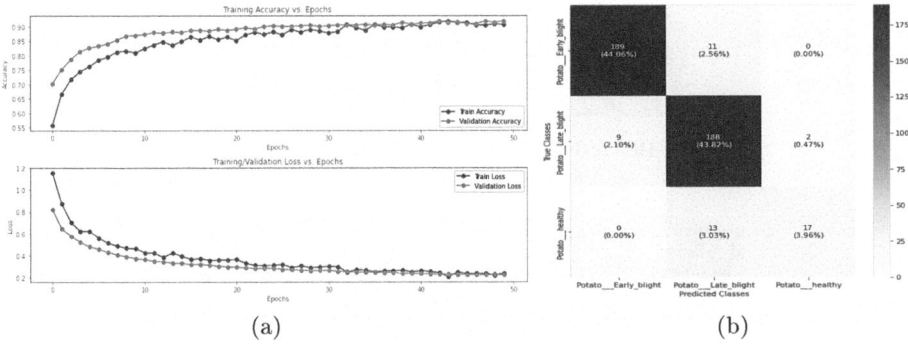

(a)   (b)

Fig. 8. (a) The Plot of Accuracy and Loss using Inception V3, (b) Confusion matrix for Inception V3

Figure 8(a) depicts the training and validation progress of an InceptionV3 model over 50 epochs. The training accuracy shows a steady increase, accompanied by a consistent decrease in training loss, indicating effective learning. While validation metrics display some fluctuations, the InceptionV3 model ultimately achieves an impressive final accuracy of 0.92, positioning itself as the top-performing model. This graph underscores the InceptionV3 model's excellence in image classification.

In Fig. 8(b), the Confusion Matrix evaluates the InceptionV3 model's classification performance. Notably, it achieves a high accuracy of 44.06% in classifying Early Blight and a precision rate of 43.82% for Late Blight. Additionally, the model displays 3.96% accuracy in identifying Healthy instances. This analysis provides technical insights into the model's adeptness in effectively discerning these categories within the dataset.

4.2 Data Testing

Following the meticulous training phase with our thoughtfully curated datasets, a critical testing process ensued. In this phase, newly acquired data, distinct from the training dataset, was utilized to evaluate the performance of a diverse set of models, including VGG16, VGG19, ResNet50, ResNet152, and Inception V3. This comprehensive testing procedure aims to rigorously assess the efficacy and generalization capabilities of each model, exposing them to the challenges presented by previously unseen data. Through this detailed testing protocol, our goal is to extract nuanced insights into the robustness and adaptability of these models, providing valuable information to guide decisions regarding their suitability for real-world applications (Figs. 9, 10 and 11).

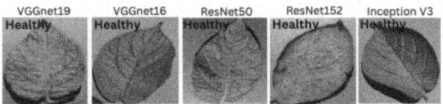

Fig. 9. Testing Results for Healthy Leaves.

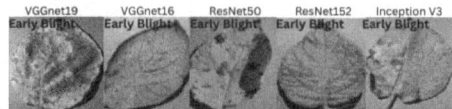

Fig. 10. Testing Results for Early Blight Leaves.

Fig. 11. Testing results for Late blight leaves

In this research study, the evaluation encompassed five distinct models: VGGNet16, demonstrating an accuracy of 0.89, VGGNet19, achieving an accuracy rate of 0.86, ResNet50, delivering an accuracy score of 0.73, ResNet152, yielding an accuracy of 0.70 and Inception V3, standing out with the highest accuracy of 0.92.

The experimentation phase involved consistent parameter configuration across all models, with hyperparameters set as follows: 50 epochs, a batch size of 32, image dimensions of 224 × 224, and the Adam optimizer with a learning rate of 0.001. The selection of 50 epochs was influenced by practical considerations, as the computational constraints of working on a CPU deemed it an optimal balance between model convergence and evaluation time. This choice proved effective, as it yielded satisfactory accuracy within the given epoch limit. The learning rate of 0.001 was chosen to facilitate a gradual descent during optimization, promoting stability in model training. The Adam optimizer, known for its adaptive learning rate and momentum, was selected for its efficiency in handling diverse datasets and achieving robust convergence. Each model underwent training with these parameters, but the time required per epoch varied, reflecting the computational demands of the respective architectures. Specifically, VGGNet16 demanded an average of approximately 800 s per epoch, translating to a total training duration of approximately 10.8 h for 50 epochs. VGGNet19, with an average epoch duration of 900 s, necessitated around 12.5 h for the same number of epochs. Meanwhile, ResNet50 required an average of 600 s per epoch, leading to a total training time of approximately 8.3 h for 50 epochs. ResNet152, with an average epoch duration of 450 s, completed 50 epochs in approximately 6.25 h. Finally, Inception V3 exhibited the shortest training time per epoch, averaging 250 s, and completed 50 epochs in roughly 3.4 h.

It's worth emphasizing that precision during the testing of these models can be influenced by factors like image noise. To optimize classification accuracy, it is advisable to implement preprocessing techniques, such as the removal of extraneous elements and ensuring that each image frame contains only one identifiable leaf object.

4.3 Performance Metrics

In evaluating the experimental results, key performance metrics, including precision (1), recall (2), accuracy score (3), and F1-score (4), were utilized. These metrics, selected purposefully for potato leaf disease classification, offer unique insights. Precision, emphasizing the accuracy of positive predictions, minimizes false positives, ensuring the identified diseases are genuinely present. Recall focuses on the model's effectiveness in capturing all instances of positive classes, providing a comprehensive evaluation and minimizing false negatives. Accuracy, offering a general measure of correctness across all classes, provides an overall assessment of the model's performance. F1-score, as a harmonic mean of precision and recall, serves as a suitable metric for the imbalanced dataset, striking a balance between precision and recall. This comprehensive set of metrics contributes to a nuanced and informative evaluation of the model's efficacy in potato leaf disease classification.

1. **Precision (P):** Precision is the ratio of correctly predicted positive observations to all positive predictions made by a classifier.

$$P = \frac{TP}{TP + FP} \tag{1}$$

2. **Recall (R):** The recall is calculated by dividing the number of correctly predicted positive observations by the total number of actual positive instances in the dataset.

$$R = \frac{TP}{TP + FN} \tag{2}$$

3. **Accuracy Score (Acc):** The accuracy score is a measure of the proportion of correctly classified instances, considering both true positives and true negatives.

$$Acc = \frac{TP + TN}{TP + TN + FP + FN} \tag{3}$$

4. **F1-Score (F1):** The F1-Score (F1) is a metric that signifies the harmonic mean of precision and recall. It provides a well-balanced evaluation of a classifier's performance, especially when dealing with imbalanced datasets.

$$F1 = 2 * \frac{P * R}{P + R} \tag{4}$$

To define each metric, the terms true positives (TP), false positives (FP), true negatives (TN), and false negatives (FN) are used to describe different aspects of the measurement.

Table 2 provides an insightful overview of the diverse performance metrics employed to ascertain the efficacy of potato image classification across various state-of-the-art models. The comprehensive roster of models encompasses VGG16, VGG19, ResNet50, ResNet152, and InceptionV3, each meticulously evaluated in the context of their classification capabilities concerning distinct class instances—namely, Early Blight, Late Blight, and Healthy. The utilization of precision, accuracy, recall, and F1 score as metrics

Table 2. Performance Measures of all the models

Model/Report	Class	P	R	Acc	F1
VGGNet16	Early Blight	0.93	0.94	0.89	0.93
	Late Blight	0.86	0.92		0.89
	Healthy	0.92	0.37		0.52
VGGNet19	Early Blight	0.90	0.95	0.86	0.92
	Late Blight	0.85	0.85		0.85
	Healthy	0.50	0.27		0.35
ResNet50	Early Blight	0.67	0.94	0.73	0.78
	Late Blight	0.85	0.63		0.73
	Healthy	0.00	0.00		0.00
ResNet152	Early Blight	0.63	0.95	0.70	0.76
	Late Blight	0.86	0.55		0.67
	Healthy	0.00	0.00		0.00
InceptionV3	Early Blight	0.95	0.94	0.92	0.95
	Late Blight	0.89	0.94		0.91
	Healthy	0.89	0.57		0.69

for potato image classification adds layers of granularity to the assessment, allowing for a nuanced understanding of the models' performance across the spectrum of categories. This systematic approach to performance evaluation not only facilitates a comparative analysis of the models but also unveils the specific strengths and areas of improvement for each in the challenging task of potato image classification.

5   Discussion

The utilization of Convolutional Neural Networks (CNNs) for classifying potato leaf diseases marks a significant stride in harnessing deep learning for agricultural issues. CNNs, such as VGG, ResNet, and Inception, showcase their effectiveness in capturing hierarchical features essential for discerning specific patterns linked to various diseases on potato leaves.

The InceptionV3 architecture, with its intricate design, emerges as the most promising model for potato leaf disease classification. Its utilization of inception modules with filters of multiple sizes enables effective detection of lesions, discolorations, and anomalies on potato leaves. The concatenation of outputs from these modules plays a crucial role in capturing complex relationships among features, contributing to a holistic understanding of the input data.

In the experimental phase, the research evaluated five models—VGGNet16, VGGNet19, ResNet50, ResNet152, and InceptionV3—using common hyperparameters.

InceptionV3 demonstrated the shortest training time per epoch, achieving an impressive accuracy of 91.84% in classifying early blight, late blight, and healthy potato plants.

Key performance metrics, including precision, recall, accuracy, and F1-score, provided a comprehensive assessment of the models' efficacy. InceptionV3's standout performance underscores the importance of selecting architectures tailored to the complexities of potato leaf disease classification.

In conclusion, this research contributes significantly to addressing the global challenge of potato disease detection, with InceptionV3 proving to be a top-performing model. The study not only advances agricultural technology but also highlights the broader potential of deep learning in enhancing food security and sustainable agriculture.

6 Conclusion and Future Work

Due to the escalating consumption of potatoes, effective disease management is imperative for ensuring global food security and sustaining agriculture. This study marks a notable stride in tackling the worldwide challenge of potato disease detection and classification. Evaluating prominent convolutional neural network architectures such as VGG16, VGG19, ResNet50, ResNet152, and InceptionV3, the research identifies InceptionV3 as the top-performing model, achieving an impressive 91.84% accuracy in classifying early blight, late blight, and healthy potato plants.Subsequent research will concentrate on enhancing the model's accuracy and resilience through advanced methodologies. Leveraging transfer learning, the study aims to capitalize on pre-existing model knowledge, fine-tuning it for heightened performance. Additionally, the exploration of transformer architectures is on the agenda to elevate the precision of potato disease classification. Acknowledging the study's current limitation of a relatively modest dataset comprising 2152 images, future efforts will focus on procuring and utilizing a more extensive and diverse dataset. This expansion is intended to bolster the model's generalization capabilities, making it proficient for real-time disease classification across diverse agricultural contexts. The envisaged future research also includes efforts to develop real-time field deployment systems, monitor disease progression, collaborate with agricultural stakeholders for large-scale implementation, and consistently refine the model's capabilities for addressing evolving challenges in potato disease management.

References

1. Beals, K.A.: Potatoes, nutrition and health. Am. J. Potato Res. **96**(2), 102–110 (2019)
2. Sholihati, R.A., Sulistijono, I.A., Risnumawan, A., Kusumawati, E.: Potato leaf disease classification using deep learning approach. In: 2020 International Electronics Symposium (IES), pp. 392–397. IEEE (2020)
3. Arora, R.K., Sharma, S.: Pre and post harvest diseases of potato and their management. In: Goyal, A., Manoharachary, C. (eds) Future Challenges in Crop Protection Against Fungal Pathogens. Fungal Biology, pp. 149–183. Springer, New York (2014). https://doi.org/10.1007/978-1-4939-1188-2_6
4. Tm, P., Pranathi, A., SaiAshritha, K., Chittaragi, N. B., Koolagudi, S.G.: Tomato leaf disease detection using convolutional neural networks. In: 2018 Eleventh International Conference on Contemporary Computing (IC3), pp. 1–5. IEEE (2018)

5. Sladojevic, S., Arsenovic, M., Anderla, A., Culibrk, D., Stefanovic, D.: Deep neural networks-based recognition of plant diseases by leaf image classification. Comput. Intell. Neurosci. **2016** (2016)
6. Gupta, M., Sinha, A.: Multi-class autoencoder-ensembled prediction model for detection of COVID-19 severity. Evol. Intel. **16**(4), 1433–1445 (2023)
7. LeCun, Y., Bengio, Y., Hinton, G.: Deep learning. Nature **521**(7553), 436–444 (2015)
8. Mandal, S.N., et al.: Image-based potato phoma blight severity analysis through deep learning. J. Inst. Eng. (India): Ser. B **104**(1), 181–192 (2023)
9. Risnumawan, A., Sulistijono, I. A., Abawajy, J.: Text detection in low resolution scene images using convolutional neural network. In: Herawan, T., Ghazali, R., Nawi, N.M., Deris, M.M. (eds) SCDM 2016. AISC, vol. 549, pp. 366–375. Springer, Cham (2017). https://doi.org/10. 1007/978-3-319-51281-5_37
10. Afakh, M.L., Risnumawan, A., Anggraeni, M.E., Tamara, M.N., Ningrum, E.S.: Aksara jawa text detection in scene images using convolutional neural network. In: 2017 International Electronics Symposium on Knowledge Creation and Intelligent Computing (IES-KCIC), pp. 77–82. IEEE (2017)
11. Arshad, F., et al.: PLDPNet: End-to-end hybrid deep learning framework for potato leaf disease prediction. Alex. Eng. J. **78**, 406–418 (2023)
12. Sulistijono, I.A., Risnumawan, A.: From concrete to abstract: Multilayer neural networks for disaster victims detection. In: 2016 International Electronics Symposium (IES), pp. 93–98. IEEE (2016)
13. Sulistijono, I.A., et al.: Implementation of victims detection framework on post disaster scenario. In: 2018 International Electronics Symposium on Engineering Technology and Applications (IES-ETA), pp. 253–259. IEEE (2018)
14. Anwar, M.K., Risnumawan, A., Darmawan, A., Tamara, M.N., Purnomo, D.S.: Deep multi-layer network for automatic targeting system of gun turret. In: 2017 International Electronics Symposium on Engineering Technology and Applications (IES-ETA), pp. 134–139. IEEE (2017)
15. Anwar, M.K., et al.:. Deep features representation for automatic targeting system of gun turret. In: 2018 International Electronics Symposium on Engineering Technology and Applications (IES-ETA), pp. 107–112. IEEE (2018)
16. Imaduddin, H., Anwar, M.K., Perdana, M.I., Sulistijono, I.A., Risnumawan, A.: Indonesian vehicle license plate number detection using deep convolutional neural network. In: 2018 International Electronics Symposium on Knowledge Creation and Intelligent Computing (IES-KCIC), pp. 158–163. IEEE (2018)
17. Talukder, M.S.H., et al.: PotatoPestNet: a CTInceptionV3-RS-based neural network for accurate identification of potato pests. Smart Agric. Technol. **5**, 100297 (2023)
18. Fujita, E., Kawasaki, Y., Uga, H., Kagiwada, S., Iyatomi, H.: Basic investigation on a robust and practical plant diagnostic system. In: 2016 15th IEEE International Conference on Machine Learning and Applications (ICMLA), pp. 989–992. IEEE (2016)
19. Rastogi, A., Arora, R., Sharma, S.: Leaf disease detection and grading using computer vision technology & fuzzy logic. In: 2015 2nd International Conference on Signal Processing and Integrated Networks (SPIN), pp. 500–505. IEEE (2015)
20. Mohanty, S.P., Hughes, D.P., Salathé, M.: Using deep learning for image-based plant disease detection. Front. Plant Sci. **7**, 1419 (2016)
21. Jia, S., Jiang, S., Lin, Z., Li, N., Xu, M., Yu, S.: A survey: Deep learning for hyperspectral image classification with few labeled samples. Neurocomputing **448**, 179–204 (2021)
22. Liu, Y., Zhang, Z., Liu, X., Wang, L., Xia, X.: Performance evaluation of a deep learning based wet coal image classification. Miner. Eng. **171**, 107126 (2021)
23. Wang, P., Fan, E., Wang, P.: Comparative analysis of image classification algorithms based on traditional machine learning and deep learning. Pattern Recogn. Lett. **141**, 61–67 (2021)

24. Korot, E., et al.: Code-free deep learning for multi-modality medical image classification. Nat. Mach. Intell. **3**(4), 288–298 (2021)
25. Arshaghi, A., Ashourian, M., Ghabeli, L.: Potato diseases detection and classification using deep learning methods. Multimed. Tools Appl. **82**(4), 5725–5742 (2023)
26. Tiwari, D., Ashish, M., Gangwar, N., Sharma, A., Patel, S., Bhardwaj, S.: Potato leaf diseases detection using deep learning. In: 2020 4th International Conference on Intelligent Computing and Control Systems (ICICCS), pp. 461–466. IEEE (2020)
27. Gené-Mola, J., et al.: Fruit detection and 3D location using instance segmentation neural networks and structure-from-motion photogrammetry. Comput. Electron. Agric. **169**, 105165 (2020)
28. Gupta, M., Singhal, Y.K., Sinha, A.: Assessing spatiotemporal transmission dynamics of COVID-19 outbreak using AI analytics. In: Gupta, D., Khanna, A., Kansal, V., Fortino, G., Hassanien, A.E. (eds) DoSCI 2021. AISC, vol. 1374, pp. 829-838. Springer, Singapore (2022). https://doi.org/10.1007/978-981-16-3346-1_67
29. García, S., Luengo, J., Herrera, F.:. Data Preprocessing in Data Mining, vol. 72, pp. 59–139. Springer, Cham (2015)
30. Jiang, H., et al.: A review of deep learning-based multiple-lesion recognition from medical images: classification, detection and segmentation. Comput. Biol. Med. 106726 (2023)
31. https://www.kaggle.com/datasets/arjuntejaswi/plant-village

Enhancing Monkeypox Disease Detection Using Computer Vision-Based Approaches and Deep Learning

Imtiaj Ahmed[✉] , Rayan , Sayma Akter Tihany , and Adnan Mahmud

Department of Computer Science and Engineering,
East West University, Dhaka 1212, Bangladesh
imtiajahmed15@gmail.com

Abstract. This research addresses the challenge of monkeypox, a contagious skin disease that may be spread from animal to human contact. Recognizing its broad reach and potential severity, the research emphasizes the critical need for early and precise detection. The project sets out to evaluate how well a computer vision-based approach and CNN-based technology can accurately spot instances of monkeypox. Various algorithms, including Googlenet, Resnet50, VGG16, VGG19, Darknet53, Mobilenetv2, Inceptionv3, Inceptionresnetv2, and Yolov7, are put to the test. Notably, Mobilenetv2 stands out with an exceptional validation accuracy of 95.2%, showcasing its effectiveness in pinpointing and identifying the disease. The study emphasizes the significance of deep learning optimizers in improving the outcomes of image classification, not only for monkeypox detection but also for making strides in diagnosing monkeypox diseases. Furthermore, the research suggests avenues for refining deep learning architectures and optimization techniques.

Keywords: Monkeypox · Skin diseases · OpenCV · Machine learning · MobileNetV2

1 Introduction

Monkeypox, a complex disease with various clinical signs, poses a real challenge in getting diagnosed and treated promptly. The usual detection methods depend heavily on human observations and lab tests, which can lead to errors and delays in critical decision-making. Our mission, however, is steering us into a new era where the fusion of image processing and deep learning is set to transform how we detect Monkeypox. Imagine a world where artificial intelligence works hand in hand with the intricate details hidden in medical images. That's the vision driving our research - a vision where spotting Monkeypox early isn't just a dream but a reality that can genuinely improve patient outcomes.

In this paper, we take you on a journey through the fascinating realms of image processing and deep learning, revealing the methods behind our innovative

A. K. Bairwa et al. (Eds.): ICCAIML 2024, CCIS 2184, pp. 346–356, 2025.
https://doi.org/10.1007/978-3-031-71481-8_27

approach. We are using advanced technologies like convolutional neural networks and sophisticated image analysis to create a sturdy framework. Our goal isn't just to replace traditional diagnostic methods but to enhance them, providing healthcare professionals with a tool that speeds up accurate diagnoses. As we dive into the details of our research, let's collectively explore the potential of this collaboration between medical expertise and cutting-edge technology. Together, we're paving the way for a future where Monkeypox detection goes beyond the usual methods, bringing us closer to a world where early intervention is the standard rather than the exception [1].

2   Related Works

The goal of this research [2] was to find a deep-learning model for monkeypox (monkeypox) detection. MobileNetV2 outperformed the other five pre-trained models with 98.16% accuracy, 0.97 recall, 0.98 precision, and 0.97 F1-score. For the early and precise identification of monkeypox in clinical settings, the model demonstrated encouraging results.

The monkeypox virus [3] raises the possibility of a global pandemic, but ML provides a remedy. We offer the "Monkeypox2022" dataset for using image analysis to detect the illness, which is freely accessible on GitHub. With the help of LIME for feature extraction, our modified VGG16 model achieves high accuracy of 97.18% and 88.88% in two trials.

A contagious virus mostly found in Africa, monkeypox, has spread around the world. Headaches, chills, fever, skin tumours, and rashes mimicking smallpox are among the symptoms. Early diagnosis is made possible by several AI models. 34 current research on monkeypox diagnostics, spread modelling, treatment, vaccine development, and risk management were examined [4].

Monkeypox is a contagious and potentially deadly disease caused by the monkeypox virus. This work proposes a secure, computer vision-based method for diagnosing monkeypox using deep learning techniques and skin lesion images, offering promising results for widespread deployment [5].

Monkeypox is the monkeypox virus, of contagious and potentially deadly disease with symptoms like skin lesions, rashes, and fever. The latest outbreak has spread globally. Traditional diagnosis is risky for medical staff, but IoT and AI offer a smart and secure method using computer vision to analyze skin lesion images. This non-invasive approach, evaluated on two datasets, shows promising results and can be a cost-effective solution for underprivileged areas lacking lab infrastructure [6].

Medical diagnosis is aided by machine learning [4], especially in dermatology when separating related disorders like skin cancer. Accurate identification is essential for efficient epidemiological management during a recent monkeypox epidemic. An open database of photos of monkeypox and control subjects was used to accomplish this. Training and test sets were created from the pre-processed photos in an 80/20 ratio. Using MiniGoggleNet, six experiments of various epochs were carried out. The best model has an accuracy of 0.9708, a loss function of 0.1442, an AUC of 0.74 for class 0 and class 1, a micro-average AUC of 0.76, and a macro-average AUC of 0.74 after 50 training iterations.

Monkeypox poses a growing global threat [7], with rising cases and transmission risks. Machine Learning (ML) offers promise in diagnosing diseases via image analysis. Here, we present a Monkeypox diagnostic model, GRA-TLA, leveraging ML techniques. Testing on ten CNN models showed accuracies of 77% to 88% for binary classification and 84% to 99% for multiclass classification. Our approach proved computationally efficient, with LIME providing valuable insights into disease indicators.

The fear of monkeypox [8], which has the potential to spark a global pandemic, is impeding the global recovery from COVID-19. Image-based diagnostics can help diagnose monkeypox by using machine learning, similar to its successful uses in cancer and COVID-19 identification. To increase the precision of monkeypox picture classification, this study suggests two techniques that combine transfer learning, meta-heuristic optimization, and a deep neural network. Evaluation utilizing a public dataset shows the suggested approaches' higher performance and efficacy, with an average classification accuracy of 98.8%.

This research [9] leverages deep learning to enhance monkeypox detection, addressing misdiagnosis challenges due to similar symptoms with chickenpox. A custom CNN model achieved superior accuracy (99.60%) compared to existing models, aiding rapid and precise diagnosis using digital skin images.

In 1970 [10], the 9th Democratic Republic of the Congo became the first human to contract monkeypox after smallpox eradication. Mostly seen in the Congo Basin's rainforests, the disease spread across 11 African nations [11]. The discovery of the varicella virus during monkeypox outbreaks suggests changes in how these diseases spread [12]. Since 2017 [13], Nigeria has faced over 500 suspected cases, with 200 confirmed and a 3% fatality rate. The 2003 U.S. outbreak, linked to pet prairie dogs, highlights monkeypox's global impact. Recent cases [14] involve Nigerian tourists in various non-endemic countries.

3   Materials and Methods

3.1   Dataset Used

Using a transfer learning method,

- Proposing the pre-trained CNN models to be developed using deep learning.
- Collecting data on paper usage is challenging, involving information on consumption, document types, processes, costs, and waste.
- Data analysis containing hospital and medical sector information, provides insights.
- Analyze paper consumption in hospitals, identifying limitations, and evaluating potential benefits of systems.

The Monkeypox Skin Lesion Dataset from IEEE DataPort was used in this word. This dataset, which is used for binary classification, concentrates on skin lesion pictures linked to measles, chickenpox, and monkeypox.

3.2   Data Analysis

Images of instances of chickenpox and measles are included in the classes titled "Monkeypox" and "Others." The three primary folders that make up the monkeypox skin lesion dataset are listed below. There are 72870 photos in the first folder, "Original Images," altogether. The remaining 42260 photos showcase Chickenpox and Measles, while 3502 of them are designated as "Monkeypox."

The "Augmented Images" folder contains enhanced photos for both classes, serving as a repository of additional images, which also serves as a collection of extra pictures. These photos were made using a variety of techniques, including scaling, rotation, translation, reflection, shear, hue, saturation, and contrast modifications. The "Fold1" directory contains one of the datasets used in three-fold cross-validation. This folder's original images have roughly been split into training, validation, and test sets with a ratio of 70:10:20. It is significant to notice that only improved copies of the images are included in the training and validation sets, but only the original photographs are included in the test set.

Mainly the Fold1 dataset was utilised in this work to train the computer vision models. It comprises 276'5410 Monkeypox pictures and 32690 Other (Chickenpox and Measles) images in the training subset, 20000 Monkeypox images and 25000 Other images in the test subset, and 16800 Monkeypox images and 25200 Other images in the validation subset. Figure 1, represents the MonkeyPox Sample Images.

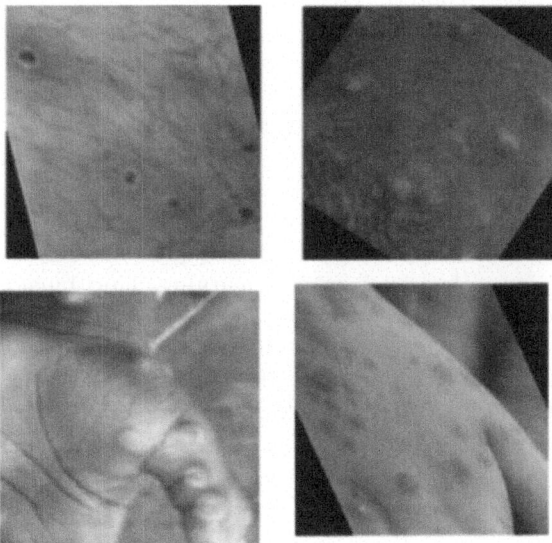

Fig. 1. MonkeyPox Sample Images

In Fig. 2, the architecture appears to be a convolutional neural network (CNN) for MonkeyPox disease detection. The Monkeypox Skin Lesion Dataset is

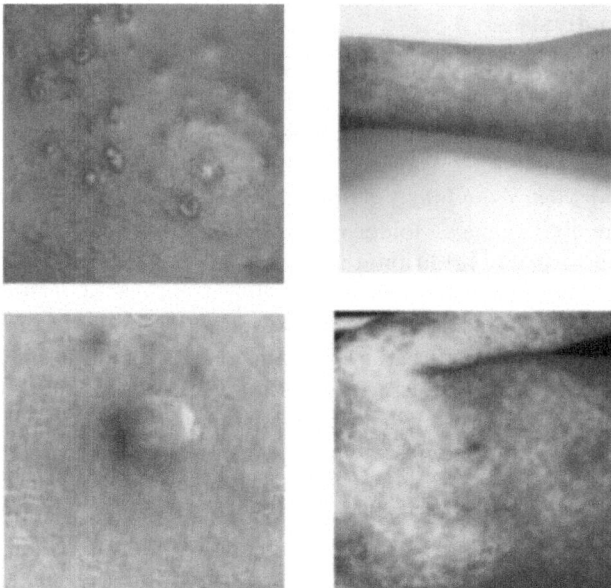

Fig. 2. Other Sample Images

a crucial tool for research and the creation of computer vision algorithms that can precisely identify and categorize skin lesions associated with monkeypox. Initial examination of a set of sample photos reveals distinctive features of monkeypox rashes, which are distinguished by their severity and localized appearance on the hands, feet, and mouth. However on the other hand, frequently cause rashes that are more evenly distributed over the entire body. The Monkeypox Skin Lesion Dataset provides a singular chance to further our comprehension of skin lesions connected to monkeypox. We can train models capable of accurately recognizing and categorizing these particular sorts of rashes by utilizing computer vision approaches, such as deep learning algorithms. By enabling earlier and more precise diagnoses of monkeypox and differentiating it from other dermatological disorders, the results of this study have the potential to make a substantial contribution to the area of dermatology. In the end, this research could support the creation of enhanced management plans and public health responses to monkeypox epidemics.

4   Proposal Model

In our paper, we provide a unique research strategy that uses the CNN models and transfer learning to get over the problems brought on by small datasets. Our goal is to create a very effective model that can correctly categorize skin lesions brought on by chickenpox, measles, and monkeypox.

We start our study by obtaining the weights for the pre-trained CNN models from the vast ImageNet dataset. We may utilize the vast knowledge and skills of the pre-trained model in picture classification tasks by designating the foundation layers as trainable while leaving the top layers non-trainable.

We can successfully utilize the CNN model's learned features and representations by using this transfer learning technique, allowing us to extract valuable information from the skin lesion photos. With the help of our study, we want to create a reliable and accurate classification model that will help recognise and treat skin lesions caused by measles, chickenpox, and monkeypox, ultimately leading to better healthcare practices in this area (Fig. 3).

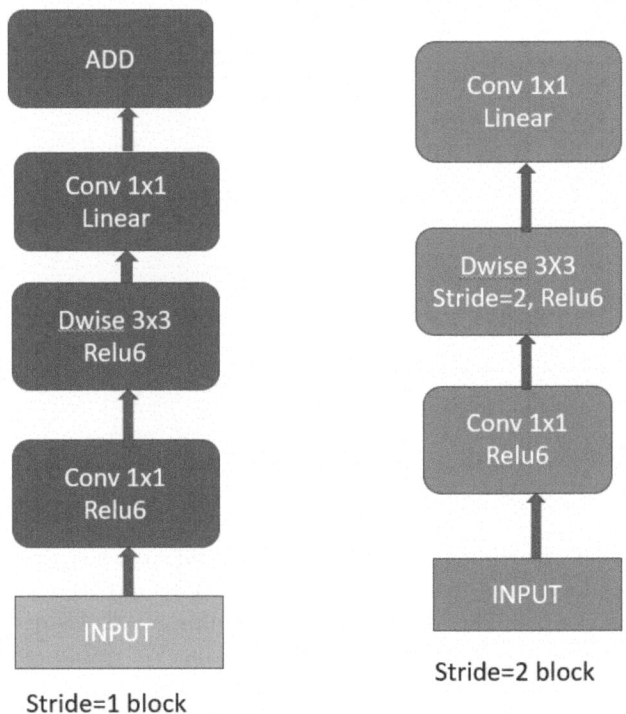

Fig. 3. Convolutional neural network Blocks

Additional layers are added to adapt the model to our particular job. Since the challenge is a two-class classification problem, binary cross-entropy is used as the loss function in the model's compilation together with accuracy as the performance metric. Callbacks are used to improve the training process.

5   Experimental Result

The model parameters are kept using the least validation loss as a criterion in the ModelCheckpoint callback. The ReduceLROnPlateau function dynamically

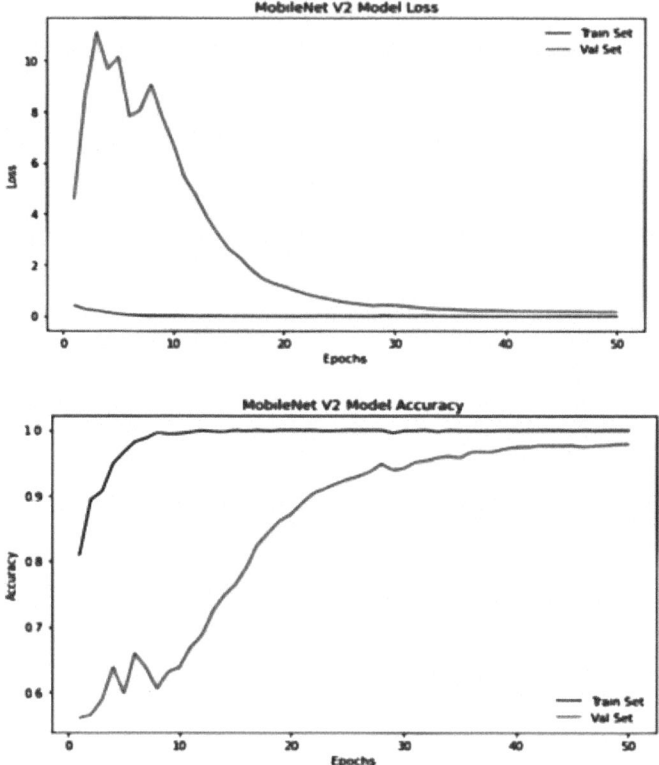

Fig. 4. Model loss and Accuracy

reduces our model's learning rate by 50% if the validation loss does not improve for two consecutive epochs. We used a batch size of 32 and 50 iterations to train the model using the training data during the training phase. Using test data, the model's performance was assessed; after 50 epochs. We examined the training history after the training was finished and found that the model's validation accuracy and loss stabilized at about 40 epochs. The weights were loaded for the model with the highest validation accuracy and lowest validation loss over training. These weights will be used to assess the model's performance according to certain standards. By adopting the suggested model, we want to solve the problems brought on by a small dataset and offer a precise and trustworthy classification tool for skin lesions associated with chickenpox, measles, and monkeypox (Fig. 4).

We looked at the model's efficiency. The model was created using pre-trained CNN models, whose weights were taken from ImageNet. The model was altered by adding new layers and making some trainable for the research. The model was learned from a dataset for 50 epochs of training. On the training data, Mobilenetv2 get a 96.10% accuracy rate. Techniques like preserving the best weights and modifying the learning rate enhanced the model. The effectiveness of the model was then evaluated using the test data and the validation data. The model's accuracy on the test data was 96.10%, which is a respectable result. However, the accuracy decreased to 95.198% when using the hidden validation data. This might be a result of the limited training dataset, which prevented the model from being trained on all the changes seen in the validation data.

Table 1, there are showing the MobileNetV2 model training accuracy, Testing accuracy, precision, and recall.

Table 1. Table of Models, % of their Accuracy, Precision and F1-score

Model	Training Accuracy	Validation Accuracy	Recall	Precision	F1-score
Googlenet	92.2	91.8	91.5	90.7	93.1
Resnet50	94.2	91.2	92.3	92.6	91.2
VGG16	96.3	91.5	90.1	89.2	92.1
VGG19	95.3	93.2	92.2	92.6	93.2
Darknet53	91.5	89.6	88.2	87.1	88.7
Mobilenetv2	96.1	95.2	95.2	94.5	95.8
Inceptionv3	80.5	78.5	79.3	78.8	78.9
Inceptionresnetv2	80.1	78.8	79.7	79.1	80.2
Yolov7	92.2	92.1	93.1	92.2	92.7

A comparative analysis of various optimization algorithms for monkeypox disease detection revealed the following validation accuracies % are: Googlenet (91.8), Resnet50 (91.2), VGG16 (91.5), VGG19 (93.2), Darknet53 (89.6), Mobilenetv2 (87.1), Inceptionv3 (77.5), Inceptionresnetv2 (75.8), Efficientnetb0 (81.2), Desnet201 (85.8), Xception (86.5), Nasnetlarge (81.8), and Yolov7 (96.1). Among these algorithms, Yolov7 exhibited the highest % age of validation accuracy of 96.1, showing its outstanding efficacy in diagnosing and localising monkeypox diseases. This suggests that, when tested on a different validation dataset, Mobilenetv2 outperformed the other algorithms in properly recognizing and localizing objects within pictures. It demonstrates that Mobilenetv2's design and training procedure were successful in obtaining higher performance in object detection tests. These findings shed light on the efficacy of various algorithms for detecting monkeypox diseases, leading to the selection of appropriate models to improve agricultural output and disease control techniques.

Figures 5 display the model loss, model accuracy, and confusion matrix of the MobileNetV2 model, respectively.

The classification report and confusion matrix are thoroughly examined, and it is clear that the model performs better than others in properly forecasting instances of monkeypox. The model has a better sensitivity, indicating an improvement in its ability to detect cases of monkeypox. This suggests that the

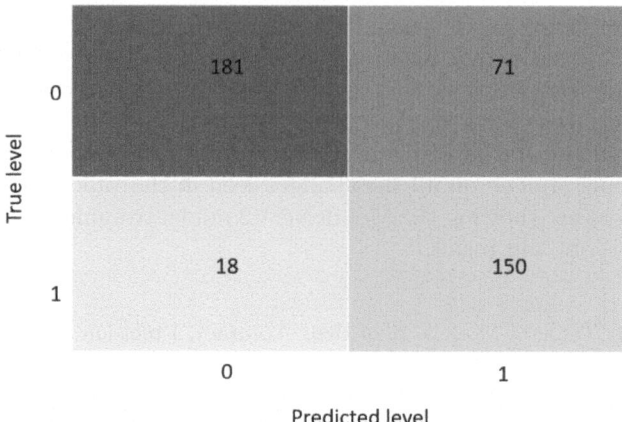

Fig. 5. Confusion Matrix

model has a strong capacity to identify instances of monkeypox. We may evaluate the model's ability to predict outcomes across several classes by using the classification report and confusion matrix. Notably, it excels in correctly identifying cases of monkeypox, demonstrating its effectiveness in this particular area of categorization. Due to the increased sensitivity, there is a lower chance that cases of monkeypox may be mistakenly labelled as negative cases. As a result, the model successfully diagnoses monkeypox, increasing the possibility of quick and efficient intervention in such situations.

In summary, the model's performance analysis, based on the classification report and confusion matrix, highlights its exceptional accuracy and capability to accurately forecast incidences of monkeypox, suggesting its strength in this particular area of classification.

6   Discussion

According to our study's findings, the model's validation accuracy was about 95.2%. This means that, on average, 95.2 out of 100 incidents were properly classified by the model as cases of monkeypox. The program also demonstrated a 93.6% sensitivity for identifying positive instances of monkeypox. We did see difficulties, though, when misclassifications occurred because the illness resembled measles or chickenpox. These challenges are caused by the similarity of the symptoms and the individual variation in monkeypox presentation. We suggest enlarging the training dataset and adding further picture augmentations to improve the model's performance. The algorithm may be trained to recognize more variants if the training data contains a wider variety of instances of monkeypox. Furthermore, our results demonstrate the potential of computer vision-based models like MobileNetV2 for locating cases of monkeypox. The model's accuracy and robustness might yet be improved, especially when it comes to instances where Monkeypox imitates other illnesses.

7   Conclusion

This research embarks on a mission to enhance our capabilities in identifying monkeypoxes through the fusion of image processing and deep learning approaches. Imagine a scenario where technology plays a pivotal role in early diagnosis, potentially saving lives and minimizing the impact of this ailment on affected communities. Our project aims to evaluate the effectiveness of cutting-edge technologies, including computer vision-based approaches and CNN-based models. We're delving into a diverse array of algorithms - Googlenet, Resnet50, VGG16, VGG19, Darknet53, Mobilenetv2, Inceptionv3, Inceptionresnetv2, and Yolov7 - to understand how well they can spot instances of monkeypox. A star performer in our evaluation is Mobilenetv2, boasting an exceptional validation accuracy of 95.2%. This not only underscores its effectiveness in pinpointing and identifying the disease but also raises hopes for a tech-driven leap in disease detection. We're not just stopping at monkeypox; our study recognizes the broader implications for monkeypox disease diagnosis, emphasizing the significance of deep learning optimizers in refining image classification outcomes.

References

1. Altun, M., Gürüler, H., Özkaraca, O., Khan, F., Khan, J., Lee, Y.: Monkeypox detection using CNN with transfer learning. Sensors **23**(4), 1783 (2023)
2. Jaradat, A.S., et al.: Automated Monkeypox skin lesion detection using deep learning and transfer learning techniques. Int. J. Environ. Res. Public Health **20**(5), 4422 (2023)
3. Ahsan, M.M., Uddin, M.R., Farjana, M., Sakib, A.N., Momin, K.A., Luna, S.A.: The image data collection and implementation of the deep learning-based model in detecting Monkeypox disease using modified VGG16. arXiv preprint arXiv:2206.01862 (2022)
4. Chadaga, K., et al.: Application of artificial intelligence techniques for Monkeypox: a systematic review. Diagnostics **13**(5), 824 (2023)
5. Almufareh, M.F., Tehsin, S., Humayun, M., Kausar, S.: A transfer learning approach for clinical detection support of Monkeypox skin lesions. Diagnostics **13**(8), 1503 (2023)
6. Alakus, T.B., Baykara, M.: Comparison of Monkeypox and wart DNA sequences with deep learning model. Appl. Sci. **12**(20), 10216 (2022)
7. Ahsan, M.M., et al.: Deep transfer learning approaches for Monkeypox disease diagnosis. Expert Syst. Appl. **216**, 119483 (2023)
8. Abdelhamid, A.A., et al.: Classification of Monkeypox images based on transfer learning and the Al-Biruni earth radius optimization algorithm. Mathematics **10**(19), 3614 (2022)
9. Uzun Ozsahin, D., et al.: Computer-aided detection and classification of Monkeypox and chickenpox lesion in human subjects using deep learning framework. Diagnostics **13**(2), 292 (2023)
10. Marin, M., Lopez, A.: Varicella (Chickenpox). https://wwwnc.cdc.gov/travel/yellowbook/2020/travel-related-infectious-diseases/varicella-chickenpox. Accessed 14 Oct 2022

11. Abed Alah, M., Abdeen, S., Tayar, E., Bougmiza, I.: The story behind the first few cases of Monkeypox infection in non-endemic countries, 2022. J. Infect. Public Health **15**, 970–974 (2022)
12. Ali, S.N., et al.: Monkeypox skin lesion detection using deep learning models. A preliminary feasibility study. arXiv arXiv:2207.03342 (2022)
13. Chickenpox Stock Photos, Pictures & Royalty-Free Images-iStock. https://www.istockphoto.com/search/2/image?page=15&phrase=chickenpox. Accessed 14 Oct 2022
14. Nafisa Ali, S., et al.: Monkeypox skin lesion detection using deep learning models: a feasibility study. arXiv, 13. https://arxiv.org/pdf/2207.03342.pdf. Accessed 24 Aug 2022

Tomato Leaf Disease Prediction Based on Deep Learning Techniques

Anirudh Singh⬤, Satyam Kumar⬤, and Deepjyoti Choudhury$^{(\boxtimes)}$⬤

Department of AIML, School of CSE, Manipal University Jaipur, Dehmi Kalan, Near GVK Toll Plaza, Jaipur 303007, Rajasthan, India
`deepjyotichoudhury05@gmail.com`

Abstract. Tomatoes, globally cherished and pivotal to diverse cuisines, face substantial threats to their quality and quantity due to various diseases. This research paper applies a deep learning approach for the precise detection of tomato leaf diseases. Our methodology incorporates a range of classifiers, including Random Forest (R.F), Inception V3, DenseNet, ResNet50, Xception, and MobileNet. Our results showcase the proficiency of these classifiers in distinguishing among nine distinct disease classes and one healthy class. Random Forest classifier exhibits an accuracy of 68.00%. Inception V3 excels with an impressive accuracy of 97%, coupled with high precision (98%), recall (96%), and an overall F1-Score of 97.50%. DenseNet demonstrates robust performance with an accuracy of 94.00%, precision of 98%, recall of 89%, and an F1-Score of 93%. ResNet50 closely follows with an accuracy of 93.30%, precision of 97%, recall of 91%, and an F1-Score of 94%. Xception maintains a well-balanced performance, achieving an accuracy of 95%, precision of 93%, recall of 99%, and F1-Score of 96%. Lastly, MobileNet, while achieving a moderate accuracy of 76%, demonstrates precision, recall, and F1-Score of 66%, 90%, and 76%, respectively. These experimental findings not only underscore the varied performance of each classifier concerning different disease classes but also emphasize the significant potential of deep learning to enhance the accuracy and efficiency of tomato leaf disease detection. This contribution supports ongoing endeavors to improve crop yield and quality through the application of advanced machine learning techniques in agriculture.

Keywords: disease · prediction · tomato · leaf · agriculture

1 Introduction

India, ranking as the third-largest global producer and exporter of tomatoes, commands a vast agricultural landscape spanning over 3,50,000 hectares, yielding a substantial annual harvest of approximately 53,00,000 tons [1]. The significance of tomatoes in the country's agricultural milieu is not only measured by sheer volume but also by the economic and nutritional impact they impart. However, the grandeur of this vital crop is persistently challenged by the looming specter of diseases that can exert a profound influence on both the quantity and quality of tomato yields. These diseases, adept at traversing from leaves to fruits, necessitate a vigilant and preemptive approach to ensure early detection

A. K. Bairwa et al. (Eds.): ICCAIML 2024, CCIS 2184, pp. 357–375, 2025.
https://doi.org/10.1007/978-3-031-71481-8_28

on the plant's foliage. Traditional disease detection methods, relying on manual observation by expert teams, present inherent challenges, manifesting as labor-intensive and time-consuming processes that often lead to delayed detection and response [2]. Recognizing the imperative for a more efficient and timely approach, researchers have been fervently engaged in the development of automated methods. These approaches, which include advanced methods like deep learning, machine learning, and digital image processing, have emerged as game-changing instruments with the potential to completely change how diseases are detected in agriculture. Beyond streamlining detection processes, these technological advancements hold the promise of reducing pesticide usage, thereby fostering sustainable agricultural practices and elevating overall crop production and quality. Against this backdrop, our research endeavors to address the critical challenge of predicting tomato leaf diseases through the innovative application of machine learning techniques. Specifically, Convolutional Neural Network (CNN) models, with ResNet (Residual Net), Inception V3, DenseNet, Xception, and MobileNet as prominent exemplars, have been strategically employed. The integration of machine learning into disease prediction protocols is motivated by its inherent ability to efficiently analyze extensive datasets, extracting intricate patterns that may elude manual inspection.

Central to our study is a meticulously curated dataset of tomato leaf images sourced from the Plant Village Dataset. This comprehensive dataset encompasses six distinct disease categories afflicting tomato leaves: Bacterial Spot, Late Blight, Leaf Mold, Septoria Leaf Spot, Tomato Mosaic, Yellow Curved, in addition to a category representing healthy leaves [3]. Our approach pivots on the strategic utilization of a CNN model, engineered to meticulously scrutinize these images with the overarching goal of not only classifying but predicting the presence of diseases with a high degree of accuracy. The selected CNN models, ResNet, Inception V3, DenseNet, Xception, and MobileNet, have proven to be robust in the context of image classification and have demonstrated notable success in various applications, including plant disease detection. These models leverage the power of deep learning to decipher complex patterns within the dataset, enabling accurate and efficient disease identification.

The subsequent sections of this research paper unfold as a journey through the nuances of our methodology and the intricacies of our findings. Section 2 initiates the discourse with an in-depth background study, providing a nuanced understanding of the prevailing challenges in tomato cultivation, considering both the historical and contemporary perspectives. Section 3 meticulously delineates the proposed methodology, offering a comprehensive exploration of the theoretical underpinnings and practical considerations guiding our approach towards disease prediction. Section 4 delves into the nuanced experimental settings and implementation strategies, shedding light on the datasets employed, the intricacies of model development, and the deliberate choices made in our research design. In conclusion, Sect. 5 provides a thorough and in-depth examination of the outcomes obtained from our research.

2 Related Work

Pivotal contribution to the literature came from Thangaraj et al. [4], who conducted a comprehensive review and discussion on artificial intelligence in tomato leaf disease detection. In 2019, Kaur and Bhatia [5] proposed an improved method for disease detection and classification, contributing to the precision agriculture landscape. The subsequent year witnessed Kibriya et al. [6] extending the capabilities of CNNs, further showcasing their effectiveness in disease detection. Chowdhury et al. [7], in 2021, focused on deep learning techniques for tomato leaf disease detection, emphasizing the applicability of these methods in agriculture. This work demonstrated a pivotal shift towards leveraging advanced technologies for automating disease identification processes in the agricultural sector.

Hasan et al. [8] explored precision farming and introduced transfer learning to enhance tomato leaf disease detection. Kaushik et al. [9], in 2020, contributed to the field by emphasizing the importance of data augmentation for improved disease detection. Durmuş et al. [10] proposed a disease detection system on tomato leaves using deep learning, underscoring the importance of advanced technologies in agriculture. The subsequent year witnessed Karthik et al. [11] introducing attention-embedded residual CNNs for disease detection in tomato leaves, demonstrating the potential of attention mechanisms in refining model accuracy.

Anandhakrishnan and Jaisakthi [12] delved into image-based disease detection, contributing to sustainable agriculture. Their work highlighted the continuous exploration of innovative technologies for addressing agricultural challenges. Chowdhury et al. [13] explored automatic and reliable leaf disease detection using deep learning techniques, showcasing the ongoing efforts to automate and streamline disease identification processes. David et al. [14] conducted a literature review in 2021, offering a comprehensive overview of disease detection in tomato leaves using deep learning.

Mim et al. [15] explored image processing techniques in 2019, presenting an alternative avenue for disease detection. Their work contributed to the diversification of methodologies employed in tomato leaf disease detection. Irmak and Saygili [16] extended the use of CNNs for tomato leaf disease detection, contributing to the ongoing advancements. Their work showcased the continued exploration and refinement of deep learning techniques for agriculture applications. The application of transfer learning for mobile applications was introduced by Bir et al. [17] in 2020, demonstrating the adaptability of advanced techniques to diverse contexts. Rahman et al. [18] proposed an image processing-based system for detection, identification, and treatment of tomato leaf diseases in 2023, showcasing the potential for holistic solutions in disease management.

Nagamani and Sarojadevi [19] explored deep learning techniques for disease detection, further diversifying the methodologies. Their work highlighted the ongoing exploration of innovative technologies to improve disease identification accuracy. Hong et al. [20] focused on disease detection and classification in 2020, emphasizing the role of deep learning in addressing complex agricultural challenges. Zhao et al. [21] introduced an improved CNN with an attention module for disease diagnosis, showcasing the ongoing efforts to enhance model architectures. Batool et al. [22] extended the classification and identification of tomato leaf diseases using deep neural networks. Their work contributed to the ongoing exploration of neural network architectures for disease detection.

Zhang et al. [23] worked on improving object detection for tomato diseases using deep learning in 2020, showcasing the potential for enhancing disease identification through advanced computer vision techniques. Wu et al. [24] explored tomato leaf disease identification based on deep CNNs in 2021, contributing to the refinement of deep learning models for agricultural applications. Paymode et al. [25] contributed to CNN-based disease detection, showcasing the continued emphasis on leveraging convolutional neural networks for accurate disease identification. Alhaj Ali et al. [26] delved into deep learning for tomato leaf diseases, contributing to the ongoing efforts to improve disease detection accuracy. Anandhakrishnan and Murugaiyan [27] focused on pretrained deep CNN models for disease detection, showcasing the potential for leveraging pre-existing knowledge in deep learning applications. Jasim and Al-Tuwaijari [28] explored plant leaf diseases detection using image processing and deep learning techniques, contributing to the diversification of methodologies. In 2021, Kodali and Gudala [29] specifically addressed tomato leaf disease detection using CNNs, providing a targeted approach to disease identification. Their work contributes to the specificity required in agricultural applications.

Adhikari et al. [30] developed a system for tomato plant diseases detection, showcasing the application of machine learning in addressing broader agricultural challenges. Gehlot et al. [31] introduced the "Tomato-Village" dataset in 2023, providing a real-world environment for end-to-end tomato disease detection. This dataset serves as a valuable resource for benchmarking and advancing the development of robust disease detection models. This comprehensive review encapsulates the evolution of tomato leaf disease detection, highlighting the diverse methodologies and approaches that collectively contribute to the advancement of precision agriculture. The continuous exploration of advanced technologies and the development of innovative datasets underscore the dynamic nature of research in this field.

3 Design and Implementation

The workflow diagram is shown in Fig. 1.

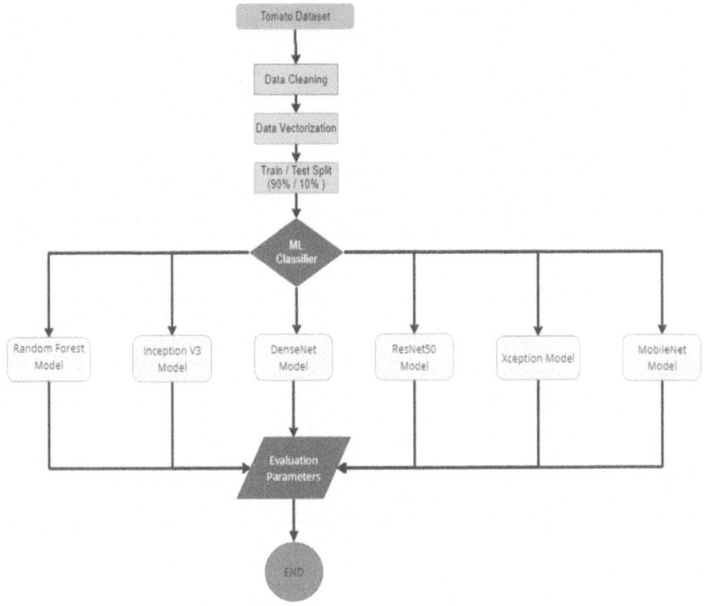

Fig. 1. Workflow diagram of our working model

3.1 Dataset Description

These datasets collectively form the foundation for our research, enabling comprehensive evaluation and comparison of deep learning models for.

- Benchmark Dataset (Kaggle): The Benchmark dataset serves as the primary dataset for evaluating the performance of our machine learning models in tomato leaf disease detection. The images in this dataset come from two different sources: the Tomato-Village dataset from Kaggle and the Tomato Leaf Disease dataset, resulting in a significant and diverse collection of data for robust model testing.

- Tomato-Village Dataset (Kaggle): This dataset emerges in response to the critical demand for authentic and varied datasets reflecting real-world scenarios of tomato diseases. Notably, existing datasets, notably Plant Village, predominantly originate from controlled laboratory environments, constraining their relevance in addressing real-world challenges. Noteworthy for its focus on diseases prevalent in the Jodhpur and Jaipur districts of Rajasthan, India, including Leaf Miner, spotted wilt virus, and nutrition deficiency diseases.

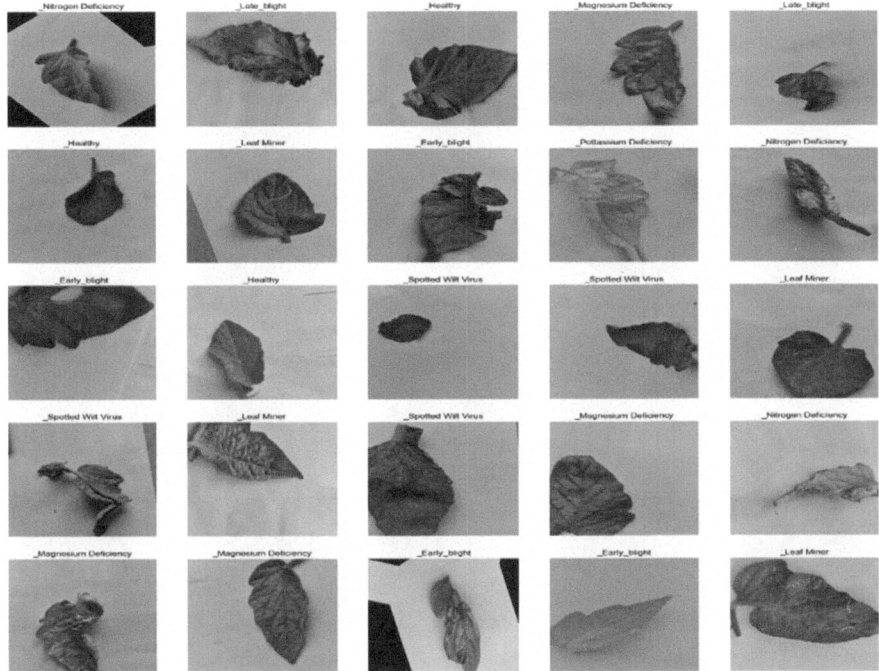

- Tomato Leaf Disease (Kaggle): This dataset focuses on the detection of diseases in tomato plant leaves using deep learning techniques. With a specific emphasis on the diverse diseases that afflict plant leaves, the project aims to create models capable of accurately identifying and classifying these diseases. Deep learning algorithms are applied to train models for disease detection. Additionally, the dataset explores the impact of transfer learning on disease detection, providing insights into the efficacy of leveraging pre-trained models for improved accuracy in tomato leaf disease identification.

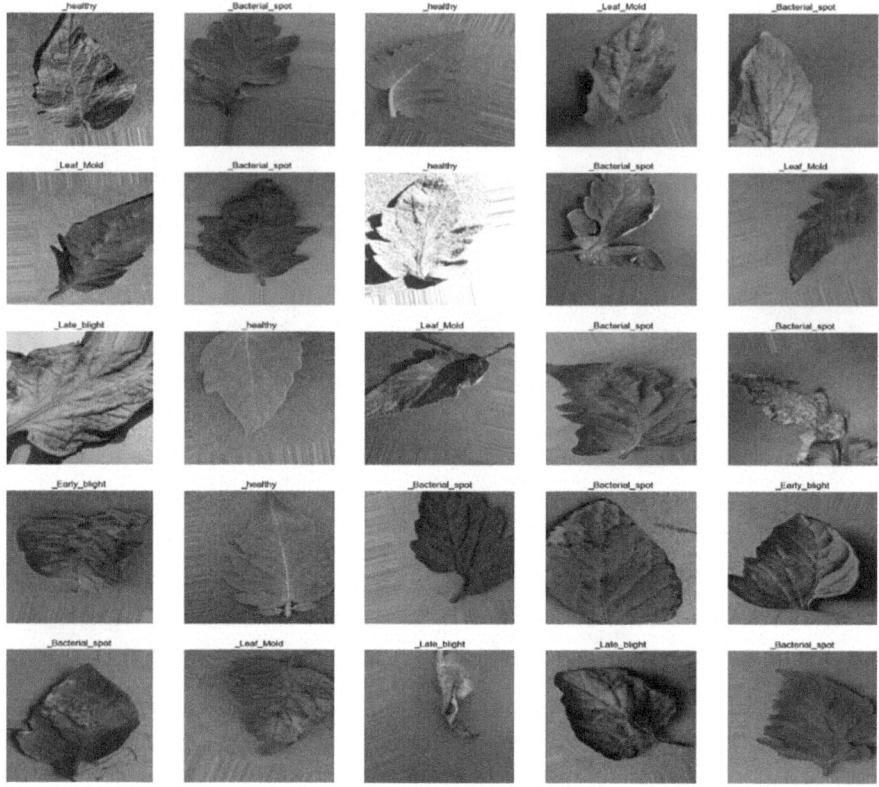

3.2   Data Vectorization and Feature Selection

3.2.1   Data Vectorization

The process of data vectorization for the purpose of predicting tomato leaf disease through deep learning begins with the gathering of tomato leaf images. Utilizing the OpenCV library, images are loaded, and a structured Data Frame is constructed to manage file paths and corresponding labels. To augment the dataset's diversity and improve model generalization, an ImageDataGenerator is employed, applying transformations such as horizontal flipping, rotation, and shift during training. This not only introduces variability but also prepares the model for real-world scenarios where leaves may exhibit different orientations.

The dimensions of the images play a crucial role in understanding the dataset's characteristics. A subset of training images is randomly selected, and their average height, width, and aspect ratio are calculated. This information provides insights into the inherent variability of leaf sizes and shapes within the dataset. The final step in data vectorization involves configuring data generators for training, validation, and testing, ensuring consistent image sizes (256 × 256 pixels) and appropriate batch sizes (30).

These generators efficiently handle the flow of data to the model during training and evaluation, setting the stage for the subsequent machine learning process.

3.2.2 Feature Selection

Feature selection is a critical aspect of building an effective tomato leaf disease prediction model. In this research, a strategic trimming process is implemented to balance the number of samples per class. The 'trim' function ensures that each class is represented by a specified number of images (both maximum and minimum), promoting model robustness and preventing bias towards classes with larger sample sizes. Convolutional Neural Network (CNN) architecture, specifically the *EfficientNetB3* model, forms the backbone of feature extraction. During training, the CNN layers automatically learn to discern relevant patterns, textures, and local spatial information from the input tomato leaf images. The subsequent layers, including Batch Normalization, Dense, and Dropout layers, contribute to further feature refinement and model adaptability. The chosen model, *EfficientNetB3*, is pre-trained on ImageNet, allowing it to capture both low-level details and high-level features during training. The fine-tuning process aligns the learned features with the specifics of tomato leaf disease patterns. The model's ability to automatically select discriminative features is enhanced through regularization techniques, such as L1 and L2 regularization in the Dense layer. In summary, the combination of meticulous data vectorization and thoughtful feature selection techniques forms the foundation for a robust machine learning model capable of accurately predicting tomato leaf diseases. The insights gained from image dimensions and the pre-trained *Efficient-NetB3* model contribute to a holistic understanding of the dataset, paving the way for effective disease classification.

3.3 Machine Learning Models

This research employs a comprehensive suite of state-of-the-art models, including Random Forest (RF), Inception V3, DenseNet, ResNet50, Xception, and MobileNet, to develop a robust and versatile predictive system. Each model brings its unique strengths and characteristics to the forefront, contributing to the overall efficacy of disease classification.

- Random Forest (RF):

   As an ensemble learning method, RF excels in handling high-dimensional data with complex relationships. It leverages a multitude of decision trees, each trained on a random subset of the data. The collective wisdom of these trees culminates in a powerful predictive model capable of capturing intricate patterns and nuances present in the tomato leaf images. RF's ability to handle diverse feature sets makes it a valuable component in the ensemble of classifiers.

- Inception v3:

   This CNN architecture is renowned for its depth and efficiency in extracting hierarchical features. Inception V3 introduces inception modules that enable the simultaneous processing of multiple receptive fields. This results in the model's capability to capture both local and global features, making it well-suited for discerning the intricate patterns indicative of various tomato leaf diseases.

- DenseNet:

Densely Connected Convolutional Networks, or DenseNet, introduces a unique connectivity pattern where each layer receives direct input from all preceding layers. This architecture promotes feature reuse and enhances gradient flow throughout the network. In the context of tomato leaf disease prediction, DenseNet's dense connections enable effective information propagation.

- ResNet50:

Residual Networks, or ResNets, address the challenge of training deep networks by introducing residual connections. ResNet50, a variant of ResNet, comprises 50 layers and is renowned for its depth. This architecture facilitates the training of deeper models without succumbing to issues like vanishing gradients. ResNet50's skip connections enhance the flow of information, enabling the model to learn intricate disease-related features.

- Xception:

Inspired by the idea of extreme depth, Xception (Extreme Inception) explores a novel architecture by replacing the standard inception modules with depthwise separable convolutions. This modification significantly reduces the number of parameters, making the model more computationally efficient. In the context of tomato leaf disease prediction, Xception's depthwise separable convolutions contribute to the model's ability to capture detailed spatial information efficiently.

- MobileNet:

In the realm of tomato leaf disease prediction, MobileNet's lightweight design allows for efficient inference without compromising on classification accuracy.

3.4 Confusion Matrix and Accuracy Curve

- Confusion matrix:

The Confusion Matrix is a tabular representation that delineates the classifier's predictions against the ground truth labels. The four quadrants of the matrix are populated with values indicating true positives, true negatives, false positives, and false negatives.

- True Positives (TP): Instances where the classifier correctly predicts the presence of a specific disease.
- True Negatives (TN): Instances correctly identified as not having the predicted disease.
- False Positives (FP): Instances where the classifier incorrectly predicts the presence of a disease (Type I error).
- False Negatives (FN): Instances incorrectly identified as not having the predicted disease (Type II error).

In our research on tomato leaf disease prediction utilizing advanced machine learning classifiers, the Confusion Matrix and Accuracy Curve play pivotal roles in assessing and interpreting the performance of our models.

- Accuracy Curve:

The Accuracy Curve is a graphical representation of the classifier's performance across different epochs or iterations during the training process. In our research, we employ machine learning models such as Random Forest, Inception V3, DenseNet, ResNet50, Xception, and MobileNet. The Accuracy Curve allows us to track how well these models are learning and adapting over time. Training Accuracy: Represents the accuracy of the model on the training dataset. Validation Accuracy: Reflects the accuracy of the model on a separate validation dataset not used for training. The trajectory of the Accuracy Curve provides valuable insights into the learning dynamics of our models. Observing steady increases in both training and validation accuracy indicates effective learning and generalization. However, significant gaps between training and validation accuracy curves could signal overfitting, necessitating adjustments to the model architecture or training strategy.

3.5 Evaluation Parameters

- Accuracy:

Accuracy assesses the correctness of predictions by comparing the sum of true positives and true negatives to the total instances.

$$\text{Accuracy} = (TP + TN)/(TP + TN + FP + FN) \tag{1}$$

- Precision:

Precision gauges the precision of positive predictions, the formula is given below:

$$\text{Precision} = TP/(TP + FP) \tag{2}$$

- Recall:

Recall measures the ability to capture positive instances, The formula is given below:

$$\text{Recall} = TP/(TP + FN) \tag{3}$$

- F1-Score:

The F1-score offers a balanced assessment, The formula is given below:

$$\text{F1-Score} = 2 * (\text{Precision} * \text{Recall})/(\text{Precision} + \text{Recall}) \tag{4}$$

Table 1. Benchmark Dataset

Classifier	Accuracy (%)	Precision (%)	Recall (%)	F1-Score (%)
R.F	68	64	71	68
Inception v3	97	98	96	97.5
DenseNet	94	98	89	93
ResNet50	93.3	97	91	94
Xception	95	93	99	96
MobileNet	76	66	90	76

4   Results Analysis

Across all datasets, Inception V3 consistently exhibits outstanding performance, boasting high accuracy, precision, recall, and F1-Score. DenseNet also demonstrates commendable results, particularly excelling in accuracy and precision. On the other hand, Random Forest and MobileNet exhibit varying degrees of performance across datasets, indicating sensitivity to dataset characteristics (Table 1).

The analysis of the BenchMark Dataset for Tomato Leaf Disease Detection from Kaggle reveals distinct performance characteristics among various classifiers, each evaluated based on key metrics such as Accuracy, Precision, Recall, and F1-Score. In this dataset, Inception V3 emerges as the standout performer, showcasing exceptional accuracy at 97%, precise disease identification with a precision of 98%, robust recall at 96%, and an overall F1-Score of 97.50%. DenseNet also demonstrates commendable results with a high accuracy of 94%, noteworthy precision of 98%, and balanced recall at 89%, resulting in an impressive F1-Score of 93%. ResNet50 and Xception exhibit a balanced performance, with ResNet50 achieving accuracy of 93.30% while Xception attains 95% accuracy, 93% precision, and an impressive 99% recall. The Random Forest (R.F) model, with a moderate accuracy of 68%, shows balanced precision and recall, indicating its competence in disease detection. However, MobileNet, while achieving a reasonable accuracy of 76%, demonstrates sensitivity to certain dataset characteristics with comparatively lower precision (66%) and high recall (90%). This comprehensive analysis underscores the efficacy of deep learning models, particularly Inception V3 and DenseNet, in accurately identifying tomato leaf diseases, and emphasizes the need for careful model selection based on dataset characteristics to optimize disease detection outcomes. The Confusion Matrix and Accuracy Curves for Inception V3 model is given below (Fig. 2 and Table 2).

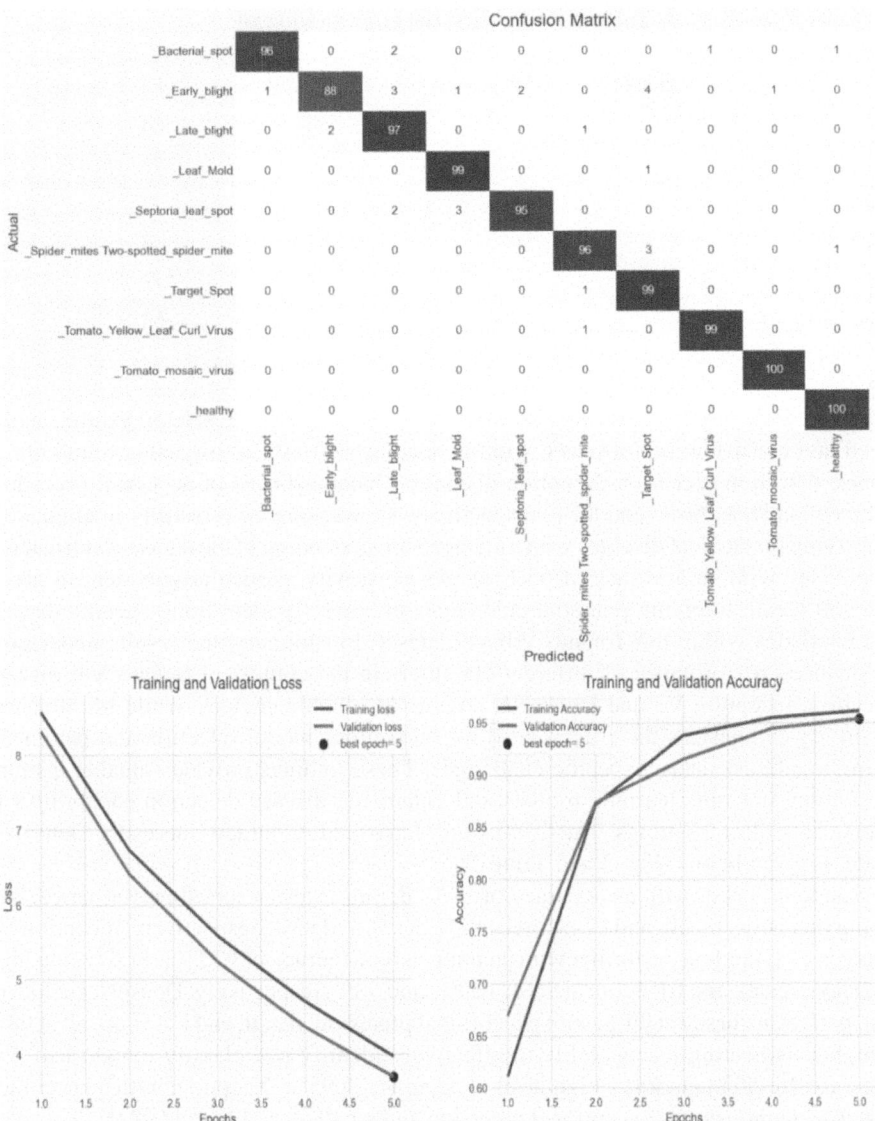

Fig. 2. Confusion Matrix and Accuracy Curve on Benchmark Dataset

Random Forest, displaying a moderate accuracy of 61%, demonstrates a balanced performance in precision and recall, suggesting its capability for correct predictions. Inception V3 emerges as a robust performer with an accuracy of 69%, demonstrating balanced precision and recall, highlighting its reliability in accurately identifying and classifying tomato leaf diseases. DenseNet achieves commendable accuracy at 74%, primarily driven by its high precision rate, while the slightly lower recall indicates

Table 2. Kaggle Dataset (Tomato Village)

Classifier	Accuracy (%)	Precision (%)	Recall (%)	F1-Score (%)
R.F	61	57	73	64
Inception v3	69	69	75	72
DenseNet	74	80	69	74
ResNet50	64	61	65	64
Xception	70	66	80	72
MobileNet	63	80	52	63

potential challenges in capturing all instances of leaf diseases. ResNet50 exhibits a balanced and consistent performance with an accuracy of 64%, showcasing reliability in disease detection scenarios. Xception displays a balanced triad of accuracy, precision, and recall at 70%, 66%, and 80%, respectively, showcasing its versatility in effectively identifying tomato leaf diseases with a nuanced understanding of the dataset's intricacies. MobileNet, with an accuracy of 63%, shows sensitivity, particularly evident in precision and recall, implying potential challenges in correctly identifying specific disease characteristics within the Tomato Village dataset. In summary, the results underscore the nuanced performance of different classifiers in the context of tomato leaf disease detection. Inception V3 and DenseNet emerge as reliable choices, while MobileNet's sensitivity emphasizes the crucial need for meticulous model selection and parameter tuning to address dataset-specific challenges. These findings provide valuable insights for refining machine learning models and enhancing disease detection capabilities in diverse agricultural settings. The Confusion Matrix and Accuracy Curves for DenseNet Model is given below (Fig. 3 and Table 3).

Random Forest, with an accuracy of 83%, demonstrates a robust performance, balancing precision, recall, and F1-Score at 82%, 84%, and 83%, respectively. Inception V3 emerges as a standout performer with an impressive accuracy of 97.5%, showcasing high precision, recall, and F1-Score at 96%, 100%, and 98%, respectively. DenseNet achieves near-perfect accuracy at 99%, with exceptional precision, recall, and F1-Score all at 98% or higher, indicating its capability to effectively identify and classify tomato leaf diseases. ResNet50 maintains a high level of accuracy at 95%. Xception, with an accuracy of 98.6%, demonstrates exceptional precision and recall at 99.42% and 98.92%, respectively, resulting in a robust F1-Score of 98.68%. MobileNet, with an accuracy of 88%, exhibits balanced precision, recall, and F1-Score metrics, all hovering around 87–89%. Inception V3, DenseNet, and Xception emerge as top-performing models, showcasing their potential for accurate identification and classification of tomato leaf diseases. These findings contribute valuable insights for the selection of classifiers based on specific requirements and performance expectations in agricultural settings. The Confusion Matrix and Accuracy Curve for DenseNet Model is Given Below (Fig. 4).

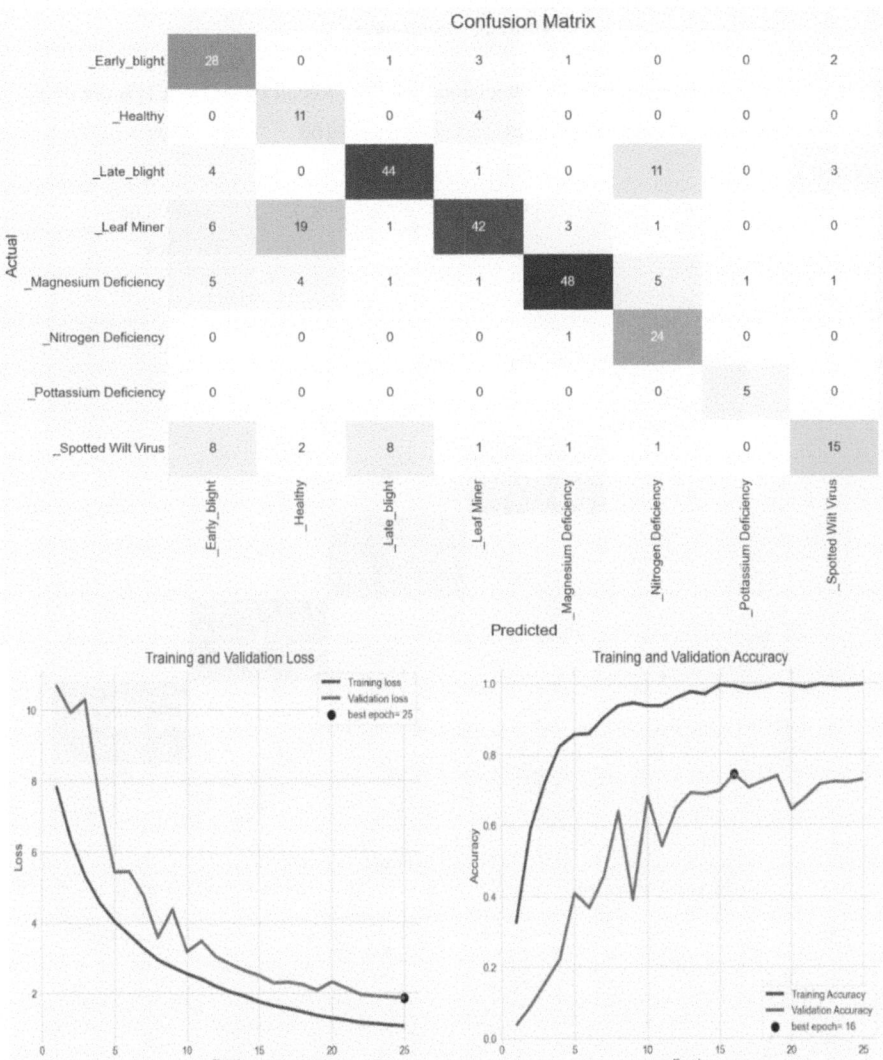

Fig. 3. Confusion Matrix and Accuracy Curve on Kaggle Dataset (Tomato Village)

Table 3. Kaggle Dataset (Tomato Leaf Disease)

Classifier	Accuracy (%)	Precision (%)	Recall (%)	F1-Score (%)
R.F	83	82	84	83
Inception v3	97.5	96	100	98

(continued)

Table 3. (*continued*)

Classifier	Accuracy (%)	Precision (%)	Recall (%)	F1-Score (%)
DenseNet	99	98	100	99
ResNet50	95	93	97	95
Xception	98.6	99.42	98.92	98.68
MobileNet	88	87	89	88

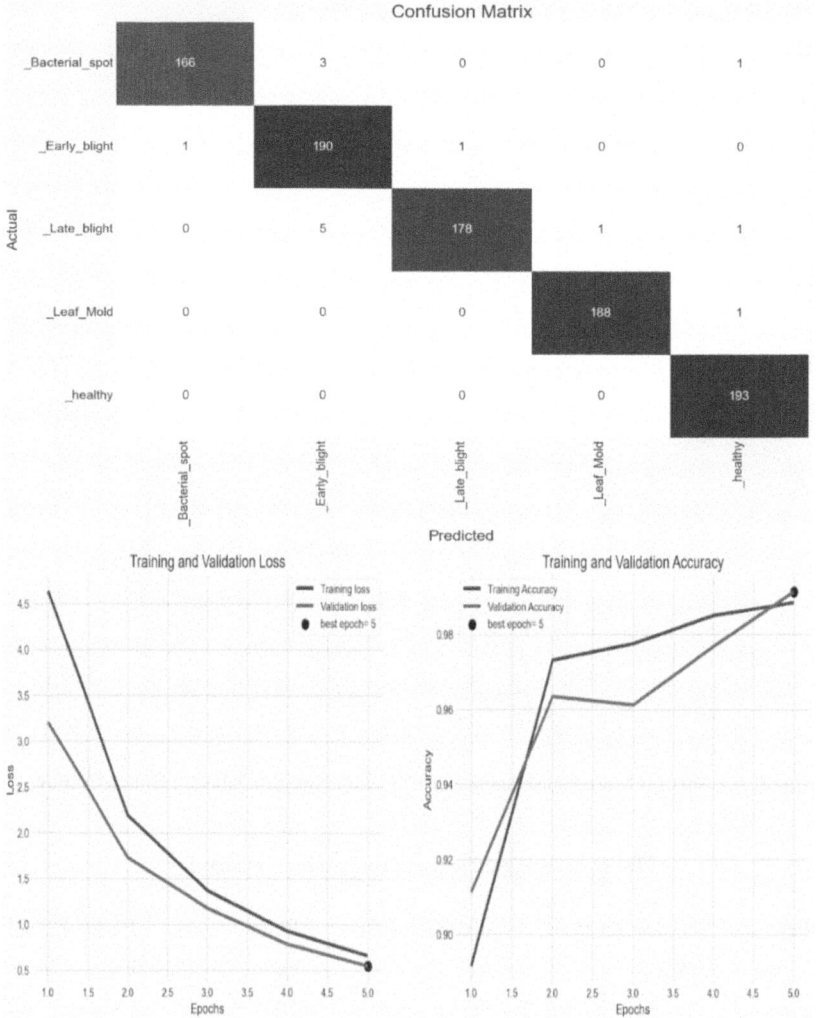

Fig. 4. Confusion Matrix and Accuracy Curve on Kaggle Dataset (Tomato Leaf Disease)

5 Conclusion and Future Work

In the culmination of our extensive exploration into the realm of tomato leaf disease prediction using deep learning, a thorough analysis of diverse classifiers – including Random Forest, Inception V3, DenseNet, ResNet50, Xception, and MobileNet – across three distinct datasets has provided nuanced insights. The BenchMark Dataset (Tomato Leaf Disease Detection Kaggle), Kaggle Dataset (Tomato Village), and Kaggle Dataset (Tomato Leaf Disease) have served as robust testing grounds, enabling a meticulous examination of the classifiers' performances. The discerning evaluation has illuminated the efficacy of specific models, with Inception V3, DenseNet, and Xception emerging as stalwart contenders, showcasing high accuracy and commendable balance in precision, recall, and F1-Score. The utilization of sophisticated machine learning methods, specifically deep learning models, holds potential to improve the precision and effectiveness of disease identification in tomato plants. Yet, the dynamic landscape of agricultural technology demands continuous refinement. Subsequent endeavors should be dedicated to fine-tuning and optimizing the proposed models, exploring the potential synergies offered by ensemble methods, and delving into transfer learning approaches to enhance model generalization across diverse datasets.

The scalability and real-world applicability of these models should be subject to thorough investigation, considering the practical constraints and exigencies of agricultural settings. Moreover, the integration of cutting-edge technologies such as edge computing and the Internet of Things (IoT) stands as a promising avenue for crafting more efficient and real-time disease detection systems tailored for tomato crops. Collaborative efforts involving domain experts, agronomists, and farmers will be pivotal in refining models to align seamlessly with the intricate needs of agricultural practices. The continuous expansion and diversification of annotated datasets, incorporating geographical variations and diverse weather conditions, will contribute substantially to fortifying the robustness and generalization capabilities of machine learning models. In summation, this research not only establishes a foundation for the application of machine learning in tomato leaf disease prediction but also underscores the potential of advanced models for precise and timely detection. The identified directions for future research encapsulate exciting prospects, promising to elevate the impact of these models on agricultural practices.

References

1. Agarwal, M., et al.: ToLeD: tomato leaf disease detection using convolution neural network. Proc. Comput. Sci. **167**, 293–301 (2020)
2. Tm, P., et al.: Tomato leaf disease detection using convolutional neural networks. In: 2018 Eleventh International Conference on Contemporary Computing (IC3). IEEE (2018)
3. Ashok, S., et al.: Tomato leaf disease detection using deep learning techniques. In: 2020 5th International Conference on Communication and Electronics Systems (ICCES). IEEE (2020)
4. Thangaraj, R., et al.: Artificial intelligence in tomato leaf disease detection: a comprehensive review and discussion. J. Plant Dis. Prot. **129**(3), 469–488 (2022)
5. Kaur, M., Bhatia, R.: Development of an improved tomato leaf disease detection and classification method. In: 2019 IEEE Conference on Information and Communication Technology. IEEE (2019)

6. Kibriya, H., et al.: Tomato leaf disease detection using convolution neural network. In: 2021 International Bhurban Conference on Applied Sciences and Technologies (IBCAST). IEEE (2021)

7. Chowdhury, M.E.H., et al.: Tomato leaf diseases detection using deep learning technique. Technol. Agric. **453** (2021)

8. Hasan, M., Tanawala, B., Patel, K.J.: Deep learning precision farming: tomato leaf disease detection by transfer learning. In: Proceedings of 2nd International Conference on Advanced Computing and Software Engineering (ICACSE) (2019)

9. Kaushik, M., et al.: Tomato leaf disease detection using convolutional neural network with data augmentation. In: 2020 5th International Conference on Communication and Electronics Systems (ICCES). IEEE (2020)

10. Durmuş, H., Güneş, E.O., Kırcı, M.: Disease detection on the leaves of the tomato plants by using deep learning. In: 2017 6th International Conference on Agro-Geoinformatics. IEEE (2017)

11. Karthik, R., et al.: Attention embedded residual CNN for disease detection in tomato leaves. Appl. Soft Comput. **86**, 105933 (2020)

12. Anandhakrishnan, T., Jaisakthi, S.M.: Deep Convolutional Neural Networks for image based tomato leaf disease detection. Sustain. Chem. Pharm. **30**, 100793 (2022)

13. Chowdhury, M.E.H., et al.: Automatic and reliable leaf disease detection using deep learning techniques. AgriEngineering **3**(2), 294–312 (2021)

14. David, H.E., et al.: Literature review of disease detection in tomato leaf using deep learning techniques. In: 2021 7th International Conference on Advanced Computing and Communication Systems (ICACCS), vol. 1. IEEE (2021)

15. Mim, T.T., et al.: Leaves diseases detection of tomato using image processing. In: 2019 8th International Conference System Modeling and Advancement in Research Trends (SMART). IEEE (2019)

16. Irmak, G., Saygili, A.: Tomato leaf disease detection and classification using convolutional neural networks. In: 2020 Innovations in Intelligent Systems and Applications Conference (ASYU). IEEE (2020)

17. Bir, P., Kumar, R., Singh, G.: Transfer learning based tomato leaf disease detection for mobile applications. In: 2020 IEEE International Conference on Computing, Power and Communication Technologies (GUCON). IEEE (2020)

18. Rahman, S.U., et al.: Image processing based system for the detection, identification and treatment of tomato leaf diseases. Multimed. Tools Appl. **82**(6), 9431–9445 (2023)

19. Nagamani, H.S., Sarojadevi, H.: Tomato leaf disease detection using deep learning techniques. Int. J. Adv. Comput. Sci. Appl. **13**(1) (2022)

20. Hong, H., Lin, J., Huang, F.: Tomato disease detection and classification by deep learning. In: 2020 International Conference on Big Data, Artificial Intelligence and Internet of Things Engineering (ICBAIE). IEEE (2020)

21. Zhao, S., et al.: Tomato leaf disease diagnosis based on improved convolution neural network by attention module. Agriculture **11**(7), 651 (2021)

22. Batool, A., et al.: Classification and identification of tomato leaf disease using deep neural network. In: 2020 International Conference on Engineering and Emerging Technologies (ICEET). IEEE (2020)

23. Zhang, Y., Song, C., Zhang, D.: Deep learning-based object detection improvement for tomato disease. IEEE Access **8**, 56607–56614 (2020)

24. Wu, Y., Xu, L., Goodman, E.D.: Tomato leaf disease identification and detection based on deep convolutional neural network. Intell. Autom. Soft Comput. **28**(2) (2021)

25. Paymode, A.S., Magar, S.P., Malode, V.B.: Tomato leaf disease detection and classification using convolution neural network. In: 2021 International Conference on Emerging Smart Computing and Informatics (ESCI). IEEE (2021)

26. Alhaj Ali, A., Chramcov, B., Jasek, R., Katta, R., Krayem, S., Awwama, E.: Tomato leaf diseases detection using deep learning. In: Silhavy, R., Silhavy, P., Prokopova, Z. (eds.) CoMeSySo 2021. LNNS, vol. 231, pp. 199–208. Springer, Cham (2021). https://doi.org/10.1007/978-3-030-90321-3_18
27. Anandhakrishnan, T., Murugaiyan, J.S.M.: Identification of tomato leaf disease detection using pretrained deep convolutional neural network models. Scalable Comput.: Pract. Exp. **21**(4), 625–635 (2020)
28. Jasim, M.A., Al-Tuwaijari, J.M.: Plant leaf diseases detection and classification using image processing and deep learning techniques. In: 2020 International Conference on Computer Science and Software Engineering (CSASE). IEEE (2020)
29. Kodali, R.K., Gudala, P.: Tomato plant leaf disease detection using CNN. In: 2021 IEEE 9th Region 10 Humanitarian Technology Conference (R10-HTC). IEEE (2021)
30. Adhikari, S., et al.: Tomato plant diseases detection system, vol. I, pp. 81–86 (2018)
31. Gehlot, M., Saxena, R.K., Gandhi, G.C.: "Tomato-Village": a dataset for end-to-end tomato disease detection in a real-world environment. Multimed. Syst. 1–24 (2023)

Predicting Stock Market Trends: Machine Learning Approaches of a Possible Uptrend or Downtrend

Raima Joseph and Mausumi Goswami[✉]

Department of Computer Science and Engineering, Christ (Deemed to Be) University, Kengeri, Bangalore, Karnataka 560074, India
mausumi.goswami@christunievrsity.in

Abstract. This paper delves into a statistical analysis of the stock market, emphasizing the significance of accuracy in stock predictions. Large data sets can be handled by machine learning algorithms, which can also forecast outcomes based on past data and spot intricate patterns in financial data. They assist control risks, automate decision-making procedures, and adjust to changing circumstances. Multisource data can be combined by ML models to provide a comprehensive picture of market circumstances. They can manage intricate, nonlinear interactions, provide impartial analysis, and lessen human bias. Models are able to adjust to shifting market conditions through ongoing learning and retraining. They must, however, exercise caution when deploying models in real-world situations and ensure that they are validated. Although machine learning has advantages for stock market analysis, it must be carefully evaluated for dangers and validated before being used in practical situations. The traditional machine learning model, Logistic Regression has been used in order to predict stock prices. It focuses on binary classification based on the trend of the stock. Through the model training and evaluation and additional analysis done on the results, this research contributes towards obtaining predictions and studying reasons of a possible uptrend or downtrend to further assist companies.

Keywords: Stock Market · Machine Learning · Logistic Regression

1 Introduction

Stock prices represent ownership or a part of a company bought, or rather invested in, and sold in the finance market. In order to benefit from investing in stocks, many researches and studies have shed a light on Machine Learning (ML) techniques to foresee stock prices. The performance of stock trends and prices is hard to predict as they are uncertain. As a lot of factors incorporate accurate analysis and prediction of the stock data, it is necessary to work forward in stock prediction using various efficient ML techniques. Stock market (SM) analysis essentially refers to the process of interpreting the factors affecting the performance of stock in order to make informed decisions. The impact of SM on the global business is profound in risk management for making strategic choices when it comes to investing. It also creates investor confidence by creating a well-analysed prediction. Stock price prediction consists a lot of scope for accuracy including its challenges. As mentioned within the background, while a company's financial performance is required for fundamental analysis, this paper focuses on technical analysis using machine learning. The global financial stability increases as the analysis of future status of the stocks can prevent financial crises.

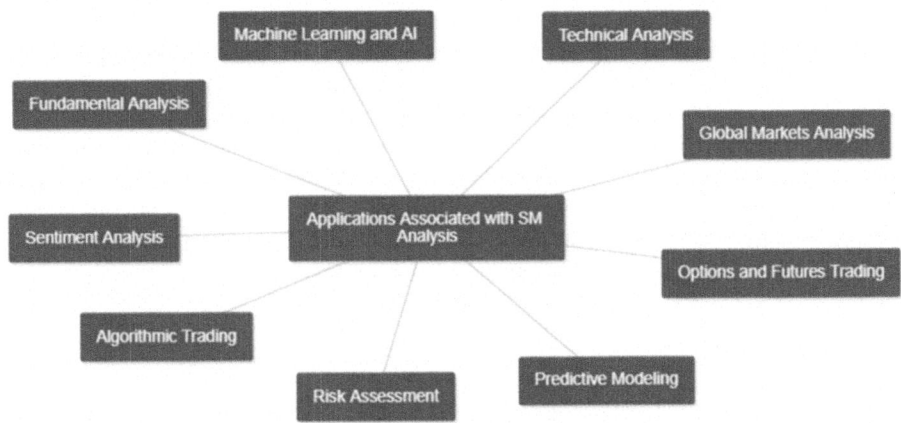

Fig. 1. Applications of ML in SMA

Various papers suggest that, [1] there can be two main approaches to stock prediction: Fundamental and technical analysis. Fundamental analysis includes evaluating a company's finances and its industry, while technical analysis relies on historical data and predictive data to forecast price trends. While these methods have been traditionally used, there are limitations. The data and the trends followed can be subjective which can lead to different interpretations by different investors. Machine learning models can identify market inefficiencies that traditional analysis methods may not account for. Another limitation is that handling large amounts of financial data manually can be very time consuming and challenging. A subset of artificial intelligence, machine learning, uses algorithms as models to analyse data. In the context of SM, machine learning concepts can be applied in different factors. The main applications include predictive modelling, risk assessment, algorithmic trading and sentiment analysis (see Fig. 1). Here, we focus on predictive modelling [2]. Various techniques in machine learning include:

1. Traditional ML analysis: Algorithms such as regression or classification when given sequential data, such as that of stock prices. These models prove to be efficient in terms of performance.
2. Neural-Network based analysis: Deep learning models that capture intricate patterns in price data.
3. Time-series based analysis: These models analyse the historical data and identify trends over time.

4. Graph-based analysis: Financial networks and creating relationships between stock prices and technical indicators in order to branch out options.

The main idea would be to analyse historical data of each company by its prices and identify patterns in order to predict what the trend would be the next day. The question this paper highlights is: Will tomorrow's price fall or rise?

Using big data techniques, stock market prediction combines fundamental and technical analysis. Whereas fundamental analysis employs sentiment analysis to analyze social media data, technical analysis uses machine learning to analyze historical stock price data [6, 7, 8, 9, 10, 11, 12, 13, 14, 15, 16]. Online user-generated textual material predicts stock prices and offers insightful analysis of investor psychology. Support vector machines are used to include sentiment analysis into machine learning techniques. Few studies use deep neural networks or other advanced machine learning techniques to forecast daily stock market returns. Support vector machines and artificial neural networks are the most often utilized algorithms for stock market prediction, according to a systematic analysis of 122 academic publications. For predicting the bitcoin market, a hybrid strategy combining natural language processing, deep learning, and machine learning is suggested group methods such as bagging, boosting.

2 Background

Accurate stock price predictions are invaluable for investors and financial organisations, as they can help make investment decisions, manage risks, and come up with better strategies. Firstly, we have some terms related to analysis and machine learning. Stock market indicators are some data points derived from historical data of the company. An example of such indicators is Moving Average, a mean price value of a particular set of days. These are very useful for identifying patterns in trends. Historical data refers to the past stock prices, volumes traded, opening and closing prices and the high as well as low prices of the stock for a specific period. This data serves as the foundation dataset for analysing trends. Machine learning models are trained using this very historical data (Fig. 2).

The methodology brings out a broader understanding of the features or factors used in the training of the ML algorithms. The algorithms that can be used for SM analysis include Regression, Classification and Deep Learning models. Regressive models bring out a relationship between the input variables (indicators and features) and the target variable (stock prices). Classification prediction involves categorising data into groups or classes. For SM, the classification required is the trend pattern. An uptrend means that the price of the stock has been increasing steadily, and downtrend indicates a decline in the trend of the stocks. This brings about the conclusion that the task would be a binary classification [3, 4]; uptrend or downtrend. Hence, we choose traditional ML analysis by using logistic regression to identify the relationship between input features and the days to predict an uptrend or a downtrend. In the context of SM analysis and prediction, accuracy refers to the performance of the model and the ability for it to make accurate forecasts, when compared to the actual prices. The accuracy of the predictive model decides the validity of the predictions which impacts the development of strategy highly. It also serves as a competitive advantage. Highly accurate predictions also help

Date	Close/Last	Volume	Open	High	Low
08/17/2023	$133.98	48,354,090	$135.46	$136.085	$133.53
08/16/2023	$135.07	41,675,900	$137.19	$137.27	$135.01
08/15/2023	$137.67	42,781,520	$140.05	$141.2778	$137.23
08/14/2023	$140.57	47,148,700	$138.30	$140.59	$137.75
08/11/2023	$138.41	42,905,830	$137.40	$139.33	$137.00
08/10/2023	$138.56	58,928,400	$139.075	$140.41	$137.49
08/09/2023	$137.85	50,017,350	$139.97	$140.32	$137.10
08/08/2023	$139.94	51,710,500	$140.62	$140.84	$138.42
08/07/2023	$142.22	71,213,110	$140.99	$142.54	$138.95

Fig. 2. Historical Quotes for AMZN from NASDAQ

in assessing the chances of loss or gain, hence risk management. We continue with technical analysis and use Machine Learning to use the pointers and the past data to predict prices. The companies chosen for this analysis are Microsoft [MSFT], Amazon [AMZN], Google [GOOG], Apple [AAPL] and Tesla [TSLA] from NASDAQ Historical quotes for Logistic Regression to determine next day predicted price trends.

3   Methodology

Data collection: Collected historical stock price data for Microsoft (MSFT), Google (GOOG), Amazon (AMZN), Tesla (TSLA), and Apple (AAPL) from the NASDAQ stock exchange. The dataset includes daily Open, High, Low, Volume, and Closing (OHLCV) prices. The dataset size ranges from August 2018–2023. Training and testing data is initially set to 80 and 20% respectively. The size of each dataset comes out to be 1005 and 251 each.

3.1   Features

These are common indicators in financial analysis and stand for:

1. Open: At the beginning of a trading session, the price at which a stock opens for trading.
2. High: The highest price at which the stock was traded during a trading session.
3. Low: The lowest price at which the stock was traded during a trading session.

4. Close: The price at which a stock closes at the end of a trading session.
5. Volume: During a given period (market activity), the number of shares or contracts traded.

These terms are fundamental in financial analysis and are widely used by traders, investors, and analysts to assess market trends.

3.2 Feature Selection

1. Next-Day Close: This feature represents the next trading day close price. It is essentially the value of the "Close" data with the next day values. It maps the data to '1' if the next day's price is greater than the previous day, else it maps '0'.
2. Open-Close: This is the variance between the opening price and the closing price of a stock during a trading period. It represents the change between the opening and closing prices and can indicate pre-market activity, i.e., the few minutes right before the trading session commences.
3. Daily Return: Daily return measures the change in the price percentage of a stock from one trading day to the next. Calculated as the following equation.

$$((Close - Previous\ Close)/Previous\ Close) \times 100 \tag{1}$$

4. Open-Open: This feature represents the difference between the open price of a trading day and the open price of the previous trading day. It can indicate how the market opens compared to the previous day, reflecting financial market conditions within the closed hours of trading sessions.
5. Moving Average: Moving averages are designed to classify the trend direction of a stock. An increasing moving average show that the stock is in an uptrend, while a falling moving average shows that it is in a downtrend. The "rolling mean" function has been used to find the simple moving average over varying ranges.

3.3 Challenges

Relevance of Open-Open vs. Open-Close. The relevance of using Open-Open over Open-Close was assessed by calculating trends along with close prices for a specific range. Results showed that a higher difference in Open-Close results in a downtrend of the close price, whereas the Open-Open gap fluctuates. Broadly, there is very little difference between the features, and Open-Open does not prove to be very relevant (Fig. 3).

Observations
Whenever there is a significant fluctuation with Open-Close, a similar spike or fall occurs for the same duration in the return prices. Graphs for Open-Close and Open-Open are similar. The presence of both features together does not seem to hinder the performance of the model. This made it necessary to verify the relevance of each with respect to the accuracy shift that occurs whenever one of the features is not present (Tables 1 and 2).

It is observed that the accuracy does not decline by a lot when Open-Open is omitted. However, the accuracy significantly reduces with removal of Open-Close, indicating that

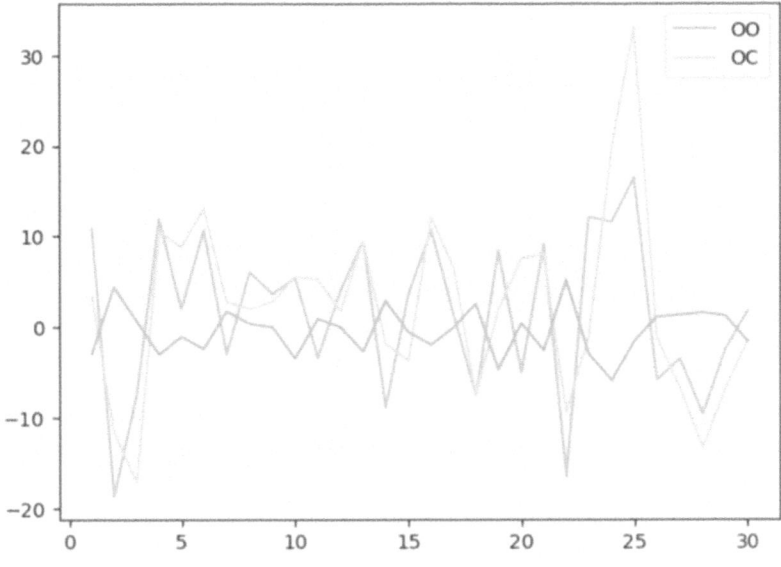

Fig. 3. Open-Open vs. Open-Close with Returns

Table 1. Results without Open-Open

	precision	recall	f1-score	support
0	0.5201612903	1	0.6843501326	129
1	1	0.02459016393	0.048	122
accuracy	0.5258964143	0.5258964143	0.5258964143	0.5258964143
macro avg	0.7600806452	0.512295082	0.3661750663	251
weighted avg	0.7533896671	0.5258964143	0.3750484745	251

Open-Close is much more relevant than Open-Open. To generalize the data to relevant features and inputs and reduce noise, Open-Open feature is removed. This also helps to reduce the chances of over-fitting, where training accuracy is higher than testing accuracy.

Appropriate Range for Moving Average. For calculating the simple moving average, the rolling mean function is used to find the average close prices over a window of trading days. The range of the window plays a crucial role in generating an apt moving average. Depending on the type of trader, the time or range of the window can be adjusted. Long-term traders prefer a weekly average and short-term traders would select a daily moving average.

Results

Table 2. Results without Open-Close

	precision	recall	f1-score	support
0	0.8424657534	0.9461538462	0.8913043478	130
1	0.9333333333	0.8099173554	0.8672566372	121
accuracy	0.8804780876	0.8804780876	0.8804780876	0.8804780876
macro avg	0.8878995434	0.8780356008	0.8792804925	251
weighted avg	0.8862704433	0.8804780876	0.8797116268	251

Due to these results, a window of 5 days gives the highest accuracy, and hence, we move forward with a window range or period within 5 days (Figs. 4, 5 and 6).

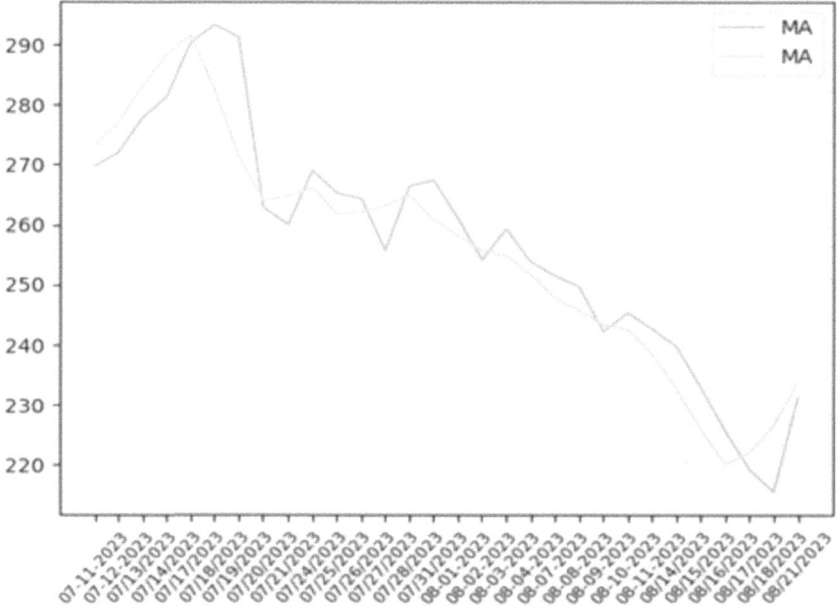

Fig. 4. Accuracy with a window size of 3 days: 88.44%

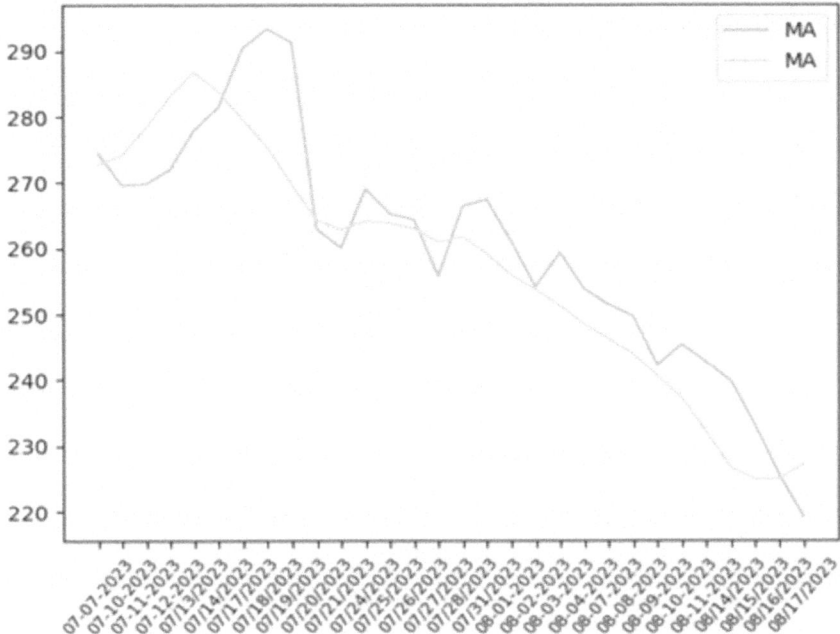

Fig. 5. Accuracy with a window size of 5 days: 88.87%

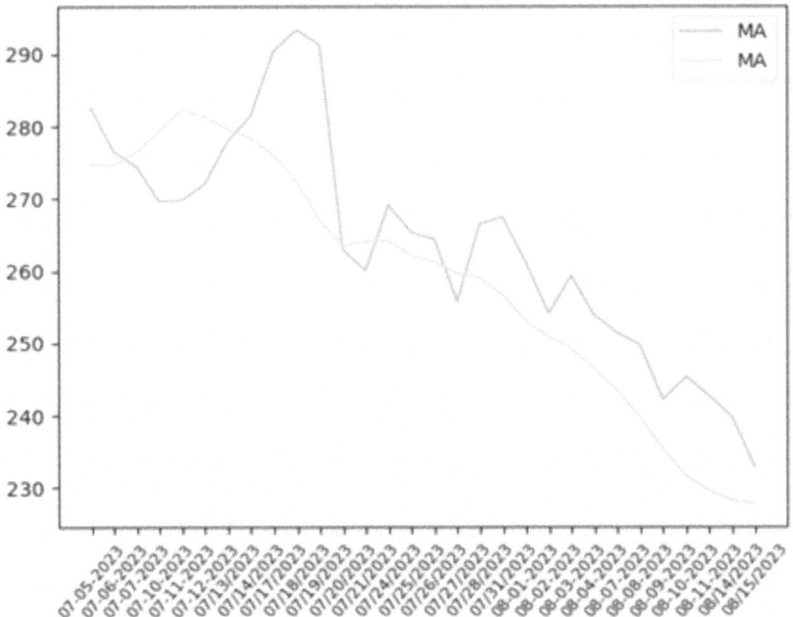

Fig. 6. Accuracy with a window size of 7 days: 88.4%

3.4 Logistic Regression

An ML algorithm that is used for binary classification tasks. For the context of SM analysis and prediction, logistic regression can be utilised to forecast the uptrend or the downtrend of the stock. It effectively makes a connection between the input and output variables, in this case the features of the stock data-set and the predicted values, and converts it into two classes; 0 or 1 (downtrend or uptrend).

A sigmoid function is used to reduce the outputs to 0 or 1 (see Fig. 7).

Here, 'x' is the linear combination calculated by the model as:

$$X = c0 + c1 * feature1 + c2 * feature2 + \ldots + cN * featureN$$

$$Feature = OHCLV + Added\ features$$

where c0, c1... cN are the coefficients that are learned by the logistic regression model. Simply put, X here gives the linear regression output and this X value is used in the above-mentioned Sigmoid function to find the binary outputs between 0 and 1. f(x) mentioned gives the value between 0 and 1 with the threshold value as 0.5. This means that the values above 0.5 are termed as uptrend (1) and below 0.5 are termed as downtrend (0)

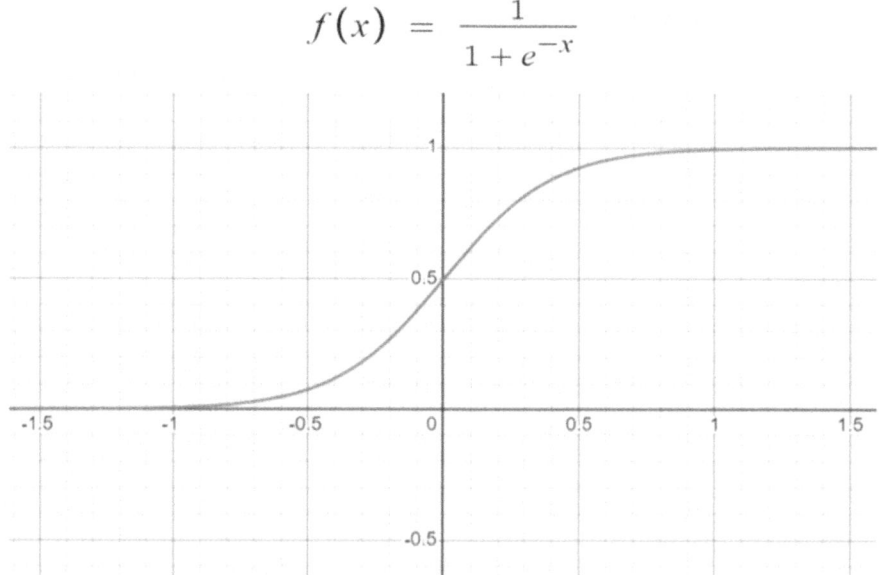

Fig. 7. Sigmoid function and plot.

3.5 Splitting and Scaling the Data

We need to split the data to train the model using a given data as well as testing it with hidden data. Hidden and given data-sets are taken as the training and testing data, which is retrieved from one data-set split in a ratio of 7:3 or 8:2. This is done to that the model created generalises to unseen data as well as it generalises to the given data.

This model takes in 80% data as training and 20% as testing. Scaling the data refers to the normalisation of features in a data-set in order to achieve consistency. Having dissimilar data or high differences between values in a data-set would affect the performance of the model. Scaling leads to more accurate predictions as it makes data uniform. For instance, the Volume in the data-set is in millions while the Close and Open prices are in hundreds. Larger values may dominate the learning princess of the model.

Min-Max scaler [5] is a commonly used scaler for the purpose of logistic regression.

This scaler does not change the relationship between data and retains the relevance of the features. The mathematical formula for scaling the data is:

$$Xscaled = (X - Xmin)/(Xmax - Xmin) \tag{2}$$

where Xscaled is the value, we get after scaling the feature X using Minmax Scaler. Xmin and Xmax are the minimum and maximum values of the X feature in the dataset. Finally, the scaled dataset is fed into the Logistic Regression model from Python's Sci-kit learn library.

4   Predictions and Accuracy

The model is trained using the training dataset, and the mentioned mathematical calculations are performed. The model calculates X values or the linear regression output and feeds it to the sigmoid function to get corresponding outputs for the predicted price of each day. The coefficients (model parameters) to be found for the optimal values are iteratively found by the function using optimization techniques. Model evaluation is needed for any machine learning model to be assessed. Certain performance metrics are used to evaluate the accurateness of the model. The first technique employed is to calculate the confusion matrix. It basically tells us the number of values the model predicts as true positives, false positives, true negatives, and false negatives. True values are predicted correctly, and the false values are predicted as the class which was incorrect. A classification report generates a summary of metrics that determine the accurate predictions and inaccuracies of the model (Table 3).

Table 3. Accuracy Results for each Company Stocks.

Ticker	Company	LR Accuracy
AMZN	Amazon	92.43
TSLA	Tesla	88.84
MSFT	Microsoft	86.85
GOOGLE	Alphabet Inc (Google)	90.03
AAPL	Apple	88.04

5 Conclusion

This paper navigated through the relevance of stock analysis and the application of machine learning in the prediction and analysis of certain stock markets with five chosen companies. The results of the Logistic Regression prediction model showed that after fine-tuning the parameters and feature engineering, accurate predictions could be made with the test datasets. Among the five companies amazon showed highest accuracy. The second highest is Google. Tesla and Apple are very close. Microsoft dataset reflects 86.85 percentage of accuracy. This research can help investors to make a better management of their portfolios.

References

1. Obthong, M., Tantisantiwong, N., Jeamwatthanachai, W., Wills, G.: A Survey on Machine Learning for Stock Price Prediction: Algorithms and Techniques, 06 May 2020. https://eprints.soton.ac.uk/437785/
2. Mintarya, L.N., Halim, J.N.M., Angie, C., Achmad, S., Kurniawan, A.: Machine learning approaches in stock market prediction: a systematic literature review. Proc. Comput. Sci. **216**, 96–102 (2023). https://doi.org/10.1016/j.procs.2022.12.115
3. Ali, S.S., Mubeen, M., Hussain, I.L.: Prediction of stock performance by using logistic regression model: evidence from Pakistan stock exchange (PSX). Asian J. Empir. Res. **8**(7), 247–258 (2018)
4. Zaidi, M., Amirat, A.: Forecasting stock market trends by logistic regression and neural networks. Int. J. Econ. Commer. Manag. (2016). https://ijecm.co.uk/wp-content/uploads/2016/06/4614.pdf. Evidence From Ksa Stock Market
5. Chatterjee, A., Bhowmick, H., Sen, J.: Stock price prediction using time series, econometric, machine learning, and deep learning models. In: 2021 IEEE Mysore Sub Section International Conference (MysuruCon), pp. 289–296, October 2021. https://doi.org/10.1109/MysuruCon52639.2021.9641610
6. Attigeri, G.V., Manohara Pai, M.M., Pai, R.M., Nayak, A.: Stock market prediction: a big data approach. In: TENCON 2015 - 2015 IEEE Region 10 Conference (2015). (IF:3)
7. Sharma, A., Bhuriya, D., Singh, U.: Survey of stock market prediction using machine learning approach. In: 2017 International Conference of Electronics, Communication (2017). (IF: 4)
8. Ren, R., Wu, D.D., Liu, T.: Forecasting stock market movement direction using sentiment analysis and support vector machine. IEEE Syst. J. (2019). (IF:4)
9. Zhong, X., Enke, D.: Predicting the daily return direction of the stock market using hybrid machine learning algorithms. Financ. Innov. (2019). (IF: 4)
10. Nti, I.K., Adekoya, A.F., Weyori, B.A.: A systematic review of fundamental and technical analysis of stock market predictions. Artif. Intell. Rev. (2019). (IF: 4)
11. Gandhmal, D.P., Kumar, K.: Systematic analysis and review of stock market prediction techniques. Comput. Sci. Rev. (2019). (IF: 4)
12. Mehtab, S., Sen, J.: A robust predictive model for stock price prediction using deep learning and natural language processing. Econom. Model.: Cap. Mark. - Forecast. eJ. (2019). (IF:3)
13. Valencia, F., Gómez-Espinosa, A., Valdés-Aguirre, B.: Price movement prediction of cryptocurrencies using sentiment analysis and machine learning. Entropy (Basel, Switzerland) (2019). (IF:4)

14. Nti, I.K., Adekoya, A.F., Weyori, B.A.: A comprehensive evaluation of ensemble learning for stock-market prediction. J. Big Data (2020). (IF:4)
15. Nabipour, M., Nayyeri, P., Jabani, H., Shahab, S., Mosavi, A.: Predicting stock market trends using machine learning and deep learning algorithms via continuous and binary data; a comparative analysis. IEEE Access (2020). (IF: 4)

A Scientific Classification-Based Survey of Intrusion Detection System Using Machine Learning Approach

Neha Srivastava$^{(\boxtimes)}$ (ID) and R. K. Singh (ID)

Kamla Nehru Institute of Technology, Sultanpur, Uttar Pradesh, India
`nehush2604@gmail.com, rksingh@knit.ac.in`

Abstract. In this paper, the prime focus is to find and explore the field of intrusion detection with supervised ML approaches. The objective is to create and publish a scientific classification for combined IDS and supervised ML techniques. To do the same, examination of the fundamentals of IDS, ML algorithms (supervised), and cyber attacks is done. In addition, the work has been explored that has recently been done in the sphere of supervised learning intrusion detection. Then, using these works, a taxonomy is provided using different publicly available datasets to classify the high and accurate performance of supervised machine learning algorithms.

Keywords: Intrusion Detection System · Cyber Attacks · Machine Learning

1 Introduction

The possibility of cyberattacks is increasing with the rise in internet usage. Now, there are new classifications of attacks, and it is necessary to recognize these types of new attacks using intelligent systems. An IDS can be described as software that examines network traffic to find malicious activity. It can be distributed into two types: one is signature-based and the other is anomaly-based. Signature-based is used to search for already available attacks and compares new attacks with old ones; if it matches, an alarm is raised. While anomaly-based detection searches the new type of traffic to see if it is showing unusual behavior and reports it as an anomaly. This detection is very critical for detecting zero-day attacks [17].

There is a need for huge amounts of data to create a model for detecting normal and abnormal behavior. Thus, it is required to create a model using supervised ML algorithms to efficiently comprehend the data and to create a prediction model for the identification of new attacks. The use of feature reduction along with feature selection helps improve classification to overcome the problem of high dimensionality in data. It also studied the best strategies for supervised machine learning applications in intrusion detection systems, which can further be used for the efficient detection of new cyberattacks.

The major aim of this investigation work is to create a survey based on supervised ML approaches, and IDS. The paper begins with the introduction of IDS and focus on its

A. K. Bairwa et al. (Eds.): ICCAIML 2024, CCIS 2184, pp. 388–396, 2025.
https://doi.org/10.1007/978-3-031-71481-8_30

significance, types, and definition. Then, it discuss the well-known supervised learning methods, the publicly available datasets in this field, and the idea of dimensionality reduction. It also discuss various types of cyber-attacks. Finally, it provides a taxonomy to find out which methods are most appropriate to use with various IDS available datasets and if feature selection could improve classification results or not.

The paper is presented in five sections which are as follows: Sect. 2 covers the information related to intrusion detection systems, supervised ML, and cyberattacks. Section 3 comes with the study of supervised ML methods for intrusion detection systems. Section 4 discusses the scientific classification provided based on the studies. Finally, Sect. 5 presents the summary and findings of this research paper.

2 Background

This particular section provides information on the following topics:

- Intrusion Detection System
- Supervised ML (Machine Learning)
- Cyber Attack

2.1 Intrusion Detection System

Computer networking is crucial for information sharing and communication. Computers, however, can be vulnerable to outside dangers that require constant monitoring and detection [1]. Such risks have the potential to breach the computer security's three pier that are: confidentiality, integrity, and availability (CIA).

An intrusion can be defined as an attempt to bypass a computer or network's security measures or breach the CIA. It is a procedure of tracking and examining network or system activity for indicators of intrusion [2]. Thus, the four basic goals of IDS is to (1) monitor the host and network (2) analyze the behavior of the user computer network, (3) raise alerts and (4) react to apprehensive behavior.

IDS systems are widely discussed in the literature. However, two issues plague a lot of IDS:

1. High false alarm rate: Many significant violations go unreported while alarms are activated for non-threatening infractions.
2. Because new attacks are crucial to spot, attention has been given to the employment of cutting-edge ML algorithms for IDS.

IDS is often be categorized in two ways: 1) based upon methodology, and 2) based upon platform being monitored.

With respect to the approach, it can divide Intrusion detection system into detection of anomaly-based and misused-based. Anomaly-based detection techniques create a model of standard behavior based on system activity.

This methodology attempts to make a model that closely resembles a system's typical behavior by monitoring the activity of a user with time. The IDS is suited for encountering zero-day attacks since it shows every event that diverges from typical behavior as malicious activity and raises an alarm. There were two problems identified as significant drawbacks: 1) unable to handle the changes in typical user behavior, and 2) a significant FPR i.e. false positive rate [3].

On the other side signature based detection compares captured events with known assaults and threats in order to identify potential intrusions. Comparing this method with the anomaly based detection method, it is much more accurate and generates minimum false alarms. However, it cannot be used to identify fresh attacks [4].

There are two categories: one is Host-based IDS (HIDS) and other one is network-based IDS (NIDS) into which IDS is classified based on the platform being monitored.

The tasks on the user's system where Host-based IDS is installed are tracked and analyzed [5]. It is possible to keep track of some of the system's dynamic conduct and state. On the other side, Network-based IDS keeps an eye on network traffic to find assaults that are transmitted via a network connection remotely [6]. It is a gadget that is dispersed among networks to examine traffic passing through the ones it is seated on. It could consist of h/w or s/w. NIDS often has two network interfaces -: one is for controlling and reporting, the other one to listen network communications [7].

2.2 Supervised Machine Learning Approaches

For automatic identification of intrusions, there are typically two methods of machine learning: supervised learning and unsupervised learning [8–11]. This survey's major objective is to use supervised machine learning methods for IDS. Labeled data must be there to train a model for detection using supervised learning or classification. The following step-by-step action can be used to summarize the categorization process:

Collection of Data , the process of gathering data to supply the knowledge needed to train the categorization model. Basically, a set of features that can differentiate between the classes is used to define the datasets. Since collecting data is a crucial task, there are numerous benchmark publicly available data sets available, including KDD'99 [14], NSL-KDD [14], and CICIDS2017 [14]. A summary of mentioned data collection is included in Table 1.

Table 1. Description of IDS datasets

Dataset	Description
CICIDS2017	It took five days to develop in a realistic environment. It has both bidirectional and packet-based traffic. There are 80 extracted features. There are various types of attacks which includes DDoS, DoS, Web attacks, BF (bruteforce) attacks as FTP and SSH, infiltration, heartbleed, botnet and many more
NSL-KDD	This data is an improvement of the KDD'99 dataset that addresses issues with duplication, redundancy, and data imbalance
KDD-99	It is produced by simulating regular and hostile traffic in a military setting (US AirForce LAN). Rat tcpdump files representing nine weeks of simulation are included. 41 intrinsic, content, and traffic-related factors are used to describe the dataset. Simulated assaults include DoS, Prob, U2R, and R2L

Reduction of Data: High dimensionality in feature spaces may be difficult and can cause problems like the "curse of dimensionality" when there is a presence of little amount of data in big-dimensional [13]. Therefore, techniques like PCA and LDA may be used to convert the data into a low-dimensional space [8]. Feature selection techniques such as the best-first algorithm [9] can also be used to select the best-performing features in the opposite way.

Classification: At this phase, one portion of the data has to be used for developing the model known as training data, while the other portion has to be used for testing the performance of the classification model known as testing data. The data divided into training set and testing set is defined as hold-out test. One other common test type is the N-fold cross-validation test, in which the data is split into N-folds, with the tenth being used to test the model and the other nine being used for training. A fresh set of 9-folds for training and a 10-fold for testing are then used to repeat the process. The average value across all folds is used to calculate accuracy.

Evaluation of Performance: The performance of classification model is evaluated with the help of accuracy and the false positive rate (FPR).

2.3 Cybersecurity Attacks

Cyberattacks are deliberate, malicious attempts to breach another person's or organization's information system. Usually, the attacker takes advantage of the victim's network outage. There are various types of cyber-attacks [16]:

Spyware, ransomware, viruses, and worms are examples of dangerous software that falls under the umbrella term "malware." Malware enters a network by taking advantage of a weakness, usually caused by clicking on a link or some attachment that installs or downloads software.

The practice of sending phony emails or other contact that appears to be from a trustworthy source is known as phishing. The goal is to either use malware to infect the victim's computer or steal private information like credit card numbers and login credentials. Phishing is a cyber-threat that is becoming more frequent.

Eavesdropping attacks, sometimes referred to as man-in-the-middle (MitM) attacks, happen when an attacker interjects oneself into a two-party transaction. After severing the connection, the attackers can filter and pilfer information.

In an effort to exhaust the bandwidth and resources of servers, networks, or systems, a denial-of-service assault floods them with traffic. As such, legitimate requests cannot be processed by the system. Attackers can even use many compromised devices to carry out this attack.

An attack known as a "SQL injection" happens when a hacker introduces malicious code into a SQL-using server, compelling it to divulge data that it would not typically. All it takes to execute a SQL injection is for an attacker to insert malicious code anywhere in the search bar of website which is vulnerable.

A zero-day exploit occurs when a network vulnerability is discovered just before a patch and/or fix has been implemented. The Attacker target the widely known susceptibility during the window of opportunity. Continuous monitoring is required for threat identification resulting from zero-day vulnerabilities.

3   Literature Review

Automating intrusion detection with supervised ML techniques has been receiving enough attention nowadays.

As per [12], the dataset NSL-KDD was utilized to address the intrusion classification problem. In order to classify numerical attributes and address missing data issues, the data set is first preprocessed. After that, the dataset is sub-divided into four groups and categorized as train data and test data. The classification, accuracy, and False positive rate are then calculated using a random forest classifier after the data has been input. In a feature selection technique, the Symmetrical Uncertainty measure is also employed. Using feature selection led to a little performance gain, according to the authors. Random forests performed better than C4.5 when the results were compared [12]. Following feature selection, the received accuracy is of 99.67 with an average FAR of 0.005.

Authors [17] compared the evaluation of 10 algorithms based on classification on the dataset NSL-KDD, similar to the authors in [12], but applying a different and various feature selection strategy. The technique of the selection of feature is dependent on the use of attribute evaluators or filters. The authors used the following algorithms: NB, Logistics, RT, RF, Bagging, J48, OneR, ZERO, PART. The random forest classification system performs best having accuracy 99.9% and a FAR as 0.001.

Similarly, the author [18] used the dataset NSL-KDD and used a hold-out test strategy with feature selection not applicable. RF, SVM, logistical regression, and the Gaussian mixture model are the four classification techniques that are being tested. Having an accuracy of 99%, the random forest performed as the best algorithm.

The authors of [19] recently stated the working of SVM and ANN for the NSLKDD dataset. The data sample taken is the 20% of the whole dataset. The author used a chi-squared-based and correlation-based approach with 35 features which were earlier 14. The data is then sent to ANN and SVM after the selection of feature. Hence it is found that ANN with correlation-based feature achieved the accuracy of 94%.

In other approach, the authors of [20] used the KDD'99 data set with the following classes: DoS, normal, U2R, Prob and R2L. CFSSubSet Eval and Best First feature selection algorithms are used along with four approaches: k-means, SVM, NB, and RF. Based on the eight best attributes, it is found that the supervised approach (random forest) performed well in comparison with k-means having accuracy of 99%.

The authors also used KDD'99 in [21] where an Algorithm selects a good representative sample of the original dataset having a record of 550. The [21] implemented a new feature reduction method called GFR (Gradual Feature Removal) for reducing the space of feature dimension to 19 features. For classifying, the reduced characteristics are then combined using SVM. Before feature selection, the received accuracy was 98.67% with no improvement and after the selection of feature, the accuracy is 98.62%.

In [22], the author reduced the dimensional space of KDD'99 to see its impact on the overall performance of classification They looked specifically at the algorithm for picking features, getting information by entropy (IG), and using BBNN with feature vector (22). They found that the accuracy dropped to 91% when the feature was removed.

Another dataset that has generated interest is CICIDS2017. The authors of [23] and [24] used sampling followed by feature reduction and then boosting strategy to create an Intrusion detection system. This dataset has a difficult part due to its imbalance feature. To overcome this issue, they preprocessed the data with the Synthetic Minority Overamplification Technique (SMOT) to reduce the number of case in certain class. They then used a reduced feature space (25) to implement a reduction strategy using Principal Component Analysis (PCA) and ensemble feature selection (EFS). Finally, they evaluated the findings using the hold-out approach based on the Adaboost algorithm having accuracy of 81.83%.

In a recent paper [25], the authors discussed the topic of assault detection using RF and the ANN approach on the dataset CICIDS2017. They used Boruta, a feature selection software, to return the top ten most significant features. Then, they fed the classifiers the feature set. Using random forest, they reported 96% accuracy, and using ANN, 96.4%.

In order to review and brief the related research, the paper presents a classification of IDS and supervised ML approach in the following section to better understand the presented final findings and conclusions.

4 Scientific Classification of IDS and Supervised Machine Learning Algorithms

Considering the useful findings presented in the mentioned part, it constructs a scientific classification of IDS and ML in this section. The scientific classification also known as taxonomy having following characteristics, which are listed in Table 2 dataset used; 2) The usage of feature selection or not: Y or N, 3) Usefulness of feature selection: Y or N, 4) The supervised ML used, 5) Method of validation, 6) The best-received accuracy and FPR outcome. 7) year of research.

To comprehend the data shown in Table 2, it shall examine the datasets individually before providing an complete summary:

1. NSL-KDD dataset: When feature selection is applied to this dataset, it improves classification performance. Furthermore, the random forest technique performs well on this dataset and produces high-quality result utilizing several methods of validation. ANN appears to perform good on NSL-KDD; however, it has tested on only 20% of entire data as sample.
2. KDD99 dataset: When feature selection is performed on this dataset, it does not improve classification results. The data set is validated using several approaches, and no immediate result can be taken regarding the top-behaving classification algorithm; however, Random forest and Support Vector Machine do quite good on KDD'99 dataset.
3. CICIDS2017 dataset: It is not possible to directly determine the impact of feature selection; however, it was useful in enhancing performance of the classification. This dataset contains the significant imbalance issue that can be remedied through sampling techniques. QDA is the most effective algorithm.

Overall, the paper can draw the following conclusions:

1. There has been a lot of discussion in the literature on IDS using machine learning methodologies.

2. The results of the categorization are pretty good and strong, as per the review of three separate publicly available datasets.
3. The selection of feature is crucial for performance enhancement and is often necessary.
4. Data imbalancy can be a thought of concern, and sampling procedures can assist in resolving it.

Table 2. IDS Scientific Classification

Dataset	Feature Selection Applied (Y/N)	Algorithm used	Used Validation	Best Method	Result	Paper	Year
KDD'99	Y	SVM	TEN-FOLD	BBNN	Acc. 98.7%	21	2012
KDD'99	Y	RF, SVM, NB, k-means	Hold-out	RF	Acc-99%	20	2015
KDD'99	Y	BBNN	TEN-FOLD	QDA	Acc-91%	22	2015
NSL-KDD	Y	RF, J-48	TEN-FOLD	RF	Acc-99.7% FPR-.005%	12	2016
NSL-KDD	N	RF, GMM, LR, SVM	HOLD OUT	RF	Acc-99%	18	2016
NSL-KDD	Y	NB	TEN-FOLD	RF	Acc-99.9%	17	2019
20% of NSL-KDD	Y	SVM, ANN	Hold-out	RF	Acc-94%	19	2019
CICIDS2017	Y	QDA, LDA, BN, RF	Hold-out	QDA	Acc = 98.8% FPR = .001%	24	2019
CICIDS2017	Y	Adaboost	Hold-out	Adaboost	Acc = 81.83%	23	2019
CICIDS2017	Y	RF, ANN	Hold-out	RF	Acc = 96.4%	25	2020

5 Conclusion

Because of the current boom in internet material, cybercrime is on the rise. The first and immediate step for detecting and raising such attacks is to employ intrusion detection systems (IDS). Anomaly detection relies on being able to identify new attacks, which can be a tricky job. In this paper, it can be seen how IDS classification works, how supervised learning works, and how cyber security attacks work.

It then presents all the work that's been done in this area and compared it to each other using three different data sets that were made available to the public: the KDD'99 dataset, the NSL-KDD dataset, and the CICIDS2017 dataset. A scientific classification i.e., taxonomy is also provided having related studies. This classification suggests that the subject of intrusion detection utilizing supervised ML approaches is gaining popularity. The study looked at three data sets and found that the performance of various classification of the supervised learning algorithms was good and promising. It's important to pick the right features, and in many cases, it's mandatory to improve performance. Data mismatches can also be a problem, and sampling procedures can help with that.

References

1. Graham, J., Olson, R., Howard, R. (eds.): Cyber Security Essentials. CRC Press (2011)
2. Liao, H.J., Lin, C.H.R., Lin, Y.C., Tung, K.Y.: Intrusion detection system: a comprehensive review. J. Netw. Comput. Appl. **36**(1), 16–24 (2013)
3. Liu, H., Lang, B.: Machine learning and deep learning methods for intrusion detection systems: a survey. Appl. Sci. **9**(20), 4396 (2019)
4. Hamid, Y., Sugumaran, M., Balasaraswathi, V.R.: IDS using machine learning-current state of the art and future directions. Curr. J. Appl. Sci. Technol. 1–22 (2016)
5. Masdari, M., Khezri, H.: A survey and taxonomy of the fuzzy signature-based intrusion detection systems. Appl. Soft Comput. 106301 (2020)
6. Milenkoski, A., Vieira, M., Kounev, S., Avritzer, A., Payne, B.D.: Evaluating computer intrusion detection systems: a survey of common practices. ACM Comput. Surv. (CSUR) **48**(1), 1–41 (2015)
7. Conrad, E., Misenar, S., Feldman, J.: CISSP Study Guide (2012)
8. Bishop, C.M.: Pattern Recognition and Machine Learning. Springer, New York (2006)
9. Witten, I.H., Frank, E.: Data mining: practical machine learning tools and techniques with Java implementations. ACM SIGMOD Rec. **31**(1), 76–77 (2002)
10. https://en.wikipedia.org/wiki/. Accessed 10 Aug 2023
11. Sahasrabuddhe, A., Naikade, S., Ramaswamy, A., Sadliwala, B., Futane, P.: Survey on intrusion detection system using data mining techniques. Int. Res. J. Eng. Technol. **4**(5), 1780–1784 (2017)
12. Farnaaz, N., Jabbar, M.A.: Random forest modeling for network intrusion detection system. Proc. Comput. Sci. **89**, 213217 (2016)
13. Rust, J.: Using randomization to break the curse of dimensionality. Econometrica: J. Econom. Soc. 487–516 (1997)
14. https://www.unb.ca/cic/datasets/index.html. Accessed 11 Aug 2023
15. www.kaggle.com. Accessed 03 Aug 2023
16. https://www.cisco.com/c/en_in/products/security/common-cyberattacks.html#~how-cyber-attacks-work
17. Malhotra, H., Sharma, P.: Intrusion detection using machine learning and feature selection. Int. J. Comput. Netw. Inf. Secur. **11**(4) (2019)
18. Belavagi, M.C., Muniyal, B.: Performance evaluation of supervised machine learning algorithms for intrusion detection. Proc. Comput. Sci. **89**, 117–123 (2016)
19. Taher, K.A., Jisan, B.M.Y., Rahman, M.M.: Network intrusion detection using supervised machine learning technique with feature selection. In: 2019 International Conference on Robotics, Electrical and Signal Processing Techniques (ICREST), pp. 643–646. IEEE, January 2019
20. El Mourabit, Y., Bouirden, A., Toumanari, A., Moussaid, N.E.: Intrusion detection techniques in wireless sensor network using data mining algorithms: comparative evaluation based on attacks detection. Int. J. Adv. Comput. Sci. Appl. **6**(9), 164–172 (2015)

21. Li, Y., Xia, J., Zhang, S., Yan, J., Ai, X., Dai, K.: An efficient intrusion detection system based on support vector machines and gradually feature removal method. Expert Syst. Appl. **39**(1), 424–430 (2012)
22. Shah, B., Trivedi, B.H.: Reducing features of KDD CUP 1999 dataset for anomaly detection using back propagation neural network. In: 2015 Fifth International Conference on Advanced Computing & Communication Technologies, pp. 247–251. IEEE, February 2015
23. Yulianto, A., Sukarno, P., Suwastika, N.A.: Improving AdaBoost-based intrusion detection system (IDS) performance on CIC IDS 2017 dataset. J. Phys.: Conf. Ser. **1192**(1), 012018 (2019)
24. Abdulhammed, R., Faezipour, M., Musafer, H., Abuzneid, A.: Efficient network intrusion detection using PCA-based dimensionality reduction of features. In: 2019 International Symposium on Networks, Computers and Communications (ISNCC), pp. 1–6. IEEE, June 2019
25. Pelletier, Z., Abualkibash, M.: Evaluating the CIC IDS-2017 dataset using machine learning methods and creating multiple predictive models in the statistical computing language R. Science **5**(2), 187–191 (2020)

Author Index

A. K. Bairwa et al. (Eds.): ICCAIML 2024, CCIS 2184, pp. 397–399, 2025.
https://doi.org/10.1007/978-3-031-71481-8

SPRINGER NATURE

GPSR Compliance

The European Union's (EU) General Product Safety Regulation (GPSR) is a set of rules that requires consumer products to be safe and our obligations to ensure this.

If you have any concerns about our products, you can contact us on ProductSafety@springernature.com

In case Publisher is established outside the EU, the EU authorized representative is:

Springer Nature Customer Service Center GmbH
Europaplatz 3
69115 Heidelberg, Germany